Studies in Choice and Welfare

For further volumes:
http://www.springer.com/series/6869

Dan S. Felsenthal • Moshé Machover
Editors

Electoral Systems

Paradoxes, Assumptions, and Procedures

Springer

Editors
Prof. Dan S. Felsenthal
University of Haifa
School of Political Sciences
Mount Carmel
31905 Haifa
Israel

and

Centre for Philosophy of Natural
and Social Science
London School of Economics
and Political Science
Houghton Street
London WC2A 2AE
United Kingdom
dfelsenthal@poli.haifa.ac.il

Prof. Moshé Machover
King's College London
Department of Philosophy
Strand, London WC2R 2LS
United Kingdom

and

Centre for Philosophy of Natural
and Social Science
London School of Economics
and Political Science
Houghton Street
London WC2A 2AE
United Kingdom
moshe.machover@kcl.ac.uk

Studies in Choice and Welfare ISSN 1614-0311
ISBN 978-3-642-20440-1 e-ISBN 978-3-642-20441-8
DOI 10.1007/978-3-642-20441-8
Springer Heidelberg Dordrecht London New York

Library of Congress Control Number: 2011943796

Printed on acid-free paper

Springer is part of Springer Science+Business Media (www.springer.com)

Preface

The essays in this volume were presented by academic experts in voting theory from seven countries, as well as by two voting practitioners, at an international workshop on "Assessing Alternative Voting Procedures", held on 30 July – 2 August 2010 at Chateau du Baffy, Normandy, France. It received generous financial support from the Leverhulme Trust (Grant # F/07 004M).

The main purpose of the workshop was to explore both the theoretical and actual vulnerability to various voting paradoxes or pathologies of voting procedures designed to elect a single candidate. The following five relatively recent election-related events served as background to the workshop deliberations:

- The phenomenon displayed (again) in the 2000 US presidential elections, where George W. Bush was elected by the Electoral College although Albert Gore received more popular votes.
- The decision by the German Federal Constitutional Court on 3 July 2008 mandating the Bundestag to amend by June 2011 the procedure by which it is elected so as to avert, or significantly decrease, the impact of non-monotonicity afflicting it.
- Several mayoral elections conducted in the US (e.g., in Burlington, VT., in March 2009) displaying the non-monotonicity paradox.
- The release by the British Academy on 10 March 2010 of a report summarizing the properties of parliamentary voting procedures currently used in the world but without mentioning the voting paradoxes (pathologies) to which they are vulnerable.[1]
- The decision of the UK parliament in May 2010 to conduct a referendum in May 2011 as to whether the UK should elect its parliament by the Alternative Vote procedure instead of the current Plurality procedure (aka "First Past the Post").

---

[1] Hix, S., Johnston, R., & McLean, I. (2010). *Choosing an electoral system: A research report prepared for the British Academy*. London: British Academy Policy Centre. ISBN 978–085672–588–3.

As outlined below, these events are explored in some of the chapters in this volume.

The volume is divided into three parts containing a total of 13 chapters.

The first part (comprising Chaps. 1–2) contains brief introductory remarks on electoral procedures of representative assemblies and decision-making rules within them. It classifies the types of representative assemblies and the manner in which they are elected, and thereafter explores possible decision rules that could be instituted within them so as to make them truly representative.

Chapter 1 formulates two main political dichotomies, each offering two alternatives. This gives rise to a fourfold political classification of voting procedures. The first main dichotomy distinguishes between legislatures based on *proportional representation*, and those based on *district representation*. The second main dichotomy distinguishes between elections employing a *deterministic voting procedure* and those using lottery. Following this fourfold classification the chapter proceeds to explore what social-choice theory has to offer in each of these four classes.

Chapter 2 argues that regardless of how a representative assembly is elected, it cannot truly be representative if (permanent) minorities in it are unable to affect decisions due to a majoritarian decision rule. This chapter briefly examines some alternative decision rules that would increase the actual voting power of minority groups.

The second part of the volume (Chaps. 3–9) surveys paradoxes afflicting single-winner voting procedures, as well as assessing the theoretical and empirical frequencies of some of these paradoxes.

Chapter 3 contains a comprehensive review and illustration of the main paradoxes that may afflict each of 18 single-winner voting procedures. It argues that in order to better assess the probability of occurrence of every paradox to which a given voting procedure is vulnerable, one must first determine what are the necessary and/or sufficient condition(s) for this paradox to occur under the given procedure. As this has so far not been achieved with respect to most paradoxes/procedures, perhaps a more reasonable way for selecting a voting procedure would be to limit the choice of a voting procedure only to those procedures that are not vulnerable to what the author considers as especially serious pathologies.

Chapters 4 and 5 investigate the probability of occurrence – in two particular settings – of the general phenomenon variously known as *majority-deficit*, or *election inversion*, or *referendum paradox*, that can occur in any two-tier electoral system, whereby the candidate (or party) that receives the largest number of votes in the entire electorate is either not elected or does not receive the largest number of parliamentary seats.

Chapter 4 investigates this phenomenon in the context of US presidential elections where it last occurred in the 2000 US presidential election. This chapter identifies the sources of election inversions by the US Electoral College, establishes logical bounds on the phenomenon, and estimates the frequency and magnitude of inversions on the basis of historical state-by-state US presidential election data.

Chapter 5 investigates this phenomenon in the context of French local (cantonal) elections. Despite the fact that the cantons are of unequal population size, each of them is represented by one representative in the region's assembly (called *conseil general*) who is elected by the plurality-with-runoff procedure. The authors' objective is to find how many representatives should be allocated to each canton as a function of its population size so as to minimize the frequency of the referendum paradox. They find that the optimal number should be very close to being proportional to the square root of each canton's population. Thus this is probably the first experimental study to support what has long been advocated in the theoretical literature on voting power and known as the "square root rule".

Chapter 6 outlines five alternative proposals to avert, or significantly decrease, the non-monotonicity of the election procedure of the German Bundestag, whereby every voter casts two votes – one for a preferred constituency representative and the other for a preferred party list. The interplay of these two votes with the Federal structure of Germany has led to instances of non-monotonicity: a vote for a party list reduced the number of seats it received in the Bundestag. At the time of writing it is not yet known which, if any of these or other proposals currently being contemplated, will be adopted by the Bundestag and sanctioned by the German Federal Constitutional Court.

The last three chapters of Part II (Chaps. 7–9) present and defend alternative methods for assessing the probabilities of various voting paradoxes.

Chapter 7 defends computer simulations designed to estimate the probability of voting paradoxes in three-candidate single-winner elections based on the models known in the literature as the Dual Culture Condition, the Impartial Culture Condition, and the Impartial Anonymous Culture Condition. Although admittedly these models do not reflect realistic scenarios, it is argued that they still add very significantly to research on the probability of occurrence of various voting paradoxes; in particular, they suggest that most extreme voting paradoxes should be expected to be rare events.

The same authors continue to investigate in Chap. 8 which of five single-winner voting procedures (Plurality, Negative Plurality, Borda, Alternative Vote, and Coombs' procedure) is more likely to maintain the social preference ordering when there are three candidates, a Condorcet Winner exists, and various degrees of group coherence in voters' preferences are introduced.

In contrast to Chaps. 7 and 8, Chap. 9 argues that any evaluation of the probability of various voting phenomena – e.g., that the social preference ordering contains a cycle, or the likelihood that a Condorcet Winner is elected when s/he exists, or that voters vote strategically rather than sincerely – needs to be based on a statistical model that describes how voters behave in actual elections. This chapter uses two sets of data, one from actual elections and the other from survey of voters, to evaluate 12 statistical models that make different assumptions regarding voters' behavior in three-candidate single-winner elections (and hence reach different conclusions) regarding voters' behavior in three-candidte single-winner elections.

The final part of the volume comprises four chapters (Chaps. 10–13). It discusses considerations other than susceptibility to paradoxes in selecting a voting procedure.

In Chap. 10 it is argued that the paradoxes afflicting single-winner voting procedures may not be the best criterion, and that selection should be based on additional criteria. Moreover, it is argued that the determination of who of the competing candidates ought to be elected should not necessarily be based on the voters' (ordinal) rankings of the candidates because voters are often capable of a much more refined expression of their preference ordering among the candidates.

Chapter 11 is an advocacy essay, supporting the replacement of the current plurality procedure by which the UK elects its members of parliament with the Alternative Vote (AV) procedure. Although it has been decided in the 5 May 2011 referendum in the UK to keep the plurality procedure for electing the UK parliament, this chapter should be of interest not only to activist proponents or opponents of AV, but also to electoral-system scholars because of its balanced and nuanced analysis of the AV procedure.

Since all the other chapters in this volume are concerned with single-winner electoral procedures, we thought it is appropriate – at least for the purpose of charting directions for possible future research – to include in this volume also one chapter which addresses multi-winner electoral procedures, e.g., procedures for electing teams or fixed-size committees. Chapter 12 describes various multi-winner procedures and proposes several properties that may be used for assessing the desirability of such procedures. It is concluded that Approval balloting is the most natural approach to multi-winner elections.

At the end of the workshop it was agreed among its 22 participants to hold a vote as to the best single-winner procedure for electing a mayor for a city or town. Each of the participants listed in his or her ballot one or more of 18 proposed single-winner procedures that s/he approved, and it was agreed that the winner would be the procedure which received the most approval votes. The last chapter in this volume (Chap. 13) describes and analyzes this election. It also contains explanations supplied by some of the participants as to why they voted the way they did. As far as we know, this is the first time that voting theorists hold a vote on voting procedures. The decision to hold the vote was spontaneous. Consequently, no one had much time to think things over, discuss them with others, or calculate. Arguably, this detracts from the significance of the result; but perhaps it adds to its significance, in that the opportunity for strategic manipulation was diminished. In either case, it seems likely that if the experiment is ever repeated, the circumstances will be different; or that knowledge of this experiment may mean that future ballots will be more carefully considered. Thus, this vote may be unique in that it may have been the first and last "naïve" vote on voting rules by voting theorists.

Of course, some assertions made in some essays included in this volume are controversial. But this is to be expected from voting theorists and practitioners, who are engaged in the highly important and sensitive issue of how to aggregate individual preferences into a binding social decision. Perhaps one of the merits of this volume is that it brings these controversies to the attention of a wider public.

Each of the essays in this volume has been revised in light of comments received from a referee and from the editors. We would like to thank the scholars who served as referees. We are also grateful to Maurice Salles for encouraging us to edit this volume and for his helpful advice on some of the editorial decisions we made; to Martina Bihn – the Economics editor at Springer – for her excellent and smooth cooperation; and to Ruth Milewski and Dagmar Kern of Springer's staff, as well as to Anitha Murugaiyan of SPi Technologies, India, who made the publication of this book in its present form possible.

Jerusalem, Israel *Dan S. Felsenthal*
London, UK *Moshé Machover*

Contents

Part I
Representative Electoral Systems: Underlying Assumptions and Decision Rules

Chapter 1
The Underlying Assumptions of Electoral Systems

Moshé Machover

1.1 Introduction

My aim in this brief paper is modest: not to present new findings, but to propose what I regard as a useful way of classifying voting procedures, and thus organizing the way we look at them. My main thesis is that we have to make a strict distinction between two kinds of consideration in choosing a voting/election procedure:

- *Political criteria.* I use this rubric in a very broad sense, including criteria ranging from the pragmatic to the philosophical. But all of them are purely a matter of opinion, not of "right" or "wrong".
- *Social-choice considerations.* I take this rubric in the narrow sense: the logico-mathematical properties of a voting procedure, the pathologies and paradoxes that afflict it.

These two kinds of consideration are not on a par with each other: political considerations are paramount in choosing a voting procedure. For example, as far as political elections are concerned, it is politicians who usually choose the voting procedure; and even when the choice is made by referendum, the question put to referendum is framed by politicians. But politicians and their advisors – and, ideally, the general public – ought to be aware of the logico-mathematical properties of the voting procedures in question; otherwise they can easily walk into a trap. So it is wrong to dismiss these matters as of interest only to geeks.

On the other hand, social-choice theorists must recognize that their professional scientific role is confined to ascertaining the technical properties of voting procedures, including the likelihood of various pathologies manifesting themselves under each procedure. However, the decision as to which pathology (with a given

M. Machover (✉)
Centre for the Philosophy of the Natural and Social Sciences, London School of Economics and Political Science, Houghton Street, London WC2A 2AE, UK
e-mail: moshe.machover@kcl.ac.uk

D.S. Felsenthal and M. Machover, *Electoral Systems*, Studies in Choice and Welfare, DOI 10.1007/978-3-642-20441-8_1, © Springer-Verlag Berlin Heidelberg 2012

likelihood) is more tolerable than another is not a scientific matter; it is political. On this, the opinion of a social-choice theorist is not more privileged than that of any well-informed member of the public.

Similarly, social-choice theory provides information about the effects of a given system of electing a legislature regarding the stability of a parliamentary government and the number of parties with realistic prospect of winning seats (Duverger's law). However, the question as to the importance of a stable government (and the desirable *degree* of stability), or the desirability of a small or large number of such parties is a purely political one.

In what follows, I will formulate two main political dichotomies, each offering two alternatives. This gives rise to a fourfold political classification of voting procedures. I will explore what social choice theory has to offer in each of these four classes.

1.2 Two Dichotomies

The first main dichotomy is relevant for electing a representative assembly such as a legislature, not a holder of an individual post, such as a president. I state it as follows:

i. Proportional Representation (PR) vs. District Representation (DR)

This dichotomy hinges on a distinction between two quite different senses of the verb *represent* and its derivatives. Who or what is being "representative", and whom or what are they supposed to "represent"?

One sense of this term – which underlies PR – is intended, for example, in statistics, when we speak of a *representative sample*. An elected assembly is representative in this sense if it is a microcosm of the entire electorate, reflecting in true proportion (or as near to it as possible) the various shades of opinion that exist in the society as a whole. Thus it can stand as proxy for a market-place meeting of the entire citizenry; and a vote taken in the assembly may be regarded as a close approximation to a referendum. Here a member of the assembly does not represent a geographically defined constituency, but reflects a like-minded section of the electorate at large, which may well be geographically dispersed. Note that being representative in this sense is primarily an attribute of the assembly as a body, not so much of each individual member: in order to ascertain whether the assembly is indeed representative, we must examine it as a whole.

Another, quite different sense of the term – which underlies DR – is similar to the one intended when we speak of a *diplomatic representative* of a country. Note that being *a representative* in this sense is an attribute of the individual member: an assembly is representative only inasmuch as it is an assembly *of* representatives. The relationship between a representative in this sense and what s/he represents is like that between agent and principal.[1] Here every member of the assembly is personally elected for representing a particular constituency, which is usually defined geographically. Accordingly, there are a large number of constituencies, each of which elects a single representative or a small number – at most a handful – of representatives. Naturally, such a constituency may be, and normally is in fact, quite heterogeneous: its voters may differ considerably from one another in their interests, preferences, tastes and opinions. The presumed aim of a DR procedure is to elect a candidate (or a small set of candidates) that is in some sense "best" or "most suitable" for representing this heterogeneous constituency.

Although the distinction between these two senses of representation is quite fundamental, I have not seen it clearly and explicitly articulated in the social-choice literature. Perhaps this is due to my ignorance; and I stand to be corrected. At any rate, the distinction is very often ignored and the two senses are conflated.[2]

However, there are some well-known "compromise" systems that blend both types of representation. One such compromise is the so-called Additional Member system used, for example, in elections to the German Bundestag and the Scottish Assembly, whereby some members of the legislature are elected by a DR method, and the rest are elected by a PR method, designed to achieve or approach overall proportionality. A second, quite different compromise consists in dividing the electorate at large into fairly large geographically-based constituencies, within each of which elections are held using PR. This compromise is used, for example, in the UK in elections to the European Parliament; it has occasionally been used in elections to the French National Assembly.

The second main dichotomy is:

ii. Deterministic Processing (DP) vs. Lottery Processing (LP)

Here "processing" refers to the way the votes cast are processed to produce the outcome of the election.

I consider a voting procedure to be DP even if it does use lottery, provided this use is confined to resolving ties, whose occurrence is extremely unlikely. Thus an LP procedure is one that relies on lottery in a major way.

Whether use of LP is acceptable is clearly a political matter (in the broad sense) and depends on social norms and on the purpose for which the election is conducted. According to current social norms, it is considered in many countries desirable to select a trial jury by lot out of a large pool of admissible candidates. But electing a legislature by LP would probably be regarded by most people as unacceptable. Electing an individual by lottery for a position such as chairman of a meeting is quite common, but electing a holder of high political office by LP would be unacceptable – although it was normal practice under Athenian democracy.

[1] I owe this observation to Iain McLean (oral communication).

[2] This goes back to John Stuart Mill. In (Mill 1861, Chap. 7) he clearly advocates PR; but then seems to take it for granted that electing a legislature must use some form of DR.

1.3 PR Procedures

Let us now see what social choice has to offer if we opt for PR.

1.3.1 PR and DP

The only electoral procedure that really implements this combination (as far as possible) is the list system. To be precise, there are two variants of this system. In the *closed* list variant, the seats are allocated to a party's candidates in the order in which they appear on its list. In the *open* list variant, voters may indicate preference for a particular candidate in the list of their choice, and seats are allocated accordingly.[3]

The STV procedure is often claimed by politicians and journalists to be a PR system. But social-choice theorists know very well that this claim is incorrect. This is not only easy to prove in theory (for example, by observing that STV is not monotonic), but can also be seen in practice by examining the results of elections conducted under STV.[4] In fact, STV is a DR system that is ingeniously designed to produce less disproportionate outcomes than the extremely pathological plurality procedure.[5] However, the approximate degree of proportionality it produces is quite erratic. In particular, STV is biased against small and radical parties.

1.3.2 PR and LP

There is one – and as far as I know only one – procedure that implements this combination of political alternatives. It is the lottery voting procedure (LVP) proposed by the American jurist and political scientist Akhil Reed Amar (1984).[6] This is how it works. The entire electorate is divided into constituencies of roughly equal size. Elections are conducted in each constituency as under the plurality system, but with the following crucial difference. Whereas under the plurality system the winner is the candidate with the greatest number of votes, under LVP a weighted lottery is conducted, with candidates' weights proportional to the respective numbers of votes cast for them.

---

[3]Note however that the aggregation of all the individual preference orderings into a single overall ordering is problematic, due to Arrow's Theorem.

[4]For example, see http://en.wikipedia.org/wiki/Dail for results of elections to the Irish Dáil.

[5]STV is therefore advocated by people who can see the virtues of PR, but are wedded to DR either on political grounds or because they simply take it for granted. Among the latter was J S Mill (1861, Chap. 7); cf. footnote 2.

[6]It is also known, somewhat misleadingly, as the "random dictator" procedure.

Using Kolmogorov's Strong Law of Large Numbers, it is not difficult to show that the overall outcome under LVP is almost certain to be extremely close to proportionality. More precisely, if the number of constituencies is fairly large (say 100 or more) then the total number of seats won by candidates representing a given party or informal trend of opinion is very highly likely to be closely proportional to the total number of votes cast at large for such candidates.[7]

This procedure shares some of the attractive political properties of both deterministic PR and DR.[8] In fact, superficially, LVP looks like a DR procedure, but this is not really so. The winner of the election in a given constituency is not supposed to be its "best" or "most suitable" representative. In fact, her or his primary allegiance is not to the constituency but to the party or trend of opinion for which s/he stands. The constituency serves primarily as a subspace of the sampling space of the electorate at large. Indeed, in principle there is no need for the constituencies to be determined geographically; they can be quite arbitrary sections, roughly equal in size, of the electorate at large. (However, this would destroy some important political advantages of LVP.)

1.4 DR Procedures

Here things will get somewhat messy. But before that, I would like to introduce a subsidiary dichotomy, singling out a particular political principle:

iii. *Majority Rule (MR) vs. Aggregation Rules (AR)*

MR systems are based on the political view that regards majority rule as a paramount principle. The meaning of MR is clear enough when there are just two candidates. The straightforward natural generalization of this is *Condorcet's Principle*:

> If candidate x dominates candidate y (i.e., x is preferred to y by a majority of the voters), then x is socially preferable to y.

Note that in order to apply this rule, it is not necessary in principle for a voter to order the candidates in a (transitive) preference ordering. Only pairwise comparisons are needed. And a voter's comparisons may contain cycles. (It is sometimes claimed that cyclic preferences are irrational. I don't find this claim persuasive. Besides, is it politically acceptable to disqualify or ignore voters whose voting behaviour is allegedly irrational? That would be extremely dangerous....)

The alternative to MR is a mixed bag of various rules for aggregating degrees of approval (or preference) that are assigned by the voters to each candidate. These "degrees" may be ordinal, cardinal or of an intermediate kind (as in grading by

---

[7]For a proof, see Machover (2009, Sect. 6.2).

[8]For a discussion of the technical properties and political advantages of this procedure, see Amar (1984) and Machover (2009, Sect. 4.4).

marks that are not merely ordinal, but are not reducible to cardinal numbers). But in any case they require or imply at least a transitive weak ordering of the candidates by each voter.[9]

Aggregation systems pose two distinct problems. First, can degrees of approval (or preference) assigned by different voters be meaningfully aggregated? This problem is familiar in relation to utilities; but it is more general.

Second, aggregation involves loss of information: in general, the voting profile contains much more information than the outcome of the election. Arrow's theorem is a particular manifestation of this: it applies only to procedures that try to aggregate ordinal preferences (preference orderings) into a single "social" ordering. However, the problem is more general.

This loss of information can be regarded as the source of all the paradoxes and pathologies that afflict voting procedures. I will not discuss these matters any further, but refer you to Dan Felsenthal's paper (Chap. 3 of this volume).

Let me just add that as far as I know the problems posed by the paradoxes and pathologies of AR procedures arise whether we insist on deterministic processing (that is, the combination AR and DP) or allow lottery processing (that is, AR and LP).

The situation regarding MR is different – which is the reason I have singled it out in the subsidiary dichotomy (iii).

The combination MR and DP needs to be supplemented by some method of aggregating preferences, in case a Condorcet winner does not exist. Thus we are back to the problems raised in the case of the combination AR and DP.

This leaves the final combination:

1.4.1 MR and LP

For this combination, if just one candidate needs to be elected, social-choice theory provides an elegant unique optimal solution, and does not need to be supplemented by any other political principle.

This solution is provided by a beautiful theorem, proved in 1991 by Laffond et al. (1993), and independently (using a quite different method) by Fisher and Ryan (1992).

Let me outline this theorem. Consider the following *tournament game*: a two-person game in which each of two players, I and U, must nominate (independently of each other) one member of the set X of candidates standing for election. Suppose I nominates x and U nominates y. If $x \succ y$ (i.e., if x dominates y), then U pays I \$1; if $y \succ x$, then I pays U \$1; and if $x = y$ no payment is made.

[9]A very rudimentary marking is used in the plurality and approval voting procedures, where the only admissible marks are 0 and 1.

The theorem states that in this game there is a unique optimal mixed strategy. In other words, there are unique probabilities $\{p_x : x \in X\}$, with $\sum_{x \in X} p_x = 1$, such that if a player uses a lottery with these probabilities to nominate a candidate, then s/he maximizes her/his expected payoff. (By symmetry, this maximal payoff is of course 0.) Clearly, the support of this probability distribution (the set $\{x \in X : p_x > 0\}$) is a subset of the top cycle of candidates. In particular, if there is a Condorcet winner, the optimal strategy is pure, and assigns that candidate probability 1. Rather surprisingly, the support always consists of an odd number of candidates.

As pointed out by Felsenthal and Machover (1992), this provides an electoral procedure based purely on MR and LP: conduct a weighted lottery, in which each candidate x is assigned weight p_x.

1.5 Conclusion

Much of social-choice literature is concerned with the perplexing problematics of selecting an acceptable election procedure out of a large number of competing ones. What I have tried to show is that if one subscribes to certain simple "grand" *political* options, or a combination of these, then social choice can provide a single optimal procedure.

Acknowledgements Valuable comments from Dan S Felsenthal and Maurice Salles are gratefully acknowledged.

References

Amar, A. R. (1984). Choosing representatives by lottery voting. *Yale Law Journal, 93*, 1283–1308.

Felsenthal, D. S. & Machover, M. (1992). After two centuries, should Condorcet's voting procedure be implemented? *Behavioral Science, 37*, 250–274.

Fisher, D. C. & Ryan, J. (1992). Optimal strategies for a generalized 'Scissors, Paper and Stone' game. *American Mathematical Monthly, 99*, 935–942.

Laffond, G., Laslier, J., & Le Breton, M. (1993). The bipartisan set of a tournament. *Games and Economic Behavior, 5*, 182–201.

Machover, M. (2009). *Collective decision-making and supervision in a communist society.* Mimeograph, downloadable from http://tinyurl.com/ydfbs5z.

Mill, J. S. (1861). *Considerations on representative government.* Downloadable from http://www.gutenberg.org/ebooks/5669

Chapter 2
Some Informal Remarks on Devising a "Fair" Decision-Making Rule for Representative Assemblies

Dan S. Felsenthal

In the previous chapter Moshé Machover distinguished between two kinds of representative assemblies, each of which can be elected by using either a deterministic or a probabilistic voting procedure:

1. A *PR assembly* which is a microcosm of the entire electorate and where every member represents an ideologically homogeneous but geographically dispersed constituency. Machover argues – and I agree – that the only way to obtain such an assembly by using a deterministic voting procedure is to use the closed list system procedure.
2. A *DR assembly* where every member represents (is an agent of) an ideologically diverse but geographically contiguous constituency, of which s/he is in some sense the "best" representative.

However, Machover did not address an important related issue, i.e., how to devise a "fair" decision-making rule to be used by the assembly. I would like to dwell on this issue.

As far as I know, it is a universal practice in democracies that the decision rule used by both types of assembly, as well as the decision rule used in popular referenda, is majoritarian.[1] The fact that democracies employ decision-making rules within representative assemblies that are based solely on the majority principle could lead to the conclusion that a decision based on this principle can always

[1] A majoritarian decision rule requires that in order to change the status quo slightly more than half the voters and fewer than all the voters must support this change. Note that since a requirement of unanimity enables every voter to veto a change of the status quo, unanimity is not considered a majoritarian voting rule.

D.S. Felsenthal (✉)
School of Political Sciences, University of Haifa, Israel

Centre for Philosophy of Natural and Social Science, London School of Economics, UK
e-mail: dfelsenthal@poli.haifa.ac.il

D.S. Felsenthal and M. Machover, *Electoral Systems*, Studies in Choice and Welfare,
DOI 10.1007/978-3-642-20441-8_2, © Springer-Verlag Berlin Heidelberg 2012

be reconciled with democratic principles. In fact, "there is nothing inherent in democracy that requires majority rule." (Guinier 1994, p. 17). Moreover, the majority rule principle implies that the minority of representatives in an assembly are unable to affect reality – even if their number is exactly proportional to the proportion of the electorate who supported them in the election.

So the universal use of the majoritarian decision rule within representative assemblies, as well as in popular referenda, causes the minority to become totally impotent in shaping public policies. Such impotence is especially serious if the minority group is relatively large and it is a permanent one, i.e., it always consists of the same (type of) voters who may belong to the same ethnic group, or the same ideological party, or the same geographic region.

In contemplating which decision rule(s) ought to be used by a representative assembly when it engages in tasks involving the selection of one out of several possible alternatives, let us first consider three alternative political/philosophical principles or goals:

1. *Majority rule*: To guarantee that the alternative preferred by the majority of voters (or representatives) will be selected.
2. *Equiprobability of success*: To let every voter (or representative) have the same probability that his/her most preferred alternative is the one selected.
3. *Equal opportunity to avoid the worst*: To provide every voter (or representative, or alliance) with an ability to prevent that his/her/its least preferred alternative is the one selected.

To achieve the first goal one must use a deterministic voting procedure and select the alternative supported by the majority of voters or representatives.

To achieve the second goal one must use a probabilistic voting procedure which assigns to every voter or representative the same chance of being selected – and the selected voter/representative will then state which alternative s/he prefers.[2]

To achieve the third goal one must enable every voter, or group of voters of some minimal size, to veto one of the alternatives under consideration, thus guaranteeing that every voter or group of voters has some minimal effect on the selected outcome.

Achieving the first goal can never make, by definition, the majority of voters very miserable, but it may make the minority of voters very miserable. On the other hand, achieving the second goal may make either the majority or the minority very miserable if a member of the other group is selected to choose the alternative to be implemented. (Of course there is a higher probability that the selected person will belong to the majority than to the minority group.)

It seems natural that we prefer the possibility that the minority may be miserable over the possibility that the majority may be miserable – and hence we prefer to realize the first goal (principle) over the realization of the second goal. Moreover, since the realization of the second goal involves the employment of a probabilistic

[2]In an assembly made up of party-blocs, the achievement of this goal would require the probability of success of each bloc to be proportional to its weight.

voting procedure, this procedure may cause additional problems such as instability and/or inconsistency in public policies and decision-making.[3]

The realization of the third goal has two advantages over the realization of the other two goals: first, it leads to the selection of an alternative that is stable regardless of whether the social preference ordering among the available alternatives contains cycles.[4] Second, if voters behave rationally, then the selected alternative is not only Pareto-optimal – that is, no other alternative is preferred by all the voters over some other alternative – but it also does not constitute any voter's least-preferred alternative.

It is quite easy to realize the third goal when the number of voters/representatives (n) is smaller than the number of policy alternatives (m) of which one or more must be selected. In this case every voter/representative, in his/her turn, can veto one or more alternatives (depending on his/her weight) – and the alternative(s) that was/were not vetoed is/are selected.[5] However, implementing this goal is difficult when the number of representatives in the assembly is larger than the number of policy alternatives. Implementing this goal in this case – which is common in actual representative assemblies – implies that more than one representative is needed to veto any given policy alternative, and the formation of the needed alliance(s) may become complicated. Moreover, a satisfactory theory as to how to analyze such cases is still lacking.

So how, if at all, is it possible to adopt one or more decision rules in representative assemblies which will provide the representatives belonging to the minority group with both a priori as well as actual voting power proportional to their weight?

I think the answer to this question is twofold:

(a) Act according to a fourth political-philosophical principle or goal, i.e.,

   4. *Proportionality of a priori voting power to weight*: Let every voter or representative have the same probability of being critical in a division. Given that the number of seats controlled by the various parties or geographical units in a representative assembly is proportional to the number of relevant

[3] Instability may be caused by the losers' demand following any given division to conduct another round of voting, whereas inconsistency may be a result of adopting contradictory policies that were selected interchangeably by voters belonging to the majority and minority groups.

[4] A stable alternative is an alternative that cannot profitably be objected to by any voter or alliance of voters. Note that, in contrast, no alternative is stable if one uses a majoritarian decision rule and the social preference ordering among the available alternatives contains a top cycle.

[5] This procedure is known as sequential voting by veto (SVV). It was proposed originally by Mueller (1978) who presented an algorithm for determining the winning alternative under SVV, given the order in which the voters/representatives cast their vetoes. Moulin (1981, 1983, pp. 138–140) extended Mueller's idea to any situation in which n voters have to select one out of $n + 1$ alternatives and they have complete information on all other voters' preference orderings among the alternatives. Felsenthal and Machover (1992) generalized the Mueller–Moulin result to a situation in which n voters/representatives must select s out of m alternatives ($s > 0; m > n \geq 2$). For laboratory experiments with small groups operating under SVV see Yuval (2002) and Yuval and Herne (2005).

voters in the electorate,[6] one looks for a decision rule to be used by the assembly such that the *a priori relative voting power* (as measured by Banzhaf's index of relative voting power) will be as close as possible to each representative's relative weight, i.e., the proportion of voters s/he represents.[7]

However, the realization of this goal too is not problem-free. To understand just one of the problems associated with it, consider the following simple example.

Suppose a 99-seat legislative assembly with three parties, each controlling 33 seats because each received an equal proportion (1/3) of the votes in an election. The *relative* a priori voting power of each party will be 1/3 – which is proportional to each party's weight – regardless of whether the quota (q) needed to pass resolutions is $34 \leq q \leq 66$ (simple or qualified majority) or $67 \leq q \leq 99$ (unanimity). However, the *absolute* a priori voting power of every party (as measured by the Penrose measure) would be 1/2 if only simple or qualified majority would be required to pass a resolution, but only 1/4 if unanimity is needed to pass a resolution. So which quota would be fairer in this case? It would seem that here unanimity would be fairer than simple/qualified majority, because under simple/qualified majority any single party may become powerless if the other two parties form a relatively long-term binding alliance. Of course the price to be paid for granting in this case veto power to each of the three parties, is not only significant loss of (a priori) absolute voting power by each of the parties, but also the possibility of total paralysis of the legislature inasmuch as the parties are unable to agree on the passage of any bill.

(b) So perhaps some milder form of de facto (proportional) voting power would be preferable than awarding each of the three parties in the above example veto power regarding all proposed bills. It should be possible to institute arrangements, at least with respect to certain kinds of decisions, e.g., budgetary decisions, or decisions regarding certain regions or policy areas, which will enhance the a posteriori (actual) voting power of representatives belonging to the minorities in legislatures. For example, if the Red and Blue parties control 40% and 60% of the seats, respectively, in a representative assembly, then one can institute an arrangement where the Blue party would be given the prerogative of determining the total size of the annual budget, as well as dividing it into parts – one containing 60% of the total planned expenditure and the other containing 40% of the total expenditure – and let the Red party have the sole prerogative to decide how the 40% part of the budget would be allocated.

Of course such power-sharing arrangements – in decision-making bodies in general and legislatures in particular – are not problem-free, and their details are crucial. Yet it seems to me that the discussion and development of such proportional

[6] The relevant electorate is either the number of voters who supported each party represented in the assembly, or the number of voters belonging to each geographical unit represented in the assembly, or the size of financial contribution of each member-country to the common fund, e.g., the financial contribution of each member of the International Monetary Fund to its fund.

[7] For the various measurements of a priori voting power see Felsenthal and Machover (1998).

power-sharing arrangements should constitute a major new area for social choice theory to be engaged in.

So far the relevant disciplines (social choice, political science, economics, mathematics, law, philosophy) focused mainly on how to elect a representative assembly, and to a lesser degree on how to measure the a priori voting power of representatives in an assembly. In my opinion the time has come for them to shift their focus to how to devise a fair and practical decision rule(s) for a representative assembly so that *all* its members will have *actual* voting power which is as close as possible to their relative weight.

Acknowledgements I wish to thank Moshé Machover and Maurice Salles for their useful comments on these remarks.

References

Felsenthal, D.S., & Machover, M. (1992). Sequential voting by veto: extending the applicability of the Mueller–Moulin algorithm. *Theory and Decision, 33*, 223–240.

Felsenthal, D.S., & Machover, M. (1998). *The measurement of voting power: theory and practice, problems and paradoxes*. Cheltenham: Edward Elgar.

Guinier, L. (1994). *The tyranny of the majority: fundamental fairness in representative democracy*. New York: Free Press.

Moulin, H. (1981). Prudence versus sophistication in voting strategy. *Journal of Economic Theory, 24*, 398–412.

Moulin, H. (1983). *The strategy of social choice. North Holland: Amsterdam.*

Mueller, D.C. (1978). Voting by veto. *Journal of Public Economics, 10*, 55–75.

Yuval, F. (2002). Sophisticated voting under the sequential voting by veto. *Theory and Decision, 53*, 343–369.

Yuval, F., & Herne, K. (2005). Sophisticated behavior under majoritarian and non-majoritarian voting procedures. *Political Behavior, 27*, 217–237.

Part II
Paradoxes Afflicting Electoral Procedures and Their Expected Probability

Chapter 3
Review of Paradoxes Afflicting Procedures for Electing a Single Candidate

Dan S. Felsenthal

3.1 Introduction

Three factors motivated me to write this chapter:

- The recent passage (25 February 2010) by the British House of Commons of the *Constitutional Reform and Governance Bill*, clause #29 of which states that a referendum will be held by 31 October 2011 on changing the current *single member plurality* (aka *first-past-the-post*, briefly FPTP) electoral procedure for electing the British House of Commons to the (highly paradoxical) *alternative vote* (AV) procedure (aka *Instant Runoff*).[1] Similar calls for adopting the alternative vote procedure are voiced also in the US.
- My assessment that both the UK and the US will continue to elect their legislatures from single-member constituencies, but that there exist, from the point of view of social-choice theory, considerably more desirable voting procedures for electing a single candidate than the FPTP and AV procedures.
- A recent report by Hix et al. (2010) – commissioned by the British Academy and entitled *Choosing an Electoral System* – that makes no mention of standard social-choice criteria for assessing electoral procedures designed to elect one out of two or more candidates.

[1]Following the general elections held in the UK on 6 May 2010, a coalition government has been formed between the Conservative and Liberal-Democratic parties in which the two parties committed to hold a referendum on the possible change of the election procedure to the House of Commons from FPTP to AV. In the referendum held on 5 May 2011 it was decided to keep the FPTP procedure.

D.S. Felsenthal (✉)
School of Political Sciences, University of Haifa, Israel

Centre for Philosophy of Natural and Social Science, London School of Economics, UK
e-mail: dfelsenthal@poli.haifa.ac.il

D.S. Felsenthal and M. Machover, *Electoral Systems*, Studies in Choice and Welfare,
DOI 10.1007/978-3-642-20441-8_3,

I therefore thought it would be well to supplement that report by reminding social choice theorists, political scientists, as well as commentators, policymakers and interested laymen – especially in the UK and the US – of the main social-choice properties by which voting procedures for the election of one out of two or more candidates ought to be assessed, and to list and exemplify the paradoxes afflicting these voting procedures.

Thus this paper should be regarded as an updated review by which to assess from a social-choice perspective the main properties of various known voting procedures for the election of a single candidate.

Of the 18 (deterministic) voting procedures analyzed in this paper, the Condorcet-consistent procedures proposed by Copeland (1951) and by Kemeny (1969) seem to me to be the most desirable from a social-choice perspective for electing one out of several candidates.

The paper is organized as follows: In Sect. 3.2 I survey 15 paradoxes, several of which may afflict any of the 18 voting procedures that are described in Sect. 3.3. Section 3.4 summarizes and presents additional *technical-administrative* criteria which should be used in assessing the relative desirability of a voting procedure. In the detailed appendix in Sect. 3.5 I exemplify most of the paradoxes to which each of the surveyed election procedures is susceptible.

3.2 Voting Paradoxes

I define a "*voting paradox*" as an undesirable outcome that a voting procedure may produce and which may be regarded at first glance, at least by some people, as surprising or as counter-intuitive.

I distinguish between two types of voting paradoxes associated with a given voting procedure:

1. "*Simple*" or "*Straightforward*" paradoxes: These are paradoxes where the relevant data leads to a "surprising" and arguably undesirable outcome. (The relevant data include, *inter alia*, the number of voters, the number of candidates, the number of candidates that must be elected, the preference ordering of every voter among the competing candidates, the amount of information voters have regarding all other voters' preference orderings, the order in which voters cast their votes if it is not simultaneous, the order in which candidates are voted upon if candidates are not voted upon simultaneously, whether voting is open or secret, the manner in which ties are to be broken).
2. "*Conditional" paradoxes*: These are paradoxes where changing one relevant datum while holding constant all other relevant data leads to a "surprising" and arguably undesirable outcome.

An array of paradoxes of one or both types are described and analyzed by McGarvey (1953), Riker (1958), Smith (1973), Fishburn (1974, 1977, 1981, 1982), Young (1974), Niemi and Riker (1976), Doron and Kronick (1977), Doron (1979),

Richelson (1979), Gehrlein (1983), Fishburn and Brams (1983), Saari (1984, 1987, 1989, 1994, 2000, 2008), Niou (1987), Moulin (1988a), Merlin and Saari (1997), Brams, Kilgour and Zwicker (1998), Scarsini (1998), Nurmi (1998a, 1998b, 1999, 2004, 2007), Lepelley and Merlin (2001), Merlin et al. (2002), Merlin and Valognes (2004), Tideman (1987, 2006), Gehrlein and Lepelley (2011), among others.

3.2.1 Simple Paradoxes

The six best-known "simple" paradoxes that may afflict voting procedures designed to elect one out of two or more candidates are the following:

3.2.1.1 The Condorcet (or Voting, or Cyclical Majorities) Paradox (Condorcet 1785; Black 1958)

Given that the preference ordering of every voter among the competing candidates is transitive, the (amalgamated) preference ordering of the majority of voters among the competing candidates may nevertheless be intransitive. A necessary condition for this to occur is that the various majorities are composed of different persons and there exist at least three candidates. Although we do not demonstrate this paradox in the Appendix, it may occur under all ranked voting procedures, as well as under the successive elimination procedure.

3.2.1.2 The Condorcet Winner Paradox (Condorcet 1785; Black 1958)

A candidate x is not elected despite the fact that it constitutes a "Condorcet Winner", i.e., despite the fact that x is preferred by a majority of the voters over each of the other competing alternatives.[2]

3.2.1.3 The Absolute Majority Paradox

This is a special case of the Condorcet winner paradox. A candidate x may not be elected despite the fact that it is the only candidate ranked first by an absolute majority of the voters.

---

[2]Fishburn (1974, p. 544) constructs an example with 101 voters and nine candidates two of whom are candidates a and w, such that w beats each of the other eight candidates by a (slim) majority of 51 to 50 (and hence is a Condorcet winner), whereas a beats each of the other seven candidates by a considerably larger majority. Fishburn states that "examples like this suggest that some cases which have a simple-majority [Condorcet] winner do *not* represent the most satisfactory social choice." We disagree with this statement and hold that a Condorcet winner, if one exists, ought *always* to be elected.

3.2.1.4 The Condorcet Loser or Borda Paradox (Borda 1784; Black 1958)

A candidate x is elected despite the fact that it constitutes a "Condorcet Loser" i.e., despite the fact that a majority of voters prefer each of the remaining candidates to x. This paradox is a special case of the violation of Smith's (1973) Condorcet principle. According to this principle, if it is possible to partition the set of candidates into two disjoint subsets, A and B, such that each candidate in A is preferred by a majority of the voters over each candidate in B, then no candidate in B ought to be elected unless all candidates in A are elected.

3.2.1.5 The Absolute Loser Paradox

This is a special case of the Condorcet loser paradox. A candidate x may be elected despite the fact that it is ranked last by a majority of voters.

3.2.1.6 The Pareto (or Dominated Candidate) Paradox (Fishburn 1974)

A candidate x may be elected while candidate y may not be elected despite the fact that *all* voters prefer candidate y to x.

3.2.2 *Conditional Paradoxes*

The nine best-known "conditional" paradoxes that may afflict voting procedures for electing a single candidate are the following:

3.2.2.1 Additional Support (or Lack of Monotonicity or Negative Responsiveness) Paradox (Smith 1973; Fishburn 1974a, Fishburn and Brams 1983)

If candidate x is elected under a given distribution of voters' preferences among the competing candidates, it is possible that, *ceteris paribus*, x may not be elected if some voter(s) *increase(s) his (their) support for* x by moving x to a higher position in his (their) preference ordering. Alternatively, if candidate x is not elected under a given distribution of voters' preferences among the competing candidates, it is possible that, *ceteris paribus*, x will be elected if some voter(s) *decrease(s)* his (their) support for x by moving x to a lower position in his (their) preference ordering.[3]

[3]Another version of the non-monotonicity paradox (which is not demonstrated in the Appendix) is a situation where x is elected in a given electorate but may not be elected if, *ceteris paribus*, additional voters join the electorate who rank x at the top of their preference ordering, or,

3.2.2.2 Reinforcement (or Inconsistency or Multiple Districts) Paradox (Young 1974)

If x is elected in each of several disjoint electorates, it is possible that, *ceteris paribus*, x will not be elected if all electorates are combined into a single electorate.

3.2.2.3 Truncation Paradox (Brams 1982; Fishburn and Brams 1983)

A voter may obtain a more preferable outcome if, *ceteris paribus*, he lists in his ballot only part of his (sincere) preference ordering among some of the competing candidates than listing his entire preference ordering among all the competing candidates.

3.2.2.4 No-Show paradox (Fishburn and Brams 1983; Ray 1986; Moulin 1988b; Holzman 1988/89; Pérez 1995)

This is an extreme version of the truncation paradox. A voter may obtain a more preferable outcome if he decides not to participate in an election than, *ceteris paribus*, if he decides to participate in the election and vote sincerely for his top preference(s).

3.2.2.5 Twin Paradox (Moulin 1988b)

This is a special version of the no-show paradox. Two voters having the same preference ordering may obtain a preferable outcome if, *ceteris paribus*, one of them decides not to participate in the election while the other votes sincerely.

3.2.2.6 Violation of the Subset Choice Condition (SCC) (Fishburn 1974b,c, 1977)

SCC requires that when there are at least three candidates and candidate x is the unique winner, then x must not become a loser whenever any of the original losers is removed and all other things remain the same. All the voting procedures discussed in this paper except the range voting (RV) and majority judgment (MJ) procedures violate SCC.[4] In the context of individual choice theory SCC is known as Chernoff's

alternatively, a situation where x is not elected in a given electorate but may be elected if, *ceteris paribus*, additional voters join the electorate who rank x at the bottom of their preference ordering.

[4]The RV and MJ procedures satisfy SCC because these procedures do not aggregate the individual voters' preference orderings into a social preference ordering in order to determine the winner. Under these procedures every candidate is ranked (on a cardinal or ordinal scale) by every voter,

condition (1954, p. 429, postulate 4) which states that if an alternative x chosen from a set T is an element of a subset S of T, then x must be chosen also from S.

3.2.2.7 Preference Inversion Paradox

If the individual preferences of each voter are inverted it is possible that, *ceteris paribus*, the (unique) original winner will still win.

3.2.2.8 Lack of Path Independence Paradox (Farquharson 1969; Plott 1973)

If the voting on the competing candidates is conducted sequentially rather than simultaneously, it is possible that candidate x will be elected under a particular sequence but not, *ceteris paribus*, under an alternative sequence.

3.2.2.9 Strategic Voting Paradox (Gibbard 1973; Satterthwaite 1975)

There are conditions under which a voter with full knowledge of how the other voters are to vote and the decision rule being used, would have an incentive to vote in a manner that does not reflect his true preferences among the competing alternatives. All known non-dictatorial voting procedures suffer from this paradox; it is not demonstrated in the Appendix.

3.3 Voting Procedures for Electing One out of Two or More Candidates

3.3.1 Non-ranked Voting Procedures

There are four main voting procedures for electing a single candidate where voters do not have to rank-order the candidates:

---

and the winner is that candidate whose average (or median) rank is highest. Thus the elimination of any losing candidate cannot affect, *ceteris paribus*, the identity of the original winner.

It may perhaps be assumed that under Approval Voting a voter will never vote for an alternative in a subset which s/he did not "approve" in the superset, and hence that Approval Voting, too, satisfies SCC. This assumption is debatable. It can easily be shown – as in Example 3.5.1.1. below – that when there are three alternatives among whom a voter has a linear preference ordering, it would always be rational for a voter under Approval Voting to vote for his/her second preference if his/her top preference is no longer available – even if originally s/he "approved" only of his/her top preference. By doing so s/he has nothing to lose but may obtain a better outcome than by abstaining – regardless of how all other voters are going to vote. Hence in our view Approval Voting may violate SCC.

3.3.1.1 Plurality (or First Past the Post, Briefly FPTP) Voting Procedure

This is the most common procedure for electing a single candidate, and is used, *inter alia*, for electing the members of the House of Commons in the UK and the members of the House of Representatives in the US. Under this procedure every voter casts one vote for a single candidate and the candidate obtaining the largest number of votes is elected.

3.3.1.2 Plurality with Runoff Voting Procedure

Under the usual version of this procedure up to two voting rounds are conducted. In the first round each voter casts one vote for a single candidate. In order to win in the first round a candidate must obtain either a special plurality (usually at least 40% of the votes) or an absolute majority of the votes. If no candidate is declared the winner in the first round then a second round is conducted. In this round only the two candidates who obtained the highest number of votes in the first round participate, and the one who obtains the majority of votes wins. This too is a very common procedure for electing a single candidate and is used, *inter alia*, for electing the President of France.

3.3.1.3 Approval Voting (Brams and Fishburn 1978, 1983)

Under this procedure every voter has a number of votes which is equal to the number of competing candidates, and every voter can cast one vote or no vote for every candidate. The candidate obtaining the largest number of votes is elected. So far this procedure has not been used in any public elections but is already used by several professional associations and universities in electing their officers.

3.3.1.4 Successive Elimination (Farquharson 1969)

This procedure is common in parliaments when voting on alternative versions of bills. According to this procedure voting is conducted in a series of rounds. In each round two alternatives compete; the one obtaining fewer votes is eliminated and the other competes in the next round against one of the alternatives which has not yet been eliminated. The alternative winning in the last round is the ultimate winner.

3.3.2 Ranked Voting Procedures That Are Not Condorcet-Consistent

Six ranked procedures under which every voter must rank-order all competing candidates – but which do not ensure the election of a Condorcet winner when one exists – have been proposed, as far as I know, during the last 250 years. These procedures are described below. Only one of these procedures (alternative vote) is used currently in public elections.

3.3.2.1 Borda's Count (Borda 1784; Black 1958)

This voting procedure was proposed by Jean Charles de Borda in a paper he delivered in 1770 before the French Royal Academy of Sciences entitled 'Memorandum on election by ballot' ('Mémoire sur les élections au scrutin'). According to Borda's procedure each candidate, x, is given a score equal to the number of pairs (V, y) where V is a voter and y is a candidate such that V prefers x to y, and the candidate with the largest score is elected. Equivalently, each candidate x gets no points for each voter who ranks x last in his preference ordering, one point for each voter who ranks x second-to-last in his preference order, and so on, and $m - 1$ points for each voter who ranks x first in his preference order (where m is the number of candidates). Thus if all n voters have linear preference orderings among the m candidates then the total number of points obtained by all candidates is equal to the number of voters multiplied by the number of paired comparisons, i.e., to $nm(m - 1)/2$.

3.3.2.2 Alternative Vote (AV); (aka Instant Runoff Voting)

This is the version of the *single transferable vote* (STV) procedure (independently proposed by Carl George Andrae in Denmark in 1855 and by Thomas Hare in England in 1857) for electing a single candidate. It works as follows. In the first step one verifies whether there exists a candidate who is ranked first by an absolute majority of the voters. If such a candidate exists s/he is declared the winner. If no such candidate exists then, in the second step, the candidate who is ranked first by the smallest number of voters is deleted from all ballots and thereafter one again verifies whether there is now a candidate who is ranked first by an absolute majority of the voters. The elimination process continues in this way until a candidate who is ranked first by an absolute majority of the voters is found. The Alternative Vote procedure is used in electing the president of the Republic of Ireland, the Australian House of Representatives, as well as the mayors in some municipal elections in the US.

3.3.2.3 Coombs' Method (Coombs 1964, pp. 397–399; Straffin 1980; Coombs et al. 1984)

This procedure was proposed by the psychologist Clyde H. Coombs in 1964. It is similar to Alternative Vote except that the elimination in a given round under Coombs' method involves the candidate who is ranked last by the largest number of voters (instead of the candidate who is ranked first by the smallest number of voters under alternative vote).

3.3.2.4 Bucklin's Method (Hoag and Hallett 1926, pp. 485–491; Tideman 2006, p. 203)

This voting system can be used for single-member and multi-member districts. It is named after James W. Bucklin of Grand Junction, Colorado, who first promoted it in 1909. In 1913 the US Congress prescribed (in the Federal Reserve Act of 1913, Sect. 4) that this method be used for electing district directors of each Federal Reserve Bank.

Under Bucklin's method voters rank-order the competing candidates. The vote count starts like in the Alternative Vote method. If there exists a candidate who is ranked first by an absolute majority of the voters s/he is elected. Otherwise the number of voters who ranked every candidate in second place are added to the number of voters who ranked him/her first, and if now there exists a candidate supported by a majority of voters s/he is elected. If not, the counting process continues in this way by adding for each candidate his/her third, fourth, ... rankings, until a candidate is found who is supported by an absolute majority of the voters. If two or more candidates are found to be supported by a majority of voters in the same counting round then the one supported by the largest majority is elected.[5]

3.3.2.5 Range Voting (Smith 2000)

According to this procedure the suitability (or level of performance) of every candidate is assessed by every voter and is assigned a (cardinal) grade (chosen from a pre-specified range) reflecting the candidate's suitability or level of performance in the eyes of the voter. The candidate with the highest average grade is the winner. This procedure is currently championed by Warren D. Smith (see http://rangevoting.org) and used to elect the winner in various sport competitions.

---

[5]However, it is unclear how a tie between two candidates, say a and b, ought to be broken under Bucklin's procedure when both a and b are supported in the same counting round by the same number of voters and this number constitutes a majority of the voters. If one tries to break the tie between a and b in such an eventuality by performing the next counting round in which all other candidates are also allowed to participate, then it is possible that the number of (cumulated) votes of another candidate, c, will exceed that of a and b.

To see this, consider the following simple example. Suppose there are 18 voters who must elect one candidate under Bucklin's procedure and whose preference orderings among four candidates, a, b, c, d are as follows: seven voters with preference ordering $a \succ b \succ c \succ d$, eight voters with preference ordering $b \succ a \succ c \succ d$, one voter with preference ordering $d \succ c \succ a \succ b$, and two voters with preference ordering $d \succ c \succ b \succ a$. None of the candidates constitutes the top preference of a majority of the voters. However, both a and b constitute the top or second preference by a majority of voters (15). If one tries to break the tie between a and b by performing the next (third) counting round in which c and d are also allowed to participate, then c will be elected (with 18 votes), but if only a and b are allowed to participate in this counting round then b will be elected (with 17 votes).

So which candidate ought to be elected in this example under Bucklin's procedure? As far as I know, Bucklin did not supply an answer to this question.

3.3.2.6 Majority Judgment (Balinski and Laraki 2007a,b, 2011)

According to this proposed procedure, the suitability (or level of performance) of every candidate is assessed by every voter and is assigned an ordinal grade (chosen from a pre-specified range) reflecting the candidate's suitability or level of performance in the eyes of the voter. The candidate with the highest median grade is the winner.

3.3.3 Ranked Voting Procedures that are Condorcet-consistent[6]

All the eight voting procedures described in this subsection require that voters rank-order all competing candidates. Under all these procedures a *Condorcet winner*, if one exists, is elected. The procedures differ from one another regarding which candidate gets elected when the social preference ordering contains a top cycle, i.e., when a Condorcet winner does not exist.

3.3.3.1 The Minimax Procedure

Condorcet specified that the Condorcet winner (whom he called 'the majority candidate') ought to be elected if one exists. However, according to Black (1958, pp. 174–175, 187) Condorcet did not specify clearly which candidate ought to be elected when the social preference ordering contains a top cycle. Black (1958, p. 175) suggests that "It would be most in accordance with the spirit of Condorcet's ... analysis ... to discard all candidates except those with the minimum number of majorities against them and then to deem the largest size of minority to be a majority, and so on, until one candidate had only actual or deemed majorities against each of the others." In other words, the procedure attributed by Black to Condorcet when cycles exist in the social preference ordering is a *minimax procedure*[7] since it chooses that candidate whose worst loss in the paired comparisons is the least bad. This procedure is also known in the literature as the *Simpson–Kramer rule* (see Simpson 1969; Kramer 1977).

---

[6] I list here only deterministic procedures. For a Condorcet-consistent probabilistic procedure see Felsenthal and Machover (1992). I also do not list here two Condorcet-consistent deterministic procedures proposed by Tideman (1987) and by Schultze (2003) because I do not consider satisfying (or violating) the independence-of-clones property, which is the main reason why these two procedures were proposed, to be associated with any voting paradox. (A phenomenon where candidate x is more likely to be elected when two clone candidates, y and y', exist, and where x is less likely to be elected when, *ceteris paribus*, one of the clone candidates withdraws, does not seem to me surprising or counter-intuitive).

[7] Young (1977, p. 349) prefers to call this procedure "The minimax function".

3.3.3.2 Dodgson's procedure (Black 1958, pp. 222–234; McLean and Urken, 1995, pp. 288–297)

This procedure is named after the Rev. Charles Lutwidge Dodgson, aka Lewis Carroll, who proposed it in 1876. It elects the Condorcet winner when one exists. If no Condorcet winner exists it elects that candidate who requires the fewest number of switches (i.e. inversions of two adjacent candidates) in the voters' preference orderings in order to make him the Condorcet winner.

3.3.3.3 Nanson's Method (Nanson 1883; McLean and Urken, 1995, ch. 14)

Nanson's method is a recursive elimination of Borda's method. In the first step one calculates for each candidate his Borda score. In the second step the candidates whose Borda score do not exceed the average Borda score of the candidates in the first step are eliminated from all ballots and a revised Borda score is computed for the uneliminated candidates. The elimination process is continued in this way until one candidate is left. If a (strong) Condorcet winner exists then Nanson's method elects him.[8]

3.3.3.4 Copeland's Method (Copeland 1951)

Every candidate x gets one point for every paired comparison with another candidate y in which an absolute majority of the voters prefer x to y, and half a point for every paired comparison in which the number of voters preferring x to y is equal to the number of voters preferring y to x. The candidate obtaining the largest sum of points is the winner.

3.3.3.5 Black's Method (Black 1958, p. 66)

According to this method one first performs all paired comparisons to verify whether a Condorcet winner exists. If such a winner exists then s/he is elected. Otherwise the winner according to Borda's count (see above) is elected.

[8] Although Nanson's procedure satisfies the strong Condorcet condition, i.e., it always elects a candidate who beats every other candidate in paired comparisons, this procedure may not satisfy the weak Condorcet condition which requires that if there exist(s) candidate(s) who is (are) unbeaten by any other candidate then this (these) candidate(s) – and only this (these) candidate(s) – ought to be elected. For an example of violation of the weak Condorcet condition by Nanson's procedure see Niou (1987).

3.3.3.6 Kemeny's Method (Kemeny 1959; Kemeny and Snell 1960; Young and Levenglick 1978; Young 1995)

Kemeny's method (aka *Kemeny–Young rule*) specifies that up to $m!$ possible social preference orderings should be examined (where m is the number of candidates) in order to determine which of these is the "most likely" true social preference ordering.[9] The selected "most likely" social preference ordering according to this method is the one where the number of pairs $(A, \ y)$, where A is a voter and y is a candidate such that A prefers x to y, and y is ranked below x in the social preference ordering is maximized. Given the voters' various preference orderings, Kemeny's procedure can also be viewed as finding the most likely (or the best predictor, or the best compromise) true social preference ordering, called the *median preference ordering*, i.e., that social preference ordering S that minimizes the sum, over all voters i, of the number of pairs of candidates that are ordered oppositely by S and by the ith voter.[10]

3.3.3.7 Schwartz's Method (Schwartz 1972; 1986)

Thomas Schwartz's method is based on the notion that a candidate x deserves to be listed ahead of another candidate y in the social preference ordering if and only if x beats or ties with some candidate that beats y, and x beats or ties with all candidates that y beats or ties with. The Schwartz set (from which the winner should be chosen) is the smallest set of candidates who are unbeatable by candidates outside the set. The Schwartz set is also called *GOCHA* (*Generalized Optimal Choice Axiom*).

3.3.3.8 Young's Method (Young 1977)

According to Fishburn's (1977, p. 473) informal description of Young's procedure "[it] is like Dodgson's in the sense that it is based on altered profiles that have candidates who lose to no other candidate under simple majority. But unlike

[9] Tideman (2006, pp. 187–189) proposes two heuristic procedures that simplify the need to examine all $m!$ preference orderings.

[10] According to Kemeny (1959) the distance between two preference orderings, R and R', is the number of pairs of candidates (alternatives) on which they differ. For example, if $R = a \succ b \succ c \succ d$ and $R' = d \succ a \succ b \succ c$, then the distance between R and R' is 3, because they agree on three pairs $[(a \succ b), \ (a \succ c), \ (b \succ c)]$ but differ on the remaining three pairs, i.e., on the preference ordering between a and d, b and d, and between c and d. Similarly, if R'' is $c \succ d \succ a \succ b$ then the distance between R and R'' is 4 and the distance between R' and R'' is 3. According to Kemeny's procedure the most likely social preference ordering is that R such that the sum of distances of the voters' preference orderings from R is minimized. Because this R has the properties of the median central measure in statistics it is called the *median preference ordering*. The median preference ordering (but not the *mean preference ordering* which is that R which minimizes the sum of the squared differences between R and the voters' preference orderings) will be identical to the possible social preference ordering W which maximizes the sum of voters that agree with all paired comparisons implied by W.

Dodgson, Young deletes voters rather than inverting preferences to obtain the altered profiles. His procedure suggests that we remain most faithful to Condorcet's principle if the choice set consists of alternatives that can become simple majority nonlosers with removal of the fewest number of voters."

3.4 Summary

As can be seen from Tables 3.1–3.3, seven procedures (Alternative Vote, Coombs, Bucklin, Majority Judgment, Minimax, Dodgson, and Young) are susceptible to the largest number of paradoxes (10), whereas the plurality (first-past-the-post) and Borda's procedures are susceptible to the smallest number of paradoxes (6).

Of the nine Condorcet-consistent procedures, six procedures (successive elimination, minimax, Dodgson's, Nanson's, Schwartz's, and Young's) are dominated by the other three procedures (Black's, Copeland's and Kemeny's) in terms of the paradoxes to which these procedures are susceptible.

However, the number of paradoxes to which each of the various voting procedures surveyed here is vulnerable may be regarded as meaningless or even misleading. This is so for two reasons.

Table 3.1 Susceptibility of non-ranked procedures to voting paradoxes

Procedure	Plurality	Plurality ϖ runoff	Approval voting	Successive elimination
Paradox				
Condorcet pdx (cyclical majorities)	−	−	−	+
Condorcet winner pdx	+	+	+	−
Absolute majority pdx	−	−	⊕	−
Condorcet loser pdx	⊕	−	+	−
Absolute loser pdx	⊕	−	⊕	−
Pareto dominated candidate	−	−	⊕	⊕
Lack of monotonicity	−	⊕	−	−
Reinforcement	−	+	−	+
No-show	−	+	−	+
Twin	−	+	−	+
Truncation	−	−	−	+
Subset choice condition (SCC)	+	+	+	+
Preference inversion	+	+	+	−
Path independence	−	−	−	+
Strategic voting	+	+	+	+
Total ⊕ signs	2	1	4	1
Total + & ⊕ signs	6	8	8	9

Notes:
A + sign indicates that a procedure is vulnerable to the specified paradox
A ⊕ sign indicates that a procedure is vulnerable to the specified paradox which seems to us an especially intolerable paradox
A − sign indicates that a procedure is not vulnerable to the specified paradox
It is assumed that all voters have linear preference ordering among all competing candidates

Table 3.2 Susceptibility of ranked non Condorcet-consistent procedures to voting paradoxes

Procedure	Borda	Alternative Vote (AV) STV	Coombs	Bucklin	Range Voting	Majority Judgment
Paradox						
Condorcet pdx (cyclical majorities)	+	+	+	+	+	+
Condorcet winner pdx	+	+	+	+	+	+
Absolute majority pdx	⊕	–	–	–	⊕	⊕
Condorcet loser pdx	–	–	–	⊕	⊕	⊕
Absolute loser pdx	–	–	–	–	⊕	⊕
Pareto dominated candidate	–	–	–	–	–	–
Lack of monotonicity	–	⊕	⊕	–	–	–
Reinforcement	–	+	+	+	–	+
No-show	–	+	+	+	–	+
Twin	–	+	+	+	–	+
Truncation	+	+	+	+	+	+
Subset choice condition (SCC)	+	+	+	+	–	–
Preference inversion	–	+	+	+	–	–
Path independence	–	–	–	–	–	–
Strategic voting	+	+	+	+	+	+
Total ⊕ signs	1	1	1	1	3	3
Total + and ⊕ signs	6	10	10	10	7	10

Notes:
A + sign indicates that a procedure is vulnerable to the specified paradox
A ⊕ sign indicates that a procedure is vulnerable to the specified paradox which seems to us an especially intolerable paradox
A – sign indicates that a procedure is not vulnerable to the specified paradox
It is assumed that all voters have linear preference ordering among all competing candidates

First, some paradoxes are but special cases of other paradoxes or may induce the occurrence of other paradoxes, as follows:

- A procedure which is vulnerable to the absolute majority paradox is also vulnerable to the Condorcet winner paradox;
- A procedure which is vulnerable to the absolute loser paradox is also vulnerable to the Condorcet loser paradox;
- Except for the range voting and majority judgment procedures, all procedures surveyed in this chapter that are vulnerable to the Condorcet loser paradox are also vulnerable to the preference inversion paradox.
- The five procedures surveyed in this chapter which may display lack of monotonicity are also susceptible to the No-Show paradox[11];

[11]Campbell and Kelly (2002) devised a non-monotonic voting rule that does not exhibit the No-Show paradox. However, as this method violates the anonymity and neutrality conditions and hence has not been considered seriously for actual use, we ignore it.

Table 3.3 Susceptibility of ranked Condorcet-consistent procedures to voting paradoxes

Procedure	Minimax	Dodgson	Black	Copeland	Kemeny	Nanson	Schwartz	Young
Paradox								
Condorcet pdx (cyclical majorities)	+	+	+	+	+	+	+	+
Condorcet winner pdx	−	−	−	−	−	−	−	−
Absolute majority pdx	−	−	−	−	−	−	−	−
Condorcet loser pdx	⊕	⊕	−	−	−	−	−	⊕
Absolute loser pdx	⊕	−	−	−	−	−	−	⊕
Pareto dominated cand.	−	−	−	−	−	−	⊕	−
Lack of monotonicity	−	⊕	−	−	−	⊕	−	−
Reinforcement	+	+	+	+	+	+	+	+
No-show	+	+	+	+	+	+	+	+
Twin	+	+	+	+	+	+	+	+
Truncation	+	+	+	+	+	+	+	+
SCC	+	+	+	+	+	+	+	+
Preference inversion	+	+	−	−	−	−	−	+
Path independence	−	−	−	−	−	−	−	−
Strategic voting	+	+	+	+	+	+	+	+
Total ⊕ signs	2	2	0	0	0	1	1	2
Total + & ⊕ signs	10	10	7	7	7	8	8	10

Notes:
A + sign indicates that a procedure is vulnerable to the specified paradox
A ⊕ sign indicates that a procedure is vulnerable to the specified paradox which seems to us an especially intolerable paradox
A − sign indicates that a procedure is not vulnerable to the specified paradox
It is assumed that all voters have linear preference ordering among all competing candidates

- All Condorcet-consistent procedures are susceptible to the no-show paradox and hence also to the twin paradox when there exist at least four candidates.[12]

Second, and more importantly, not all the surveyed paradoxes are equally undesirable. Although assessing the severity of the various paradoxes is largely a subjective matter, *there seems to be a wide consensus that a voting procedure which is susceptible to an especially serious paradox (denoted by ⊕ in Tables 3.1–3.3), i.e., a voting procedure which may elect a pareto-dominated candidate, or elect a Condorcet (and absolute) loser, or display lack of monotonicity, or not elect an absolute winner, should be disqualified as a reasonable voting procedure regardless of the probability that these paradoxes may occur.* On the other hand, the degree of severity that should be assigned to the remaining paradoxes should depend, *inter alia*, on the likelihood of their occurrence under the procedures that are vulnerable

[12] Although all Condorcet-consistent procedures are also susceptible to the Reinforcement paradox, there is no logical connection between this paradox and the no-show paradox. As mentioned by Moulin (1988b, pp. 54–55), when there are no more than three candidates there exist Condorcet-consistent procedures which are immune to both the no-show and twin paradoxes, e.g., the minimax procedure which elects the candidate to whom the smallest majority objects.

to them. Thus, for example, a procedure which may display a given paradox only when the social preference ordering is cyclical – as is the case for most of the paradoxes afflicting the Condorcet-consistent procedures – should be deemed more desirable (and the paradoxes it may display more tolerable) than a procedure which can display the same paradox when a Condorcet winner exists.[13]

Additional criteria which should be used in assessing the relative desirability of a voting procedure are what may be called *administrative-technical criteria*. The main criteria belonging to this category are the following:

- *Requirements from the voter*: some voting procedures make it more difficult for the voter to participate in an election by requiring him/her to rank-order all competing candidates, whereas other procedures make it easier for the voter by requiring him/her to vote for just one candidate or for any candidate(s) s/he approves.
- *Ease of understanding how the winner is selected*: In order to encourage voters to participate in an election a voting procedure must be transparent, i.e., voters must understand how their votes (preferences) are aggregated into a social choice. Thus a voting procedure where the winner is the candidate who received the plurality of votes is easier to explain – and considered more transparent – than a procedure which may involve considerable mathematical calculations (e.g., Kemeny's) in order to determine the winner.
- *Ease of executing the elections*: Election procedures requiring only one voting (or counting) round are more easily executed than election procedures that may require more than one voting (or counting) round. Similarly, election procedures requiring to count only the number of votes received by each candidate are easier to conduct than those requiring the conduct of all $m(m-1)/2$ paired contests between all m candidates, or those requiring the examination of up to $m!$ possible social preference orderings in order to determine the winner.
- *Minimization of the temptation to vote insincerely*: Although all voting procedures are vulnerable to manipulation, i.e., to the phenomenon where some voters may benefit if they vote insincerely, some voting procedures (e.g., Borda's count, Range voting) are susceptible to this considerably more than others.
- *Discriminability*: One should prefer a voting procedure which is more discriminate, i.e., it is more likely to select (deterministically) a unique winner than produce a set of tied candidates – in which case the employment of additional means are needed to obtain a unique winner. Thus, for example, when the social preference ordering is cyclical then, *ceteris paribus*, Schwartz's and Copeland's methods are considerably less discriminating than the remaining Condorcet-consistent procedures surveyed in this chapter.

[13]However, in order to be able to state conclusively which of several voting procedures that are susceptible to the same paradox is more likely to display this paradox, one must know what are the necessary and/or sufficient conditions for this paradox to occur under the various compared procedures. Such knowledge is still lacking with respect to most voting procedures and paradoxes.

Of course there may exist conflicts between some of these technical-administrative criteria. For example, a procedure like Kemeny's which, on the one hand, is more difficult to execute in practice and to explain to prospective voters (and hence less transparent), is, on the other hand, more discriminate and less vulnerable to insincere behavior.

So in view of all the above criteria, which of the 18 surveyed voting procedures do I think should be preferred? Since the weakest extension of the majority rule principle when there are more than two candidates is the Condorcet winner principle, I think that the electoral system which ought to be used for electing one out of $m \geq 2$ candidates should be Condorcet-consistent.

But as one does not know before an election is conducted whether a Condorcet winner will exist or whether the social preference ordering will contain a top cycle, which of the nine Condorcet-consistent procedures surveyed and exemplified in this paper should be preferred in case a top cycle exists? In this case I think that the Successive Elimination procedure and Schwartz's procedure should be readily disqualified because of their vulnerability to electing a pareto-dominated candidate, Dodgson's and Nanson's procedures should be readily disqualified because of their lack of monotonicity, and the minimax and Young's procedures should be readily disqualified because of their vulnerability to electing an absolute or a Condorcet loser. Although Black's procedure cannot elect a Condorcet loser, it may nevertheless come quite close to it because, as demonstrated in Example 3.5.13.3 below, it violates Smith's (1973) Condorcet principle, so this procedure too seems to me not considerably more desirable than the minimax and Young's procedures.

This leaves us with a choice between the remaining two Condorcet-consistent procedures – Copeland's and Kemeny's. The choice between them depends on the importance one assigns to the above-mentioned technical-administrative criteria. Both these procedures require voters to rank-order all candidates. However, Copeland's method is probably easier than Kemeny's to explain to lay voters, as well as, when the number of candidates is large, may involve considerably fewer calculations in determining who is (are) the ultimate winner(s). Kemeny's procedure, on the other hand, is more discriminate than Copeland's when the number of candidates is relatively small, and is probably also – because of its increased complexity in determining the ultimate winner – less vulnerable to insincere voting. So if I would have to choose between these two procedures I would choose Kemeny's because most elections where a single candidate must be elected usually involve relatively few contestants – in which case Kemeny's procedure seems to have an advantage over Copeland's procedure. Moreover, as I mentioned in the description of Kemeny's procedure and as argued by Young (1995, pp. 60–62), Kemeny's procedure has also the advantage that it can be justified not only from Condorcet's perspective of the maximum likelihood rule, but also as choosing for the entire society the "median preference ordering" – which can be viewed from the perspective of modern statistics as the best compromise between the various rankings reported by the voters.

3.5 Appendix: exemplifying the Various Paradoxes That Afflict the Various Procedures

3.5.1 Demonstrating Paradoxes Afflicting the Plurality Procedure

Except for being vulnerable to strategic voting, the plurality procedure is vulnerable to the Condorcet winner paradox, the Condorcet loser paradox, the absolute loser paradox, the preference inversion paradox, and to SCC. The following example demonstrates the vulnerability of the plurality procedure to all these paradoxes simultaneously.

3.5.1.1 Example

Suppose there are nine voters who must elect one out of three candidates, a, b, and c, and whose preference orderings among these candidates are as follows:

No. of voters	Preference ordering
4	$a \succ b \succ c$
3	$b \succ c \succ a$
2	$c \succ b \succ a$

Here b is the Condorcet winner and a is not only a Condorcet loser but also an absolute loser. Nevertheless, if all voters vote for their top preference then a will be elected. Note that if c drops out of the race then b will be elected – thus demonstrating violation of SCC. Note also that if all voters invert their preference orderings then a becomes an absolute winner and hence will be elected – thus demonstrating the Preference Inversion paradox.

3.5.2 Demonstrating Paradoxes Afflicting the Plurality with Runoff Procedure

Except for being vulnerable to strategic voting, the plurality with runoff procedure is vulnerable to the Condorcet winner, lack of monotonicity, reinforcement, no-show, twin, preference inversion, and to the SCC paradoxes.

Example 3.5.2.1 below demonstrates the vulnerability of the plurality with runoff procedure to the Condorcet winner, to lack of monotonicity, and to the SCC paradoxes.

3.5.2.1 Example

Suppose there are 43 voters whose preference orderings among three candidates, a, b, and c, are as follows:

No. of voters	Preference ordering
7	$a \succ b \succ c$
9	$a \succ c \succ b$
14	$b \succ c \succ a$
13	$c \succ a \succ b$

Here the social preference ordering is $c \succ a \succ b$, i.e., c is the Condorcet winner. But if all voters vote sincerely then under the plurality with runoff procedure c will be eliminated in the first round and a will beat b in the second round and thus become the ultimate winner. (Note that if c would have withdrawn from the race prior to the first round then, *ceteris paribus*, a would have been elected already in the first round, thereby demonstrating this procedure's vulnerability to SCC).

Now suppose that, *ceteris paribus*, five of the 14 voters whose preference ordering is $b \succ c \succ a$ (who are not very happy with the prospect that a may be elected) change it to $a \succ b \succ c$ thereby increasing a's support. As a result of this change b (rather than c) will be eliminated in the first round, and c (the Condorcet winner) will beat a in the second round – thereby demonstrating the vulnerability of the plurality with runoff procedure to non-monotonicity.

Example 3.5.2.2 demonstrates the vulnerability of the plurality with runoff procedure to the reinforcement paradox.

3.5.2.2 Example

Suppose there are two districts, I and II. In district I there are 17 voters whose preference orderings among three candidates, a, b, and c, are as follows:

No. of voters	Preference ordering
4	$a \succ b \succ c$
1	$b \succ a \succ c$
5	$b \succ c \succ a$
6	$c \succ a \succ b$
1	$c \succ b \succ a$

and in district II there are 15 voters whose preference orderings among the three candidates are as follows:

No. of voters	Preference ordering
6	$a \succ c \succ b$
8	$b \succ c \succ a$
1	$c \succ a \succ b$

If all voters vote sincerely then no candidate is ranked first by an absolute majority of the voters in district I. Consequently candidate a is deleted from the race after the first round and candidate b beats candidate c in this district in the second round.

In district II candidate b, who is ranked first by the majority of voters, is elected in the first round.

However if, *ceteris paribus*, the two districts are amalgamated into a single district, we obtain the following distribution of preference orderings of the 32 voters:

No. of voters	Preference ordering
4	$a \succ b \succ c$
6	$a \succ c \succ b$
1	$b \succ a \succ c$
13	$b \succ c \succ a$
7	$c \succ a \succ b$
1	$c \succ b \succ a$

If all voters vote sincerely then no candidate is ranked first by an absolute majority of the voters. Consequently c is deleted after the first round and a beats b and is elected in the second round – in violation of the reinforcement postulate.

Example 3.5.2.3 demonstrates the vulnerability of the plurality with runoff procedure to the no-show and to the twin paradoxes.

3.5.2.3 Example

Suppose there are 11 voters whose preference orderings among three candidates, a, b, and c, are as follows:

No. of voters	Preference ordering
4	$a \succ b \succ c$
3	$b \succ c \succ a$
1	$c \succ a \succ b$
3	$c \succ b \succ a$

If all voters vote sincerely then no candidate is ranked first by an absolute majority of the voters. Consequently b is deleted after the first round and c beats a in the second round and is elected. Since the election of c is the worst outcome for the voters whose preference ordering is $a \succ b \succ c$, suppose that, *ceteris paribus*, two of them decide not to participate in the election (no-show). We thus obtain the following distribution of preference orderings:

No. of voters	Preference ordering
2	$a \succ b \succ c$
3	$b \succ c \succ a$
1	$c \succ a \succ b$
3	$c \succ b \succ a$

Here a (rather than b) is eliminated in the first round, and b beats c in the second round. Thus the $a \succ b \succ c$ voters obtained, *ceteris paribus*, a better outcome when two of them did not participate in the election than when all of them participated in the election thereby demonstrating the no-show paradox.

This example demonstrates also the vulnerability of the plurality with runoff procedure to the (weak form) of the twin paradox. Suppose that, *ceteris paribus*, there are only two voters with preference ordering $a \succ b \succ c$. One would expect these voters to welcome another "twin" voters having identical preference ordering to theirs thereby presumably giving an increased weight to their common preference ordering. Yet as we saw, the addition of these twins to the electorate results in the election of c, their worst alternative – thereby demonstrating the twin paradox.

Example 3.5.2.4 demonstrates the vulnerability of the plurality with runoff procedure to the preference inversion paradox.

3.5.2.4 Example

Suppose there are 11 voters whose preference orderings among three candidates, a, b, and c, are as follows:

No. of voters	Preference ordering
5	$a \succ b \succ c$
4	$b \succ c \succ a$
2	$c \succ a \succ b$

If all voters vote sincerely for their top preference in the first round, then c will be eliminated at the end of the first round and thereafter a will beat b in the second round. However, if all voters invert their preference orderings then b will be eliminated at the end of the first round and a will beat c in the second round – thus demonstrating the Preference inversion paradox.

3.5.3 Demonstrating the Paradoxes Afflicting the Approval Voting Procedure

Except for being vulnerable to strategic voting, the approval voting procedure is vulnerable to the Condorcet winner paradox, the Condorcet loser paradox, the absolute majority and absolute loser paradoxes, to the pareto-dominated paradox, to the Preference Inversion paradox, and to SCC.

Example 3.5.3.1 demonstrates the vulnerability of the approval voting procedure to the Condorcet winner paradox.

3.5.3.1 Example

This example is due to Felsenthal and Maoz (1988, p. 123, Example 3.2). Suppose there are 47 voters whose preference orderings among three candidates, a, b, and c, are as follows:

No. of voters	Preference ordering
18	$(a) \succ b \succ c$
6	$(b \succ c) \succ a$
8	$(b \succ a) \succ c$
2	$(c \succ a) \succ b$
13	$(c) \succ b \succ a$

The social preference ordering is $b \succ a \succ c$, i.e., b is the Condorcet winner. However, if all voters approve (and vote for) the candidates denoted between parentheses then a would get the largest number of approval votes (28) and will thus be elected.

Example 3.5.3.2 demonstrates the vulnerability of the approval voting procedure to the pareto-dominated paradox.

3.5.3.2 Example

This example is due to Felsenthal and Maoz (1988, p. 123, Example 3.4). Suppose there are three voters whose preference orderings among four candidates, a, b, c, and d, are as follows:

No. of voters	Preference ordering
1	$a \succ b \succ c \succ d$
1	$c \succ a \succ b \succ d$
1	$d \succ a \succ b \succ c$

The social preference ordering is $a \succ b \succ c \succ d$, i.e., a is the Condorcet winner. However, if each voter approves (and votes for) his top three preferences then a tie would occur between the number of votes (3) obtained by candidates a and b, and if this tie were to be broken randomly then there is a 0.5 probability that b would be elected. So if b were to be elected it would demonstrate not only that the Condorcet winner (a) was not elected but also that a pareto-dominated candidate can be elected under the approval voting procedure. (Note that *all* voters prefer a to b).

Example 3.5.3.3 demonstrates the vulnerability of the approval voting procedure to the absolute majority paradox.

3.5.3.3 Example

Suppose there are 100 voters whose preference orderings among three candidates, a, b, and c, are as follows:

No. of voters	Preference ordering
51	$a \succ b \succ c$
48	$b \succ c \succ a$
1	$c \succ b \succ a$

The social preference ordering is $a \succ b \succ c$, i.e., a is the Condorcet winner who is ranked first by an absolute majority of the voters. However, if only one candidate must be elected and if each voter approves (and votes for) his top two preferences, then b will be elected despite the fact that a is ranked first by an absolute majority of the voters.

Example 3.5.3.4 demonstrates the vulnerability of the approval voting procedure to the absolute loser and to the Condorcet loser paradoxes.

3.5.3.4 Example

Suppose there are 15 voters whose preference orderings among three candidates, a, b, and c, are as follows:

No. of voters	Preference ordering
6	$(a) \succ b \succ c$
4	$(b) \succ c \succ a$
1	$(c \succ a) \succ b$
4	$(c) \succ b \succ a$

The social preference ordering is $b \succ c \succ a$, i.e., a is not only the Condorcet loser but also the Absolute Loser because this candidate is ranked last by an absolute majority of the voters. However, if only one candidate must be elected and if all voters approve (and vote for) the candidate(s) denoted between parentheses then a will be elected.

This example can also be used to demonstrate the susceptibility of the Approval Voting procedure to the preference inversion paradox. If in the above example all voters invert their preference ordering and decide to vote, as before, either only for their top preference or for their top two preferences, then we obtain the following distribution of votes:

No. of voters	Preference ordering
6	$(c) \succ b \succ a$
4	$(a) \succ c \succ b$
1	$(b \succ a) \succ c$
4	$(a) \succ b \succ c$

Here a is not only the Condorcet winner but also the absolute winner and is elected – thereby demonstrating the susceptibility of approval voting to the preference inversion paradox.

When all voters are assumed to approve of (and vote for) originally only their top preference (as under the plurality procedure) – and subject to what we said in footnote 4 above – Example 3.5.1.1 can be used to also demonstrate the susceptibility of the Approval Voting procedure to SCC. Thus, for instance, it would be worthwhile, *ceteris paribus*, for the two voters in Example 3.5.1.1 whose original preference ordering is $c \succ b \succ a$ to vote for b if alternative c were no longer available even though they did not "approve" originally of b – because by voting for b they lose nothing but may avert the election of a, their least preferable alternative, which may be elected if they abstain.

3.5.4 Demonstrating the Paradoxes Afflicting the Successive Elimination Procedure

Except for being vulnerable to cyclical majorities and to strategic voting, the successive elimination procedure is vulnerable to pareto-dominated, reinforcement, no-show, twin, truncation, SCC, and path independence paradoxes.

Example 3.5.4.1 demonstrates the vulnerability of the successive elimination procedure to the election of a pareto-dominated candidate. A necessary condition for this to happen is that the social preference ordering is cyclical and there are at least four candidates (Fishburn, 1982, p. 131).

3.5.4.1 Example

Suppose there are 11 voters whose preference orderings among four candidates, a, b, c, and d, are as follows:

No. of voters	Preference ordering
3	$a \succ b \succ c \succ d$
2	$c \succ a \succ b \succ d$
1	$c \succ d \succ a \succ b$
5	$d \succ a \succ b \succ c$

Thus the social preference ordering is cyclical ($b \succ c \succ d \succ a \succ b$). Suppose further that all the voters always vote sincerely for their preferred candidate in each round, and that the order in which the divisions are carried out is as follows:

In round 1: d against a;
In round 2: the winner of round 1 against c;
In round 3: the winner of round 2 against b;

Given this order d beats a (6:5) in the first round, c beats d (6:5) in the second round, and b beats c (8:3) in the third round and becomes the ultimate winner. Note, however, that b is a Pareto-dominated candidate because *all* the voters prefer a to b.

This example can also be used to demonstrate the vulnerability of the successive elimination procedure to SCC.

If, *ceteris paribus*, d is deleted, then in the first round a will beat c (8:3), and in the second round a will beat b (11:0) and thus a will become the ultimate winner – in violation of SCC.

Similarly, this example can also be used to demonstrate the vulnerability of the successive elimination procedure to the no-show paradox.

If, *ceteris paribus*, two of the voters whose top preference is d decide not to participate, then a becomes the Condorcet winner and hence will be elected under the successive elimination procedure. Note that this outcome is preferred over the election of b by the two $d \succ a \succ b \succ c$ voters who decided not to participate – thus demonstrating the vulnerability of the successive elimination procedure to the no-show paradox.

This example can also be used to demonstrate the vulnerability of the successive elimination procedure to lack of path independence when the social preference ordering is cyclical.

Given the above preference orderings of the 11 voters, if the order of the divisions in each round were changed such that:

In round 1: a against b
In round 2: the winner of round 1 against c
In round 3: the winner of round 2 against d

Then in the first round a would beat b (11:0), in the second round a would also beat c (8:3), but in the third round d would beat a (6:5) and become the ultimate winner.

Example 3.5.4.2 demonstrates the vulnerability of the successive elimination procedure to the reinforcement paradox.

3.5.4.2 Example

Suppose there are two districts, I and II. In district I there are three voters whose preference orderings among four candidates are as follows:

No. of voters	Preference ordering
1	$a \succ b \succ d \succ c$
1	$b \succ d \succ c \succ a$
1	$d \succ c \succ a \succ b$

and in district II there are four voters whose preference ordering among the four candidates are as follows:

No. of Voters	Preference Ordering
3	$c \succ b \succ d \succ a$
1	$d \succ a \succ b \succ c$

If the order of divisions in each district is:

b vs. d in round 1
Winner of first round against a in round 2
Winner of second round against c in round 3

Then in each district c will be the ultimate winner.

However if, *ceteris paribus*, the two districts are amalgamated into a single district of seven voters, then d becomes the Condorcet winner and will therefore be elected under the successive elimination procedure – in violation of the reinforcement postulate.

Example 3.5.4.3 demonstrates the vulnerability of the successive elimination procedure to the Twin paradox.

3.5.4.3 Example

This example is due to Moulin (1988b, p. 54). Suppose there are six voters whose preference orderings among three candidates, a, b, and c, are as follows:

No. of voters	Preference ordering
2	$a \succ b \succ c$
2	$b \succ c \succ a$
1	$c \succ a \succ b$
1	$c \succ b \succ a$

Suppose further that the order in which the divisions are conducted is as follows:

a vs. b in round 1
Winner of round 1 vs. c in round 2

and that if there is a tie between two candidates in any of the divisions it is broken lexicographically, i.e., in favor of the candidate who is denoted by the letter that is closer to the beginning of the alphabet.

Accordingly, there is a tie between a and b in the first round which is broken in favor of a, and in the second round c beats a and becomes the ultimate winner.

In view of this result one could expect that, *ceteris paribus*, the single $c \succ b \succ a$ voter should welcome if an additional "twin" voter would join the electorate thereby providing more weight to their common preferences. However, an addition of a second $c \succ b \succ a$ voter would result, *ceteris paribus*, in a net loss to the first $c \succ b \succ a$ voter because b would become the Condorcet winner and hence also the ultimate winner under the successive elimination procedure – thus demonstrating the twin paradox.

Example 3.5.4.4 demonstrates the vulnerability of the successive elimination procedure to the Truncation paradox.

3.5.4.4 Example

Suppose there are six voters with the following preference orderings:

No. of voters	Preference ordering
1	$a \succ b \succ c \succ d$
1	$c \succ b \succ a \succ d$
2	$c \succ d \succ b \succ a$
2	$d \succ a \succ b \succ c$

Suppose further that the order in which the divisions are conducted is as follows:

First round: b vs. c
Second round: winner of first round vs. d
Third round: winner of second round vs. a

Additionally, suppose that if a tie occurs between two candidates it is broken in favor of the one denoted by a letter closer to the beginning of the alphabet.

Accordingly, in the first round there is a tie between b and c which is broken in favor of b. In the second round d beats b, and in the third round d beats a and hence becomes the ultimate winner. This is of course a very bad outcome for the single voter whose preference ordering is $a \succ b \succ c \succ d$. So suppose that, *ceteris paribus*, this voter would truncate his preferences between b, c, and d, and indicate just his top preference, a, i.e., this voter will participate only in the third round in which a will compete against the winner from the second round. As a result of such truncation c would beat b in the first round, c would beat also d in the second round, but in the third round there would be a tie between a and c – which will be broken in favor of a, a much better result for the $a \succ b \succ c \succ d$ voter, thus demonstrating the truncation paradox.

3.5.5 Demonstrating Paradoxes Afflicting Borda's Procedure

Except for being vulnerable to cyclical majorities and to strategic voting, Borda's procedure is vulnerable to the Condorcet winner, absolute majority, truncation, and SCC paradoxes. And as I shall show in Example 3.5.13.3, it also violates Smith's Condorcet principle.

Example 3.5.5.1 demonstrates simultaneously the vulnerability of Borda's procedure to the absolute majority paradox (and thus also to the Condorcet winner paradox).

3.5.5.1 Example

Suppose there are 100 voters who have to elect one out of three candidates, *a, b, c*, under Borda's procedure, and whose preference orderings are as follows:

No. of voters	Preference ordering
51	$a \succ b \succ c$
48	$b \succ c \succ a$
1	$c \succ b \succ a$

The number of Borda points awarded to candidates *a, b*, and *c*, are 102, 148, and 50, respectively, so candidate *b* is elected. However, note that candidate *a* is not only the Condorcet winner but also an absolute winner because an absolute majority of the voters rank candidate *a* as their top preference.

Example 3.5.5.2 demonstrates the vulnerability of Borda's procedure to the truncation paradox.

3.5.5.2 Example

This example is due to Fishburn (1974, p. 543). Suppose that seven voters have to elect one out of four candidates $a - d$ under Borda's procedure, and that their preference orderings among the candidates are as follows:

No. of voters	Preference ordering
3	$a \succ b \succ c \succ d$
1	$b \succ c \succ a \succ d$
1	$b \succ c \succ d \succ a$
2	$c \succ d \succ a \succ b$

Suppose further that under Borda's procedure with k candidates one assigns k points to the top-ranked candidate, $k-1$ points to the second-ranked candidate, ..., 1 point to the kth ranked candidate, and 0 points to any non-ranked candidate.

Given the above preference orderings and Borda-point assignment, the number of points awarded to candidates *a, b, c*, and *d*, are 19, 19, 20, and 12, respectively, so candidate *c* is elected. However, if the first three voters (who are not very happy with the election of candidate *c*) decide not to rank (i.e., truncate) candidate *c*, then the number of Borda points awarded to candidates *a, b, c*, and *d*, are 16, 16, 14, and 12, respectively, so candidates *a* and *b* are tied and one of them will be eventually elected depending on the rule employed for braking ties. This result is of course preferred by the first three voters to the election of candidate *c*, thereby demonstrating the truncation paradox.

Example 3.5.5.3 demonstrates the vulnerability of Borda's procedure to SCC.

3.5.5.3 Example

Suppose that 11 voters have to elect one out of three candidates, *a, b*, or *c*, under Borda's procedure and that their preference orderings among these candidates are as follows:

No. of voters	Preference ordering
3	$a \succ c \succ b$
3	$b \succ a \succ c$
5	$c \succ b \succ a$

Accordingly, the number of Borda points awarded to candidates *a, b*, and *c*, are 9,11, and 13, respectively – so candidate *c* is elected.

Now suppose that, *ceteris paribus*, candidate *b* drops out of the race. In this case the number of Borda points awarded to candidates *a* and *c* are 6 and 5, respectively, so candidate *a* would be elected – in violation of SCC.

3.5.6 Demonstrating Paradoxes Afflicting the Alternative Vote Procedure

Except for being vulnerable to cyclical majorities and to strategic voting, the Alternative Vote procedure is vulnerable to the Condorcet winner, lack of monotonicity, reinforcement, no-show, twin, truncation, preference inversion, and SCC paradoxes.

The same examples that were used to demonstrate the vulnerability of the plurality with runoff procedure to all these paradoxes (except the truncation paradox), can also be used to demonstrate the vulnerability of the alternative vote procedure to these paradoxes.

Specifically, Example 3.5.2.1 above can be used to demonstrate the vulnerability of the Alternative Vote procedure to the Condorcet winner, to lack of monotonicity,[14] and to the SCC paradoxes; Example 3.5.2.2 above can be used to demonstrate the vulnerability of the Alternative Vote procedure to the reinforcement paradox, Example 3.5.2.3 above can be used to demonstrate the vulnerability of the alternative vote procedure to the no-show and twin paradoxes, and Example 3.5.2.4 above can be used to demonstrate the vulnerability of the alternative vote procedure to the preference inversion paradox.

Example 3.5.6.1 demonstrates the vulnerability of the alternative vote procedure to the truncation paradox.

[14]A display of negative responsiveness (or lack of monotonicity) under the alternative vote procedure has actually occurred recently in the March 2009 mayoral election in Burlington, Vermont. Among the three biggest vote getters, the Republican got the most first-place votes, the Democrat the fewest, and the Progressive won after the Democrat was eliminated. Yet if many of those who ranked the Republican first had ranked the Progressive first, the Republican would have been eliminated and the Progressive would have lost to the Democrat. In March 2010 Burlington replaced the Alternative Vote procedure for electing its mayor with the Plurality with Runoff procedure – which is also susceptible to negative responsiveness. See detailed report in http://rangevoting.org/Burlington.html.

3.5.6.1 Example

This example is due to Nurmi (1999, p. 63). Suppose there are 103 voters whose preference orderings among four candidates, *a, b, c*, and *d*, are as indicated below and who must elect one of these candidates under the alternative vote procedure.

No. of voters	Preference ordering
33	$a \succ b \succ c \succ d$
29	$b \succ a \succ c \succ d$
24	$c \succ b \succ a \succ d$
17	$d \succ c \succ b \succ a$

Since none of the four candidates is ranked first by an absolute majority of the voters, candidate *d* (who is ranked first by the smallest number of voters) is eliminated. As this does not yet lead to a winner, *b* is eliminated, whereupon *a* wins.

Suppose now that, *ceteris paribus*, those 17 voters who rank *a* last decide to truncate their preference ordering and list only their top preference, *d*. In this case *d* will be eliminated first (as before), but since these 17 voters did not indicate their preference ordering among the remaining candidates, candidate *c* (rather than *b*) will be eliminated thereafter – whereupon *b* wins. This result is preferred by these 17 voters to the election of *a*, thereby demonstrating the truncation paradox.

3.5.6.2 Remark

As stated at the outset of this chapter, the UK conducted a referendum in May 2011 regarding whether to replace its plurality voting procedure in parliamentary elections with the alternative vote procedure. It may therefore be interesting to note that when there are only three competing candidates (as is usually the case in parliamentary elections in England), the alternative vote procedure is more Condorcet-efficient than the plurality procedure. This is so because, by definition, a necessary and sufficient condition for a Condorcet winner (or any other candidate) to be elected under the plurality procedure is that s/he will constitute the top preference of a plurality of the voters, whereas for a Condorcet winner to be elected under the alternative vote procedure when there are three candidates it is sufficient (but not necessary) that the Condorcet winner constitutes the top preference of a plurality of the voters. This is so because if there exist three candidates, *a, b*, and *c*, such that the social preference ordering is $a \succ b \succ c$ and *a* constitutes the top preference of the plurality of voters, then either *b* or *c* (but not *a*) must be eliminated in the first counting round, and as *a* is the Condorcet winner s/he must necessarily beat the remaining alternative in the second counting round.

So while it is a sufficient condition for a Condorcet winner to be elected under the alternative vote procedure when there are three candidates and the Condorcet winner constitutes the top preference of a plurality of the voters, it is not a necessary condition because, as can be ascertained from Example 3.5.1.1, a Condorcet winner can be elected under the Alternative Vote procedure when there are three candidates even though it does not constitute the top preference of a plurality of the voters.

However, it is no longer a sufficient condition for a Condorcet winner who is ranked first by a plurality of the voters to be elected under the alternative vote procedure once there are more than three candidates. This is demonstrated in Example 3.5.6.3.

3.5.6.3 Example

This example is due partly to Moshé Machover who provided me general guidance in its construction (private communication 13.12.2010). Suppose there are 85 voters whose preference orderings among four candidates, *a, b, c,* and d, are as indicated below and who must elect one of these candidates under the alternative vote procedure.

No. of voters	Preference ordering
15	$a \succ b \succ c \succ d$
10	$a \succ c \succ b \succ d$
13	$b \succ a \succ c \succ d$
10	$b \succ c \succ a \succ d$
14	$c \succ a \succ b \succ d$
10	$c \succ b \succ a \succ d$
6	$d \succ c \succ a \succ b$
7	$d \succ b \succ a \succ c$

The social preference ordering here is $a \succ b \succ c \succ d$, i.e., candidate a is the Condorcet winner who is ranked first by a plurality of the voters. However, as none of the candidates is ranked first by an absolute majority of the voters, one deletes first candidate d according to the alternative vote procedure, and thereafter one deletes candidate a, whereupon candidate b becomes the winner.

3.5.7 Demonstrating Paradoxes Afflicting Coombs' Procedure

Except for being vulnerable to cyclical majorities and to strategic voting, Coombs' procedure is vulnerable to the same paradoxes afflicting the alternative vote procedure, i.e., the Condorcet winner, monotonicity, reinforcement, no-show, twin, truncation, preference inversion, and the SCC paradoxes.

Example 3.5.7.1 demonstrates the vulnerability of Coombs' procedure to the Condorcet winner paradox.

3.5.7.1 Example

This example is due to Nicolaus Tideman (private communication 8.9.2010). Suppose that 45 voters have to elect under Coombs' procedure one out of three candidates, *a*, *b*, or *c*, and that their preference orderings among these three candidates are as follows:

No. of voters	Preference ordering
1	$a \succ b \succ c$
10	$a \succ c \succ b$
11	$b \succ a \succ c$
11	$b \succ c \succ a$
10	$c \succ a \succ b$
2	$c \succ b \succ a$

The social preference ordering is $b \succ c \succ a$, i.e., *b* is the Condorcet winner. However, since none of the candidates is ranked first by an absolute majority of the voters, one deletes according to Coombs' procedure the candidate who is ranked last by the largest number of voters. In the above example this candidate is *b*, the Condorcet winner. (After deleting *b* candidate *c* is ranked first by an absolute majority of the voters and is elected.)

3.5.7.2 Remark

It is not clear whether Coombs' procedure is more Condorcet-efficient than either the plurality or the alternative vote procedures. As we have already proved in Remark 3.5.6.2, a necessary and sufficient condition for a Condorcet winner to be elected under the plurality procedure is that the Condorcet winner constitutes the top preference of a plurality of the voters. This condition is sufficient (but not necessary) for a Condorcet winner to be elected under the alternative vote procedure when there are three candidates. However, as is demonstrated in Example 3.5.7.1 above, this condition is neither necessary nor sufficient for a Condorcet winner to be elected under Coombs' procedure. On the other hand, as argued by Coombs (1964, p. 399), a sufficient condition for a Condorcet winner to be elected under his proposed procedure is that the voters' preferences are single-peaked along s single dimension. But under both the plurality and alternative vote procedures a Condorcet winner may not be elected when the voters' preferences are single-peaked along a single dimension. To see this consider Example 3.5.7.3.

3.5.7.3 Example

Suppose there are 13 voters who must elect one out of three candidates, *a*, *b*, or *c*, and whose preference orderings among these candidates are as follows:

No. of voters	Preference ordering
1	$a \succ b \succ c$
2	$a \succ c \succ b$
4	$b \succ a \succ c$
6	$c \succ a \succ b$

Here a is the Condorcet winner, the voters' preferences are single-peaked, and a is elected under Coombs' procedure. However, under the plurality and alternative vote procedures c is elected.

Example 3.5.7.4 demonstrates the vulnerability of Coombs' procedure to non-monotonicity.

3.5.7.4 Example

In Example 3.5.7.1 above candidate c was elected under Coombs' procedure although candidate b is the Condorcet winner. Now suppose that, *ceteris paribus*, the 11 voters whose preference ordering is $b \succ a \succ c$ (who are not happy with the prospect that c will be elected) decide to *increase* c's support by changing their preference ordering to $b \succ c \succ a$. Candidate b is still the Condorcet winner but as a result of this change a (rather than b) will first be eliminated under Coombs' procedure, and thereafter b will be elected – in violation of the monotonicity postulate.

Example 3.5.7.5 demonstrates the vulnerability of Coombs' procedure to the no-show, truncation, and preference inversion paradoxes.

3.5.7.5 Example

Suppose there are 15 voters who must elect one out of three candidates, *a, b*, or c, under Coombs' procedure, and whose preference orderings among these candidates are as follows:

No. of voters	Preference ordering
4	$a \succ b \succ c$
4	$b \succ c \succ a$
5	$c \succ a \succ b$
2	$c \succ b \succ a$

Here no candidate is ranked first by an absolute majority of the voters. Hence, according to Coombs' procedure, a is eliminated in the first round and thereafter b is elected.

Now suppose that, *ceteris paribus*, the two voters with preference ordering $c \succ b \succ a$ decide not to participate in the election. In this case b is eliminated according to Coombs' procedure in the first round and thereafter c (the abstainers' top preference!) is elected thereby demonstrating the no-show paradox.

This example can also be used to demonstrate the vulnerability of Coombs' procedure to the truncation paradox: if the two voters with preference ordering $c \succ b \succ a$ decide to list only their top preference then, *ceteris paribus*, b would be eliminated according to Coombs' procedure and thereafter c would be elected!

If, *ceteris paribus*, all voters invert their preference orderings, then we obtain the following distribution of votes:

No. of voters	Preference ordering
4	$c \succ b \succ a$
4	$a \succ c \succ b$
5	$b \succ a \succ c$
2	$a \succ b \succ c$

As no candidate obtains an absolute majority of the votes in the first counting round, c is eliminated and thereafter b is elected in the second counting round – thus demonstrating the vulnerability of Coombs' procedure to the preference inversion paradox.

Example 3.5.7.6 demonstrates the vulnerability of Coombs' procedure to the Reinforcement paradox.

3.5.7.6 Example

Suppose there are two districts, I and II. In district I there are 34 voters whose preference orderings among three candidates, *a, b*, and *c*, are as follows:

No. of voters	Preference ordering
9	$a \succ b \succ c$
9	$b \succ c \succ a$
11	$c \succ a \succ b$
5	$c \succ b \succ a$

and in district II there are seven voters whose preference orderings among the three candidates are as follows:

No. of voters	Preference ordering
1	$a \succ b \succ c$
6	$b \succ a \succ c$

Since no candidate is ranked first by an absolute majority of the voters in district I, candidate a is eliminated under Coombs' procedure in the first round, and thereafter candidate b is elected. In district II candidate b is ranked first by an absolute majority of the voters and is elected right away.

However, if, *ceteris paribus*, the two districts are amalgamated into a single district of 41 voters then one obtains the following distribution of preferences:

No. of voters	Preference ordering
10	$a \succ b \succ c$
6	$b \succ a \succ c$
9	$b \succ c \succ a$
11	$c \succ a \succ b$
5	$c \succ b \succ a$

Since none of the three candidates is ranked first in the amalgamated district, candidate c is eliminated according to Coombs' procedure in the first round, and candidate a (rather than b) is elected thereafter – thus demonstrating the reinforcement paradox.

Example 3.5.7.7 demonstrates the vulnerability of Coombs' procedure to the Twin paradox.

3.5.7.7 Example

Suppose there are 20 voters who have to choose one out of four candidates, a, b, c, or d, under Coombs' procedure and whose preference orderings among these candidates are as follows:

No. of voters	Preference ordering
5	$a \succ b \succ d \succ c$
5	$b \succ c \succ d \succ a$
1	$b \succ a \succ d \succ c$
6	$c \succ a \succ d \succ b$
1	$c \succ b \succ a \succ d$
2	$c \succ b \succ d \succ a$

Since no voter is ranked first by an absolute majority of the voters, candidate a is eliminated according to Coombs' procedure in the first round and thereafter b is elected.

Now suppose that, *ceteris paribus*, two more voters with preference ordering $b \succ a \succ d \succ c$ join the electorate thereby apparently increasing the chances of candidate b to be elected. However, as result of this increase of the electorate candidate c (rather than a) will be eliminated in the first round under Coombs' procedure, and thereafter a tie will be created between candidates a and b – thereby *decreasing* the chances of candidate b to be elected if the tie is to be broken randomly.

Example 3.5.7.8 demonstrates the vulnerability of Coombs' procedure to SCC.

3.5.7.8 Example

Suppose that there are 29 voters having to elect under Coombs' procedure one out of four candidates, a, b, c, or d, and whose preference orderings among the four candidates are as follows:

No. of voters	Preference ordering
11	$a \succ b \succ c \succ d$
12	$b \succ c \succ d \succ a$
2	$b \succ a \succ d \succ c$
4	$c \succ a \succ d \succ b$

Since none of the candidates is ranked first by an absolute majority of the voters, one deletes according to Coombs' procedure the candidate who is ranked last by the largest number of voters. In the above example this candidate is a. After deleting a candidate b is ranked first by an absolute majority of the voters and is elected.

Now suppose that, *ceteris paribus*, candidate c drops out of the race. As a result candidate a is ranked first by an absolute majority of the voters and is elected – contrary to SCC.

3.5.8 Demonstrating Paradoxes Afflicting Bucklin's Procedure

Except for being vulnerable to cyclical majorities and to strategic voting, Bucklin's procedure is vulnerable to the Condorcet winner, Condorcet loser, reinforcement, no-show, twin, truncation, preference inversion, and SCC paradoxes.

Example 3.5.7.1 above can be used to demonstrate the susceptibility of Bucklin's procedure to the Condorcet winner paradox. In this example b is the Condorcet winner but under Bucklin's procedure a is elected because the number of voters (32) who rank a first or second exceeds the number of voters (25) who rank b first or second.

Example 3.5.8.1 demonstrates that a Condorcet loser may be elected under Bucklin's procedure.

3.5.8.1 Example

This example is due to Tideman (2006, p. 197, Example 13.13). Suppose there are 29 voters whose preference orderings among four candidates, w, x, y, z, are as follows:

No. of voters	Preference ordering
5	$w \succ x \succ y \succ z$
3	$w \succ z \succ x \succ y$
5	$x \succ y \succ w \succ z$
2	$x \succ z \succ y \succ w$
3	$y \succ w \succ x \succ z$
2	$y \succ z \succ w \succ x$
4	$z \succ w \succ x \succ y$
2	$z \succ x \succ y \succ w$
3	$z \succ y \succ w \succ x$

The social preference ordering here contains a top cycle ($w \succ x \succ y \succ w$), but since each of the three candidates w, x, y beats candidate z in pairwise contests, candidate z is a Condorcet loser. However, under Bucklin's procedure candidate z will be elected because the number of voters (16) who rank z first or second in their preference ordering exceeds the number of voters who rank any of the other three candidates in first or second place in their preference ordering.

Example 3.5.8.2 demonstrates the vulnerability of Bucklin's procedure to the Reinforcement paradox.

3.5.8.2 Example

This example is due to Tideman (2006, p. 205, Example 13.19). Suppose the are two districts, I and II. In District I there are 15 voters whose preference ordering among three candidates, a, b, c, are as follows:

No. of voters	Preference ordering
6	$a \succ c \succ b$
5	$b \succ a \succ c$
4	$c \succ b \succ a$

and in district II there are nine voters whose preference orderings among the same three candidates are as follows:

No. of Voters	Preference Ordering
5	$a \succ b \succ c$
4	$c \succ b \succ a$

Given these data a will be elected under Bucklin's procedure in district I (in the second counting round with 11 votes), as well as in district II (in the first counting round with five votes).

However if, *ceteris paribus*, the two districts are amalgamated into a single district, we obtain a district of 24 voters with the following preference orderings among the three candidates:

No. of voters	Preference ordering
5	$a \succ b \succ c$
6	$a \succ c \succ b$
5	$b \succ a \succ c$
8	$c \succ b \succ a$

In this amalgamated district candidate b will be elected under Bucklin's procedure (in the second counting round with 18 votes) – in violation of the reinforcement axiom.

Example 3.5.8.3 demonstrates the susceptibility of Bucklin's procedure to the no-show, twin, and truncation paradoxes.

3.5.8.3 Example

Suppose there are 101 voters whose preference orderings among four candidates, a, b, c, and d, are as follows:

No. of voters	Preference ordering
43	$a \succ b \succ c \succ d$
26	$b \succ c \succ d \succ a$
15	$c \succ d \succ b \succ a$
17	$d \succ a \succ b \succ c$

If one of the four candidates must be elected under Bucklin's procedure then candidate b would be elected because the number of voters (69) who rank b first or second in their preference ordering exceeds the number of voters who rank any of the other candidates in first or second place in their preference ordering.

Now suppose that *ceteris paribus*, 16 of the 17 voters whose top preference is d decide not to participate in the election. As a result candidate a would be elected because an absolute majority of the voters (43) rank a as their top preference. This result is preferable for all the voters whose top preference is d who thus obtain their second preference (instead of their third preference) – thereby demonstrating simultaneously both the No-Show and twin paradoxes.

To demonstrate the vulnerability of Bucklin's procedure to the truncation paradox suppose that in the above example the 43 voters whose top preference is a decide to list only their top preference. In this case a would be listed first or second by 60 voters – which is more than any other voter is listed first or second – thereby elected under Bucklin's procedure, an outcome which these 43 voters prefer to the election of b.

Example 3.5.8.3 above can also be used to demonstrate the vulnerability of Bucklin's procedure to SCC. We just saw that in this example candidate b is elected (with 69 votes in the second counting round) under Bucklin's procedure when all four candidates and 101 voters participate in the election. However, *ceteris paribus*,

a is elected under Bucklin's procedure (in the first counting round with 60 votes) if candidate d drops out of the race – thereby demonstrating the violation of the SCC postulate.

Example 3.5.8.4 demonstrates the vulnerability of Bucklin's procedure to the preference inversion paradox.

3.5.8.4 Example

Suppose that there are four voters whose preference orderings among five candidates, a, b, c, d, and e, are as follows:

No. of voters	Preference ordering
1	$a \succ b \succ c \succ d \succ e$
1	$e \succ d \succ c \succ b \succ a$
1	$b \succ e \succ c \succ a \succ d$
1	$d \succ a \succ c \succ e \succ b$

If one of the four candidates must be elected under Bucklin's procedure then candidate c would be elected because the number of voters (3) who rank c first, second, or third in their preference ordering exceeds the number of voters who rank any of the other candidates in first, second or third place in their preference ordering.

Now suppose that, *ceteris paribus*, all voters invert their preference orderings among the four candidates. In this case c, who is placed in the middle of all candidates' preference orderings, would still be elected – thus demonstrating the vulnerability of Bucklin's procedure to the preference inversion paradox.

3.5.9 Demonstrating the Paradoxes Afflicting the range Voting (RV) Procedure

Except for being vulnerable to cyclical majorities and to strategic voting, the Range Voting (RV) procedure is vulnerable to the Condorcet winner paradox, the Condorcet loser paradox, the absolute winner paradox, the absolute loser paradox, and to the Truncation paradox.

In contrast to all other voting procedures except majority judgment (MJ) where a necessary condition to demonstrate the paradoxes afflicting them is that there exist at least three candidates, it is possible to demonstrate most of the paradoxes afflicting the RV (and MJ) procedure when there are just two candidates. The paradoxes afflicting the MJ procedure will be demonstrated in the next subsection.

Example 3.5.9.1 demonstrates simultaneously the vulnerability of the RV procedure to the first four paradoxes listed above.

3.5.9.1 Example

Suppose there are five voters, V_1, V_2, V_3, V_4, and V_5, who award the following (cardinal) grades (on a scale of 1–10) to two candidates, *x* and *y*:

Candidates /voters	V_1	V_2	V_3	V_4	V_5	Mean grade
x	2	2	2	3	10	3.8
y	1	1	1	10	7	4.0

As the mean grade of candidate *y* is higher than that of candidate *x*, candidate *y* is elected by the RV procedure. However, note that an absolute majority of the voters (V_1, V_2, V_3, V_5) awarded candidate *x* a higher grade than they awarded to candidate *y*, and an absolute majority of the voters (V_1, V_2, and V_3) awarded *y* the lowest grade. Hence candidate *x* is not only a Condorcet winner but also an absolute winner, whereas candidate *y* is not only a Condorcet loser but also an absolute loser.

Example 3.5.9.2: demonstrates the vulnerability of the RV procedure to the truncation paradox.

3.5.9.2 Example

Suppose there are seven voters, V_1–V_7, who award the following (cardinal) grades (on a scale of 1–10) to two candidates, *x* and *y*:

Candidates/voters	V_1	V_2	V_3	V_4	V_5	V_6	V_7	Mean grade
x	1	1	1	10	5	4	7	4.143
y	2	2	2	3	8	5	8	4.286

As the mean grade of candidate *y* is higher than that of candidate *x*, candidate *y* is elected by the RV procedure. However, as voter V_4 grades candidate *x* higher than *y* he is not satisfied with this result and will be better off if he does not grade candidate *y* at all, thereby demonstrating the truncation paradox. (*Ceteris paribus*, if voter V_4 does not grade candidate *y* then this candidate will be deemed to have been awarded the lowest grade (1) by voter V_4 and, as a result, the average grade of candidate *y* will drop to 4.0 thus electing candidate *x*.)

3.5.10 Demonstrating Paradoxes Afflicting the Majority Judgment (Mj) Procedure

The paradoxes afflicting the majority judgment (MJ) procedure are discussed at length in Felsenthal and Machover (2008).

Except for being vulnerable to cyclical majorities and to strategic voting, the MJ procedure is vulnerable to the Condorcet winner paradox, the Condorcet loser paradox, the Absolute Winner paradox, the absolute loser paradox, the Truncation paradox, the Reinforcement paradox, the no-show paradox, and to the Twin paradox.

Example 3.5.10.1 demonstrates the vulnerability of the MJ procedure to the absolute winner, Condorcet winner, the absolute loser and Condorcet loser paradoxes.

3.5.10.1 Example

This example is due to Felsenthal and Machover (2008, p. 330). Suppose there are three voters, V_1, V_2, and V_3, who award the following (ordinal) grades (on a scale of A-H) to two candidates, x and y:

Candidate/voter	V_1	V_2	V_3	Median grade
x	B	C	H	C
y	A	F	G	F

As the median grade of candidate y is higher than that of candidate x, candidate y is elected by the MJ procedure. However, note that an absolute majority of the voters (V_1 and V_3) awarded candidate y a lower grade than they awarded candidate x – hence candidate x is not only a Condorcet winner but also an absolute winner, whereas candidate y is not only a Condorcet loser but also an absolute loser.

Example 3.5.10.2 demonstrates the vulnerability of the MJ procedure to the reinforcement paradox.

3.5.10.2 Example

This example is due to Felsenthal and Machover (2008, p. 327).

Suppose there are three regions, I, II, and III, in each of which 101 voters grade each of two candidates, x and y, on an ordinal scale A-D. The following lists show the distributions of grades. The figure next to a grade is the number of voters awarding that grade.

	Region I			
x:	21A	31B	48C	1D
y:	40A	11B	48C	2D
	Region II			
x:	1A	46B	14C	40D
y:	1A	45B	33C	22D
	Region III			
x:	40B	20C	41D	
y:	48B	3C	50D	

In all three elections the two candidates have equal median grades (median grade B in region I and median grade C in regions II, III), so the tie-breaking algorithm proposed by Balinski and Laraki (2007, 2011) must be used. The number of iterations required for breaking the tie in each of the three regions are 2, 7, and 2, respectively, whereupon y wins in each of the three regions.[15]

However, if the three regions are amalgamated into a single region we obtain the following distribution of grades awarded to candidates x and y by the 303 voters:

	Amalgamated region:			
x:	22A	117B	82C	82D
y:	41A	104B	84C	74D

Here again candidates x and y obtain the same median grade (C), but when one breaks this tie (after 13 iterations) x wins – in violation of the Reinforcement postulate.

Example 3.5.10.3 demonstrates the vulnerability of the MJ procedure to the no-show and twin paradoxes.

3.5.10.3 Example

This example is due to Felsenthal and Machover (2008, p. 329).

Suppose that seven voters, V_1–V_7, grade two candidates, x and y, on an ordinal scale ranging between A and F, as follows:

Candidate/voter	V_1	V_2	V_3	V_4	V_5	V_6	V_7	Median grade
x	A	A	A	D	E	E	F	D
y	B	B	B	C	F	F	F	C

Here x wins. But now suppose that voters V_1 and V_2, both of whom awarded the same grades as voter V_3, and who prefer candidate y, abstain from voting. Then we get:

Candidate/voter	V_3	V_4	V_5	V_6	V_7	Median grade
x	A	D	E	E	F	E
y	B	C	F	F	F	F

Here y wins. Thus by abstaining voters V_1 and V_2 cause their favorite candidate to win – thereby demonstrating the vulnerability of the MJ procedure to the no-show

[15] To break a tie between two leading candidates who have the same median grade, one performs one or more iterations in each of which the equal median grade of the two candidates is dropped. This process continues until one reaches a situation where the candidates' median grades are no longer the same. If no such situation is reached then the tie is broken randomly. With an even number of grades Balinski and Laraki take the median to be the lower of the two middle grades.

paradox. Similarly, since V_3 prefers candidate y to x, one could expect that if, *ceteris paribus*, the two "twins" (V_1 and V_2) – who grade the two candidates in the same way as V_3 – would join the electorate, then y would certainly be elected. However, as can be seen from the first table, in this case x would be elected, thereby demonstrating the vulnerability of the MJ procedure to the twin paradox.

Example 3.5.10.4 demonstrates the vulnerability of the MJ procedure to the truncation paradox.

3.5.10.4 Example

Suppose there are seven voters, V_1–V_7, who award the following (ordinal) grades (on a scale of A- J) to two candidates, x and y:

Candidate/voter	V_1	V_2	V_3	V_4	V_5	V_6	V_7	Median grade
x	A	A	A	J	E	D	G	D
y	B	B	B	C	H	E	H	C

Here x is elected because his median grade is higher than that of y. Voter V_6 does not like this result so if, *ceteris paribus*, he decides to grade only candidate y, then candidate x would be deemed to have been awarded the lowest grade (A) by V_6 and, consequently, candidate x's median grade would drop from D to A – causing candidate y to be elected. Voter V_6 of course prefers this result – thereby demonstrating the truncation paradox.

3.5.11 Demonstrating Paradoxes Afflicting the Minimax Procedure (aka Simpson–Kramer or Condorcet's Procedure)

Except for being vulnerable to cyclical majorities and to strategic voting, the Minimax procedure (aka Condorcet's procedure or Simpson–Kramer rule) is vulnerable to the Condorcet loser, absolute loser, no-show, twin, truncation, reinforcement, preference inversion, and SCC paradoxes.

When the social preference ordering contains a top cycle it is possible that the minimax procedure will elect a Condorcet loser which may also be an absolute loser. Example 3.5.11.1 demonstrates this.

3.5.11.1 Example

Suppose there are 11 voters whose preference orderings among four candidates, a, b, c, d, are as follows:

No. of voters	Preference ordering
2	$d \succ a \succ c \succ b$
3	$d \succ b \succ a \succ c$
3	$c \succ b \succ a \succ d$
1	$b \succ a \succ c \succ d$
2	$a \succ c \succ b \succ d$

This preference profile can be depicted as the following *matrix of paired comparisons*. In such a matrix, the entry in row i and column j is the number of voters who rank candidate i ahead of candidate j.

	a	b	c	d
a	–	4	8	6
b	7	–	4	6
c	3	7	–	6
d	5	5	5	–

As an example of how the numbers in this paired comparisons matrix are calculated, the number 4 in the second column of the first row derives from the two voters in the first row of the example plus the two voters in the last row of the example who rank a ahead of b. Similarly, the number 7 in the first column of the second row derives from the three voters in the second row of the example, plus the three voters in the third row of the example, plus the one voter in the fourth row of the example, who rank b ahead of a.

As can be seen from the paired comparisons matrix, the social preference ordering in Example 3.5.11.1 contains a top cycle $[b \succ a \succ c \succ b] \succ d$, i.e., d is the Condorcet loser which happens to be also an absolute loser. However, the minimax procedure will elect d because d's worst loss margin (6) is smaller than the worst loss margin of each of the other three candidates (7, 7, 8 for a, b, c, respectively).

This example can also be used to demonstrate the vulnerability of the minimax procedure to the preference inversion paradox. If all voters invert their preference orderings then d becomes an Absolute Winner and hence is elected under the minimax procedure.

Example 3.5.11.2 demonstrates the vulnerability of the minimax procedure to the no-show and twin paradoxes.

3.5.11.2 Example

This example is due to Hannu Nurmi (private communication 22.2.2010; this example appears also in section 10.5.5 in this volume). Suppose there are 19 voters who must elect one out of four candidates, a, b, c, d and whose preference orderings among these candidates are as follows:

No. of voters	Preference ordering
5	$d \succ b \succ c \succ a$
4	$b \succ c \succ a \succ d$
3	$a \succ d \succ c \succ b$
3	$a \succ d \succ b \succ c$
4	$c \succ a \succ b \succ d$

These preference orderings can be depicted as the following paired comparisons matrix:

	a	b	c	d
a	–	10	6	14
b	9	–	12	8
c	13	7	–	8
d	5	11	11	–

Here the social preference ordering is cyclical ($c \succ a \succ d \succ b \succ c$). So according to the Minimax procedure one should elect that candidate whose worst loss is smallest. From the paired comparisons matrix it is seen that the worst loss of candidates a, b, c, d, is 13, 11, 12, and 14, respectively, so candidate b is elected.

Now suppose that, *ceteris paribus*, three of the four voters with preference ordering $c \succ a \succ b \succ d$ decide not to participate in the election. In this case the paired comparisons matrix changes as follows:

	a	b	c	d
a	–	7	6	11
b	9	–	12	5
c	10	4	–	5
d	5	11	11	–

The social preference ordering is still cyclical but the worst losses of the four candidates are now 10, 11, 12, 11 for candidates a, b, c, d, respectively, so according to the minimax procedure candidate a is elected – which is preferable from the point of view of the absent voters – thereby demonstrating the vulnerability of the minimax procedure to the no-show paradox.

We also have here an instance of the twin paradox. We have just seen that if, *ceteris paribus*, only one of the four voters whose preference orderings are c $\succ a \succ b \succ d$ participates in the election then according to the minimax procedure candidate a is elected. But if this voter's three twin brothers join the electorate then, as we have seen at the beginning of Example 3.5.11.2, candidate b is elected according to the minimax procedure – thereby demonstrating this procedure's vulnerability to the twin paradox.

Example 3.5.11.3 demonstrates the vulnerability of the minimax procedure to the truncation paradox.

3.5.11.3 Example

This example is due to Hannu Nurmi (private communication 23.2.2010; this example appears also in section 10.5.5 in this volume). As we have seen in the first part of Example 3.5.11.2, candidate b would be elected under the minimax procedure. Now suppose that, *ceteris paribus*, the four voters whose preference ordering is $c \succ a \succ b \succ d$ would decide to state only their top two preferences, c and a. This would lead to the assumption that the probability that these voters prefer b to d is equal to the probability that they prefer d to b, which would result, in turn, in the following paired comparisons matrix:

	a	b	c	d
a	–	10	6	14
b	9	–	12	6
c	13	7	–	8
d	5	13	11	–

From this paired comparisons matrix it is easy to see that candidate c's largest loss (12 against candidates b) is smallest, hence this candidate will be elected under the minimax procedure – which is certainly preferable for the voters whose top preference is c – thus demonstrating the vulnerability of the minimax procedure to the truncation paradox.

Example 3.5.11.4 demonstrates the vulnerability of the minimax procedure to the reinforcement paradox.

3.5.11.4 Example

Suppose there are two districts, one with 11 voters whose preference orderings among four candidates are as in Example 3.5.11.1 and a second district with three voters two of whom have preference ordering $d \succ a \succ b \succ c$ and the third voter has preference ordering $b \succ a \succ c \succ d$.

As we have seen in Example 3.5.11.1, candidate d will be elected in the first district, and as candidate d is the absolute winner in the second district s/he will also be elected in the second district under the minimax procedure.

Now suppose that, *ceteris paribus*, these two districts are amalgamated into one district of 14 voters having the following paired comparisons matrix:

	a	b	c	d
a	–	6	11	7
b	8	–	7	7
c	4	7	–	7
d	7	7	7	–

From this paired comparisons matrix it is easy to see that there is a tie between candidates b and d because the largest loss of both of them is smallest (7), thus according to the minimax procedure a lottery should be conducted between them – thereby demonstrating the vulnerability of the minimax procedure to the reinforcement paradox.

Example 3.5.11.5 demonstrates the vulnerability of the minimax procedure to SCC.

3.5.11.5 Example

This example is due to Fishburn (1974, p. 540). Suppose there are seven voters who are divided into three groups who have to select under the Minimax procedure one out of four candidates, a, b, c, or d, and whose preference orderings among these candidates are as follows:

Group	No. of voters	Preference ordering
G1	3	$d \succ c \succ b \succ a$
G2	2	$a \succ d \succ c \succ b$
G3	2	$b \succ a \succ d \succ c$

From this preference list we see that the social preference ordering is cyclical ($a \succ d \succ c \succ b \succ a$). It can be depicted as a (cyclical) paired comparisons matrix as follows:

	a	b	c	d
a	–	2	4	4
b	5	–	2	2
c	3	5	–	0
d	3	5	7	–

From this matrix we can see that the worst loss of candidate a is 5 (against candidate b), the worst loss of candidate b is also 5 (against candidates c, d), the worst loss of candidate c is 7 (against candidate d) and the worst loss of candidate d is 4 (against candidate a). As candidate d's loss is the smallest, this candidate would be elected under the minimax procedure.

Now suppose that, *ceteris paribus*, candidate b drops out of the race. In this case candidate a becomes the absolute winner and will be elected under the minimax procedure – in violation of SCC.

3.5.12 Demonstrating Paradoxes Afflicting Dodgson's Procedure

Except for being vulnerable to cyclical majorities and to strategic voting, Dodgson's procedure is vulnerable to the Condorcet loser, lack of monotonicity, reinforcement, no-show, twin, truncation, preference inversion, and SCC paradoxes.

Example 3.5.12.1 demonstrates the vulnerability of Dodgson's procedure to the Condorcet loser paradox.

3.5.12.1 Example

This example is due to Fishburn (1977, p. 477). Suppose there are seven voters whose preference orderings among eight candidates, a, b, c, d, e, f, g, x, are as follows:

No. of voters	Preference ordering
1	$a \succ b \succ c \succ d \succ x \succ e \succ f \succ g$
1	$g \succ a \succ b \succ c \succ x \succ d \succ e \succ f$
1	$f \succ g \succ a \succ b \succ x \succ c \succ d \succ e$
1	$e \succ f \succ g \succ a \succ x \succ b \succ c \succ d$
1	$d \succ e \succ f \succ g \succ x \succ a \succ b \succ c$
1	$c \succ d \succ e \succ f \succ x \succ g \succ a \succ b$
1	$b \succ c \succ d \succ e \succ x \succ f \succ g \succ a$

The social preference ordering contains a top cycle $[a \succ b \succ c \succ d \succ e \succ f \succ g \succ a] \succ x$. It can be presented by the following paired comparisons matrix:

	a	b	c	d	e	f	g	x
a	–	6	5	4	3	2	1	4
b	1	–	6	5	4	3	2	4
c	2	1	–	6	5	4	3	4
d	3	2	1	–	6	5	4	4
e	4	3	2	1	–	6	5	4
f	5	4	3	2	1	–	6	4
g	6	5	4	3	2	1	–	4
x	3	3	3	3	3	3	3	–

As can easily be seen from this matrix, candidate x is a Condorcet loser as this candidate is beaten in pairwise comparisons by each of the other seven candidates. Nevertheless, candidate x will be elected in this case by Dodgson's procedure because for x to become a Condorcet winner only four preference inversions are needed (e.g., it is sufficient for any of the voters to move candidate x from fifth to first place in his preference ordering), whereas for any of the other candidates to become a Condorcet winner at least six preference inversions are needed.

This example can also be used to demonstrate the vulnerability of Dodgson's procedure to the Preference Inversion paradox. If all voters invert their preference orderings in this example then x becomes a Condorcet winner and hence is elected under Dodgson's procedure.

Example 3.5.12.2 demonstrates the vulnerability of Dodgson's procedure to lack of monotonicity. At least four candidates must exist for this to occur (Fishburn, 1982, p. 132).

3.5.12.2 Example

This example was adapted by Hannu Nurmi (private communication 15.2.2010) from Fishburn (1977, p. 478). Suppose there are 100 voters who are divided into four groups, who must elect one out of five candidates a, b, c, d, e, under Dodgson's procedure, and whose preference orderings among the candidates are as follows:

Group	No. of voters	Preference ordering
G1	42	$b \succ a \succ c \succ d \succ e$
G2	26	$a \succ e \succ c \succ b \succ d$
G3	21	$e \succ d \succ b \succ a \succ c$
G4	11	$e \succ a \succ b \succ d \succ c$

The social preference ordering has a top cycle: $[b \succ a \succ e \succ b] \succ c \succ d$. It can be depicted as the following paired comparisons matrix:

	a	b	c	d	e
a	–	37	100	79	68
b	63	–	74	79	42
c	0	26	–	68	42
d	21	21	32	–	42
e	32	58	58	58	–

For candidate a to become the Condorcet winner at least 14 voters in group G1 must change $b \succ a$ in their preference ordering to $a \succ b$, i.e., a total of 14 changes.

For candidate b to become the Condorcet winner at least nine voters from group G4 must first change $a \succ b$ to $b \succ a$ and thereafter $e \succ b$ to $b \succ e$ in their preference ordering, i.e., a total of 18 changes.

For candidate e to become the Condorcet winner at least 19 voters in group G2 must change $a \succ e$ in their preference ordering to $e \succ a$, i.e., a total of 19 changes.

Since the number of changes needed in the voters' preference orderings in order for a to become the Condorcet winner is the smallest, a would be elected under Dodgson's procedure.

Now suppose that, *ceteris paribus*, the 11 voters in group G4 *increase* their support of candidate a by changing their preference orderings from $e \succ a \succ b \succ d \succ c$ to $a \succ e \succ b \succ d \succ c$. This change can be depicted by the following paired comparisons matrix:

	a	b	c	d	e
a	–	37	100	79	79
b	63	–	74	79	42
c	0	26	–	68	42
d	21	21	32	–	42
e	21	58	58	58	–

From this matrix it is possible to see that despite the increase in a's support it would still take at least 14 persons from group G1 to change in their preference orderings $b \succ a$ to $a \succ b$ in order for a to become the Condorcet winner, whereas now for b to become the Condorcet winner only nine voters in G4 would have to change $e \succ b$ to $b \succ e$ in their preference orderings. So since the number of changes needed for b to become the Condorcet winner is smallest, b would be elected under Dodgson's procedure – thereby demonstrating lack of monotonicity.

The first part of Example 3.5.12.2 can also be used to demonstrate the vulnerability of Dodgson's procedure to the no-show and twin paradoxes. If 20 of the 21 voters in group G3 decide not to participate in the election then b becomes the Condorcet winner and will be elected according to Dodgson's procedure. The election of b is of course preferred by the members of group G3 over the election of a thus demonstrating the vulnerability of Dodgson's procedure to the no-show paradox. Adding those 20 twins back to retrieve the original profile shows that Dodgson's procedure is also vulnerable to the Twin paradox.

Example 3.5.12.2 can also be used to demonstrate the vulnerability of Dodgson's procedure to SCC. As we have seen from the paired comparisons matrix of the first part of Example 3.5.12.2 candidate a is selected by Dodgson's procedure. However if, *ceteris paribus*, candidate e drops out of the race then candidate b becomes the Condorcet winner and is elected by Dodgson's procedure – in violation of SCC.

Example 3.5.12.3 demonstrates the vulnerability of Dodgson's procedure to the Reinforcement paradox.[16]

3.5.12.3 Example

This example is due to Fishburn (1977, p. 484). Suppose there are two districts, I and II, in each of them one of four candidates, w, x, y, z, must be elected.

In district I there are seven voters, four with preference ordering $x \succ y \succ z \succ w$ and three with preference ordering $y \succ x \succ z \succ w$. Since x is here the Condorcet winner, x is elected according to Dodgson's procedure.

[16]Note that by increasing a's support, the 11 voters of group G4 obtained the election of b which for them is a less preferable alternative than the election of a. In demonstrating the non-monotonicity paradox under the other four procedures surveyed in this chapter that are susceptible to this paradox (Plurality with Runoff, Alternative Vote, Coombs, Nanson), it is exemplified not only that an original winner, w, loses after one or more voters, V_i, increase their support of w by moving w upwards in their preference ordering, but also that the voters belonging to V_i *benefit* from this because the new winner, y, is ranked higher than w in V_i's original preference ordering. However, under Dodgson's procedure it is impossible to construct such an example because when w rises in V_i's ranking, the indirect benefit, if any, goes to the candidates ranked below w in V_i's preference ordering who now find the candidates who had been ranked above w more accessible. But if V_i's initial ranking is assumed to be sincere, then it follows, by definition, that the members of V_i prefer w over any of the candidates ranked below w. So if some candidate ranked below w is elected then the members of V_i are harmed. Hence non-monotonicity under Dodgson's procedure cannot arise from considerations of strategic voting. I am grateful to Nicolaus Tideman for this insight (private communication 3.8.2011).

In district II there are 12 voters whose preference orderings are as follows:

No. of voters	Preference ordering
1	$x \succ y \succ z \succ w$
2	$y \succ x \succ z \succ w$
3	$w \succ y \succ x \succ z$
3	$z \succ w \succ y \succ x$
3	$x \succ z \succ w \succ y$

These orderings can be presented as the following paired comparisons matrix:

	w	x	y	z
w	–	6	9	3
x	6	–	4	9
y	3	8	–	6
z	9	3	6	–

Here the social preference ordering is cyclical $[w \succ y \succ x \succ z \succ w]$. For x to become the Condorcet winner only four preference inversions are needed (the two voters whose top preference is y should change their top preference to x, and one of the three voters whose top preference is w should change his top preference to x), whereas for any of the other candidates to become a Condorcet winner more than four preference inversions are needed. So according to Dodgson's procedure candidate x is elected also in district II.

Now suppose that, *ceteris paribus*, the two districts are amalgamated into a single district with 19 voters. In this case candidate y becomes the Condorcet winner and is elected according to Dodgson's procedure – thereby demonstrating its vulnerability to the Reinforcement paradox.

Example 3.5.12.4 demonstrates Dodgson's vulnerability to the Truncation paradox.

3.5.12.4 Example

Suppose there are 49 voters whose preference orderings among five candidates, a, b, c, d, e are as follows:

No. of voters	Preference ordering
11	$b \succ a \succ d \succ e \succ c$
10	$e \succ c \succ b \succ d \succ a$
10	$a \succ c \succ d \succ b \succ e$
2	$e \succ c \succ d \succ b \succ a$
2	$e \succ d \succ c \succ b \succ a$
2	$c \succ b \succ a \succ d \succ e$
1	$d \succ c \succ b \succ a \succ e$
1	$a \succ b \succ d \succ e \succ c$
10	$e \succ d \succ a \succ b \succ c$

These orderings can be presented as the following paired comparisons matrix:

	a	b	c	d	e
a	–	21	32	24	25
b	28	–	22	24	25
c	17	27	–	24	13
d	25	25	25	–	25
e	24	24	36	24	–

Here candidate d is the Condorcet winner so this candidate is elected according to Dodgson's procedure. However, if the 10 voters whose preference ordering is $e \succ d \succ a \succ b \succ c$ decide to reveal only their top preference (e) – in which case one assumes that these voters prefer candidate e over all the other four candidates and that all possible preference orderings among these candidates are equiprobable – then one obtains the following paired comparisons matrix:

	a	b	c	d	e
a	–	16	27	29	25
b	33	–	17	29	25
c	22	32	–	29	13
d	20	20	20	–	25
e	24	24	36	24	–

From this matrix we see that the social preference ordering is cyclical ($d \succ e \succ c \succ b \succ a \succ d$). So according to Dodgson's procedure candidate e is elected in this situation because for candidate e to become a Condorcet winner only three preference inversions are needed (if one of the 11 $b \succ a \succ d \succ e \succ c$ voters will change his preference ordering to $e \succ b \succ a \succ d \succ c$), whereas for any of the other candidates to become a Condorcet winner more than three preference inversions are needed – thereby demonstrating the vulnerability of Dodgson's procedure to the truncation paradox.

3.5.13 Demonstrating the Paradoxes Afflicting Black's Procedure

Since Black's procedure is a hybrid procedure (when a Condorcet winner exists it elects the Condorcet winner, and when a Condorcet winner does not exist it elects the Borda winner), it is vulnerable to the no-show, twin, truncation, reinforcement, and SCC paradoxes. Although Black's procedure is not vulnerable to the Condorcet loser paradox, it may violate Smith's (1973) Condorcet principle.

Example 3.5.13.1 demonstrates the vulnerability of Black's procedure to the no-show, twin, and truncation paradoxes.

3.5.13.1 Example

This example is partly due to Hannu Nurmi (private communication, 15.2.2010; this example appears also in section 10.5.1 in this volume). Suppose there are 16 voters whose preference orderings among five candidates, a, b, c, d, e, are as follows:

No. of voters	Preference ordering
3	$d \succ e \succ a \succ b \succ c$
3	$e \succ a \succ c \succ b \succ d$
4	$c \succ d \succ e \succ a \succ b$
3	$d \succ e \succ b \succ c \succ a$
3	$e \succ b \succ a \succ d \succ c$

Here d is the Condorcet winner and hence is elected under Black's procedure. Suppose now that, *ceteris paribus*, two of the voters whose preference ordering is $e \succ b \succ a \succ d \succ c$ decide not to participate in the election. As a result the social preference ordering becomes cyclical ($a \succ b = c = d \succ e \succ a$) and e emerges as the Borda winner and is therefore elected under Black's procedure. Since e is ranked first by the two absent voters, it turns out that they obtained a better outcome by not participating in the election – thereby demonstrating the vulnerability of Black's procedure to the no-show paradox.

We also have here an instance of the Twin paradox: if, *ceteris paribus*, the two absent voters decide to participate in the election and join their twin brother, then d becomes the Condorcet winner and will be elected under Black's procedure – thereby demonstrating the vulnerability of Black's procedure to the twin paradox.

Obviously, not voting at all is an extreme version of truncation and hence the above example can also be used to show that Black's procedure is vulnerable to the truncation paradox. Thus if, *ceteris paribus*, all three voters whose preference ordering is $e \succ b \succ a \succ d \succ c$ truncate their preference ordering after a, i.e., if they do not express their preferences between c and d – which would automatically be considered to mean that they prefer each of the three ranked alternatives over c and d and are indifferent between c and d – then the social preference ordering will become cyclic ($d \succ e \succ a \succ b \succ c \succ d$) and e will emerge as the Borda winner to be elected under Black's procedure – which is a preferable outcome for these voters.

The vulnerability of Borda's procedure (and hence also Black's) to the truncation paradox when a Condorcet winner does not exist initially is demonstrated in Example 3.5.5.2 above.

Example 3.5.13.2 demonstrates the vulnerability of Black's procedure to the reinforcement paradox.

3.5.13.2 Example

Suppose there are two districts, I and II. In district I there are five voters whose preference orderings among three candidates, a, b, and c, are as follows:

No. of voters	Preference ordering
2	$a \succ b \succ c$
2	$b \succ c \succ a$
1	$c \succ a \succ b$

and in District II there are nine voters whose preference orderings among these three candidates are as follows:

No. of **voters**	Preference **ordering**
5	$b \succ c \succ a$
4	$c \succ a \succ b$

The social preference ordering in district I is cyclical ($a \succ b \succ c \succ a$), so according to Borda's (and Black's) procedure candidate b, whose Borda score (6) is largest, is elected in this district. In district II candidate b is the Condorcet winner, so according to Black's procedure b is elected in this district too.

Now suppose that, *ceteris paribus*, the two districts are amalgamated into a single large district of 14 voters whose preference ordering among the three candidates are as follows:

No. of **voters**	Preference **ordering**
2	$a \succ b \succ c$
7	$b \succ c \succ a$
5	$c \succ a \succ b$

As the social preference ordering in the amalgamated district is cyclical ($c \succ a = b \succ c$) candidate c is elected in this district because his Borda score (17) is largest – thus demonstrating the vulnerability of Black's procedure to the reinforcement paradox.

The vulnerability of Black's procedure to SCC is demonstrated in Example 5.11.5 above. When all four candidates compete the social preference ordering is cyclical ($a \succ d \succ c \succ b \succ a$) so according to Black's procedure candidate d is elected because this candidate has the highest Borda score (15). But if, *ceteris paribus*, candidate b drops out of the race then candidate a becomes the Condorcet winner and is therefore elected according to Black's procedure – contrary to SCC.

Example 3.5.13.3 demonstrates the violation of Smith's (1973) Condorcet principle by Black's procedure.

3.5.13.3 Example

This example is due to Fishburn (1977, p. 480). Suppose there are five voters whose preference orderings among eight candidates a, b, c, d, e, x, y, z, are as follows:

No. of voters	Preference ordering
1	$a \succ b \succ c \succ x \succ y \succ z \succ d \succ e$
1	$e \succ a \succ b \succ x \succ y \succ z \succ c \succ d$
1	$d \succ e \succ a \succ x \succ y \succ z \succ b \succ c$
1	$c \succ d \succ e \succ x \succ y \succ z \succ a \succ b$
1	$b \succ c \succ d \succ x \succ y \succ z \succ e \succ a$

These preference orderings can be depicted as the following paired comparisons matrix:

	a	b	c	d	e	x	y	z
a	–	4	3	2	1	3	3	3
b	1	–	4	3	2	3	3	3
c	2	1	–	4	3	3	3	3
d	3	2	1	–	4	3	3	3
e	4	3	2	1	–	3	3	3
x	2	2	2	2	2	–	5	5
y	2	2	2	2	2	0	–	5
z	2	2	2	2	2	0	0	–

The social preference ordering here has a top cycle $[a \succ b \succ c \succ d \succ e \succ a] \succ x \succ y \succ z$, so according to Black's procedure one must use Borda's procedure in order to determine which of the eight candidates will be deemed the winner. The Borda counts of each of the candidates $a - e$ is 19, that of candidate x is 20, and those of candidates y and z are 15 and 10, respectively. So according to Black's procedure candidate x is elected because he has the highest Borda score. However, since Borda's procedure violates here Smith's (1973) Condorcet principle, so does Black's procedure.[17]

3.5.14 Demonstrating Paradoxes Afflicting Copeland's Procedure

Except for being vulnerable to cyclical majorities and to strategic voting, Copeland's procedure is vulnerable to the no-show, twin, truncation, reinforcement and SCC paradoxes.

[17]As noted above in Sect. 3.2.1.4, Smith's (1973) Condorcet principle states that if the set of candidates can be partitioned into two disjoint subsets, A and B, such that each candidate belonging to A can beat in paired comparisons each of the candidates belonging to B, then none of the candidates belonging to B ought to be elected unless all candidates in A are elected. In Example 3.5.13.3 each of candidates $a - e$ beats in paired comparisons each of the candidates x, y, z. However, Borda's procedure (and Black's) elects here candidate x although only a single candidate must be elected – in violation of Smith's Condorcet principle.

Example 3.5.14.1 demonstrates the vulnerability of Copeland's procedure to the no-show, twin, and truncation paradoxes.

3.5.14.1 Example

Suppose there are 33 voters who must select one out of four candidates, a, b, c, or d, and whose preference orderings among these four candidates are as follows:

No. of voters	Preference ordering
11	$a \succ b \succ c \succ d$
2	$b \succ c \succ a \succ d$
12	$b \succ c \succ d \succ a$
4	$c \succ a \succ d \succ b$
2	$d \succ a \succ b \succ c$
2	$d \succ b \succ a \succ c$

This preference list can be depicted as the following paired comparisons matrix:

	a	b	c	d
a	–	17	15	17
b	16	–	29	25
c	18	4	–	29
d	16	8	4	–

From this paired comparisons matrix we see that the social preference ordering has a top cycle $[a \succ b \succ c \succ a] \succ d$, so according to Copeland's procedure there is a tie between a, b and c.

Now suppose that, *ceteris paribus*, one of the two voters whose preference ordering is $b \succ c \succ a \succ d$ decides not to participate in the election. This change will result in the following paired comparisons matrix:

	a	b	c	d
a	–	17	15	16
b	15	–	28	24
c	17	4	–	28
d	16	8	4	–

From this matrix we can see that according to Copeland's procedure each of candidates b and c gets two points (since each of these two candidates beats two other candidates), while candidates a and d get 1.5 and 0.5 points, respectively. This result is certainly preferable from the point of view of the voter who decided not to participate, thus demonstrating the vulnerability of the Copeland's procedure to the no-show paradox.

The same example can also be used to demonstrate the vulnerability of Copeland's procedure to the twin paradox.

We have just seen that in the second part of this example one obtains a tie between candidates b and c. So one could expect, presumably, that if a twin brother of the voter with preference ordering $b \succ c \succ a \succ d$ joins the electorate (instead of abstaining), the chances of candidate b to get elected would increase. But as we have seen from the first part of this example when, *ceteris paribus*, two voters with preference ordering $b \succ c \succ a \succ d$ exist in the electorate then the chances of candidate b to get elected according to Copeland's procedure *decrease* because in this case one obtains a tie between b and two other candidates (a and c), whereas one obtains a tie between b and just one other candidate (c) when only one voter with preference ordering $b \succ c \succ a \succ d$ exists in the electorate – thus demonstrating the vulnerability of Copeland's procedure to the twin paradox.

To demonstrate the truncation paradox suppose that, *ceteris paribus*, in the first part of the above example the two voters with preference ordering $b \succ c \succ a \succ d$ would decide to reveal only their top preference. In this case one would have to assume that all the six possible preference orderings of these voters among candidates a, c, d are equiprobable (or, equivalently, that they are indifferent among them) and, consequently, one would obtain the following paired comparisons matrix:

	a	b	c	d
a	–	17	16	16
b	16	–	29	25
c	17	4	–	28
d	17	8	5	–

From this paired comparisons matrix it is easy to see that according to Copeland's procedure there would be a tie between candidates b and c (each obtaining two points) – which is a preferable result from the point of view of the two $b \succ c \succ a \succ d$ voters over a tie among candidates a, b, c which was obtained, *ceteris paribus*, when these voters revealed their entire preference ordering among all four candidates.

Example 3.5.14.2 demonstrates the vulnerability of Copeland's procedure to the reinforcement paradox.

3.5.14.2 Example

Suppose there are two districts, I and II. In district I there are three voters whose preference orderings among four candidates, a, b, c, and d, are as follows:

No. of voters	Preference ordering
1	$a \succ b \succ c \succ d$
1	$b \succ d \succ c \succ a$
1	$d \succ c \succ a \succ b$

and in district II there are two voters, one with preference ordering $b \succ d \succ c \succ a$, and the other with preference ordering $d \succ b \succ c \succ a$.

According to Copeland's procedure there is a tie between candidates b and d in each of the two districts.

However, *ceteris paribus*, if the two districts are amalgamated into a single district of five voters then one obtains the following preference list:

No. of **voters**	Preference **ordering**
1	$a \succ b \succ c \succ d$
2	$b \succ d \succ c \succ a$
1	$d \succ b \succ c \succ a$
1	$d \succ c \succ a \succ b$

This preference list can be depicted as the following paired comparisons matrix:

	a	b	c	d
a	–	2	1	1
b	3	–	4	3
c	4	1	–	1
d	4	2	4	–

From this paired comparisons matrix it is clear that candidate b is the Condorcet winner and hence is elected according to Copeland's procedure – contrary to the reinforcement axiom.

Example 3.5.11.5 can be used to demonstrate the vulnerability of Copeland's procedure to the SCC paradox. According to that example there is a tie according to Copeland's procedure between candidates a and d. However if, *ceteris paribus*, candidate b is eliminated then candidate a becomes the Condorcet winner and is elected by Copeland's procedure – in violation of the SCC postulate.

3.5.15 Demonstrating the Paradoxes Afflicting Kemeny's Procedure

Except for being vulnerable to cyclical majorities and to strategic voting, Kemeny's procedure is vulnerable to the reinforcement, no-show, twin, truncation, and SCC paradoxes.

Example 3.5.15.1 demonstrates the vulnerability of Kemeny's procedure to the reinforcement paradox. It can also be used to demonstrate the vulnerability of Dodgson's procedure to this paradox.

3.5.15.1 Example

This example is due to Fishburn (1977, p. 484). Suppose there are two districts, I and II.

In district I there are two voters whose preference orderings among nine candidates are as follows: $x \succ y \succ a \succ b \succ c \succ d \succ e \succ f \succ g$. Here x is the Condorcet winner and hence will be elected according to Kemeny's procedure.

In district II there are seven voters whose preference orderings among the nine candidates are as follows:

No. of voters	Preference ordering
1	$y \succ x \succ a \succ b \succ c \succ d \succ e \succ f \succ g$
1	$y \succ x \succ g \succ a \succ b \succ c \succ d \succ e \succ f$
1	$y \succ x \succ f \succ g \succ a \succ b \succ c \succ d \succ e$
1	$e \succ f \succ g \succ a \succ b \succ c \succ d \succ y \succ x$
1	$d \succ e \succ f \succ g \succ a \succ b \succ c \succ y \succ x$
1	$c \succ d \succ e \succ f \succ g \succ a \succ b \succ y \succ x$
1	$x \succ b \succ c \succ d \succ e \succ f \succ g \succ a \succ y$

These preference orderings can be depicted as the following paired comparisons matrix:

	a	b	c	d	e	f	g	x	y
a	–	6	5	4	3	2	1	3	4
b	1	–	6	5	4	3	2	3	4
c	2	1	–	6	5	4	3	3	4
d	3	2	1	–	6	5	4	3	4
e	4	3	2	1	–	6	5	3	4
f	5	4	3	2	1	–	6	3	4
g	6	5	4	3	2	1	–	3	4
x	4	4	4	4	4	4	4	–	1
y	3	3	3	3	3	3	3	6	–

The social preference ordering here is cyclical: x beats each of the seven candidates $a - g$, whereas y beats x but is beaten by each of the seven candidates $a - g$. So it is clear that according to Kemeny's procedure the closest (non-cyclical) social preference ordering here is one in which x is the top-ranked candidate. (Note that x here has also the largest Borda score). So in district II too x is elected according to Kemeny's procedure.

However, in the amalgamated district (consisting of districts I and II), we obtain the following paired comparisons matrix:

	a	b	c	d	e	f	g	x	y
a	–	8	7	6	5	4	3	3	4
b	1	–	8	7	6	5	4	3	4
c	2	1	–	8	7	6	5	3	4
d	3	2	1	–	8	7	6	3	4
e	4	3	2	1	–	8	7	3	4
f	5	4	3	2	1	–	8	3	4
g	6	5	4	3	2	1	–	3	4
x	6	6	6	6	6	6	6	–	3
y	5	5	5	5	5	5	5	6	–

According to this matrix y is the Condorcet winner and hence elected under Kemeny's procedure – thereby demonstrating its vulnerability to the reinforcement paradox.

Example 3.5.15.2 demonstrates the vulnerability of Kemeny's procedure to the No-Show, Twin, and Truncation paradoxes.

3.5.15.2 Example

This example is due to Hannu Nurmi (private communication 27.2.2010 and 17.7.2011; this example appears also in section 10.5.7 in this volume). Suppose there are 19 voters whose preference orderings among four candidates, a, b, c, d, are as follows:

No. of voters	Preference ordering
5	$d \succ b \succ c \succ a$
4	$d \succ a \succ b \succ c$
4	$b \succ c \succ a \succ d$
3	$a \succ d \succ c \succ b$
3	$a \succ d \succ b \succ c$

Here a is the Condorcet winner and is therefore elected under Kemeny's procedure.

Now suppose that, *ceteris paribus*, the four $d \succ a \succ b \succ c$ voters decide not to participate in the election. As a result we obtain that the social preference ordering is cyclical $[d \succ b \succ c \succ a \succ d]$, so according to Kemeny's procedure the most likely (transitive) social preference ordering is $d \succ b \succ c \succ a$ because the sum (57) associated with the pairwise comparisons of this social preference ordering is highest. So according to Kemeny's procedure d will now be elected – which the four absentee $d \succ a \succ b \succ c$ voters certainly prefer to the election of a, thereby demonstrating the vulnerability of Kemeny's procedure to the No-Show paradox.

We also have here an instance of the twin paradox. To show Kemeny's procedure vulnerability to the Twin paradox start with the 16-voter profile:

No. of **voters**	Preference **ordering**
5	$d \succ b \succ c \succ a$
1	$d \succ a \succ b \succ c$
4	$b \succ c \succ a \succ d$
3	$a \succ d \succ c \succ b$
3	$a \succ d \succ b \succ c$

Here the social preference ordering is cyclical $[d \succ b \succ c \succ a \succ d]$ and according to Kemeny's procedure the two most likely (transitive) social preference orderings are $d \succ b \succ c \succ a$ and $a \succ d \succ b \succ c$ because the sum (61) associated with the pairwise comparisons of these social preference orderings is highest. So according to Kemeny's procedure there is a tie between a and d (to be broken randomly).

Now suppose that, *ceteris paribus*, one twin brother of the $d \succ a \succ b \succ c$ voter joins the electorate, thereby, presumably, strengthening the position of d to be elected under Kemeny's procedure. But if this twin joins the electorate then a will be elected under Kemeny's procedure – thus demonstrating its vulnerability to the twin paradox. (*Ceteris paribus*, if one twin brother of the $d \succ a \succ b \succ c$ voter join the electorate then the social preference ordering will still be cyclical but according to Kemeny's procedure the most likely transitive social preference ordering will be topped by a, not by d, thereby demonstrating the vulnerability of Kemeny's procedure to the Twin paradox. We also have here an instance of the Truncation paradox. To show Kemeny's procedure vulnerability to this paradox suppose that the four voters with preference ordering $d \succ a \succ b \succ c$ list only their top preference (d). In this case one assumes that these voters are indifferent among a, b, and c, and as a result the social preference ordering becomes cyclical ($d \succ b \succ c \succ a \succ d$) and the most likely transitive social preference ordering will be topped by d, not by a, thereby demonstrating the vulnerability of Kemeny's procedure to the Truncation paradox).

Example 3.5.11.5 demonstrates the vulnerability of Kemeny's procedure to the SCC paradox. In that example candidate d is elected according to Kemeny's procedure (because the "most likely" social preference ordering according to this procedure is $d \succ c \succ b \succ a$) but if, *ceteris paribus*, candidate b is eliminated then candidate a becomes the Condorcet winner and is elected according to Kemeny's procedure – in violation of the SCC postulate.

3.5.16 Demonstrating Paradoxes Afflicting Nanson's Procedure

Except for being vulnerable to cyclical majorities and to strategic voting, Nanson's procedure may display non-monotonicity, as well as being vulnerable to the Reinforcement, no-show, twin, truncation, and SCC paradoxes.

Example 3.5.16.1 demonstrates the vulnerability of Nanson's procedure to lack of monotonicity.

3.5.16.1 Example

This example is due to Nicolaus Tideman (private communication, 3.8.2011). Suppose there are 36 voters who must elect one out of four candidates, a, b, c, or d, under Nanson's procedure and whose preference orderings among these candidates, as well as the resultant Borda scores of the four candidates, are as follows:

No. of voters	Preference ordering	No. of voters	Preference ordering
1	$a \succ b \succ c \succ d$	2	$c \succ a \succ b \succ d$
1	$a \succ b \succ d \succ c$	1	$c \succ a \succ d \succ b$
2	$a \succ c \succ b \succ d$	3	$c \succ b \succ a \succ d$
2	$a \succ c \succ d \succ b$	2	$c \succ b \succ d \succ a$
1	$a \succ d \succ b \succ c$	1	$c \succ d \succ a \succ b$
2	$a \succ d \succ c \succ b$	2	$c \succ d \succ b \succ a$
2	$b \succ a \succ c \succ d$	1	$d \succ a \succ b \succ c$
2	$b \succ a \succ d \succ c$	1	$d \succ a \succ c \succ b$
1	$b \succ c \succ a \succ d$	0	$d \succ b \succ c \succ a$
1	$b \succ c \succ d \succ a$	2	$d \succ b \succ a \succ c$
2	$b \succ d \succ a \succ c$	1	$d \succ c \succ a \succ b$
1	$b \succ d \succ c \succ a$	2	$d \succ c \succ b \succ a$

The Borda scores of the candidates can be derived from the sum of the lines in the following paired comparisons matrix:

	a	b	c	d	Sum
a	–	16	19	20	55
b	20	–	15	20	55
c	17	21	–	20	58
d	16	16	16	–	48
Total					216

The sum of Borda scores of all four candidates is 218,[18] hence the average Borda score is 54 (216:4). According to Nanson's procedure one eliminates at the end of every counting round those candidates whose Borda score is equal to or smaller than the average score of all candidates participating in this round. Hence only candidate d is eliminated after the first round. So in the second counting round we have:

[18]Note that the sum of the Borda scores of all candidates can also be obtained by multiplying the number of voters (36 in this example) by the number of paired comparisons among the candidates (six in this example).

No. of **voters**	Preference **ordering**
4	$a \succ b \succ c$
7	$a \succ c \succ b$
8	$b \succ a \succ c$
3	$b \succ c \succ a$
5	$c \succ a \succ b$
9	$c \succ b \succ a$

which can be depicted as the following paired comparisons matrix cum Borda scores:

	a	b	c	Sum
a	–	16	19	35
b	20	–	15	35
c	17	21	–	38
Total				108

Here the sum of Borda scores of all three candidates is 108, hence their average Borda score is 36 (108:3). So according to Nanson's procedure one eliminates at the end of the second counting round both candidates a and b – thus candidate c becomes the ultimate winner.

Now suppose that, *ceteris paribus*, the voter whose preference ordering is $a \succ b \succ c \succ d$ – who is not happy with the prospect that candidate c may be elected – is motivated to *increase* his support of candidate c by changing his preference ordering to $a \succ c \succ b \succ d$. As a result of this change the Borda scores of candidates b and c change to 54 and 59, respectively, while the Borda scores of the remaining two candidates, as well as the sum of all Borda scores and the average Borda score, remain the same. So now both candidates b and d are eliminated after the first counting round. In the second counting round one obtains that the (revised) Borda scores of candidates a and c are 19 and 17, respectively, so candidate a becomes the ultimate winner – thus demonstrating that Nanson's procedure is susceptible to lack of monotonicity.

Example 3.5.12.3, which demonstrates the vulnerability of Dodgson's procedure to the reinforcement paradox, can also be used to demonstrate the vulnerability of Nanson's procedure to this paradox.

Example 3.5.16.2 demonstrates the vulnerability of Nanson's procedure to the truncation paradox.

3.5.16.2 Example

Suppose there are 43 voters divided into six groups whose preference orderings among four candidates a, b, c, d are as follows:

Group	No. of voters	Preference ordering
G1	9	$a \succ b \succ d \succ c$
G2	5	$a \succ c \succ b \succ d$
G3	2	$a \succ c \succ d \succ b$
G4	5	$b \succ a \succ c \succ d$
G5	9	$b \succ d \succ c \succ a$
G6	13	$c \succ b \succ a \succ d$

Suppose further that under Nanson's procedure with k candidates one assigns k points to the top-ranked candidate, $k-1$ points to the second-ranked candidate, ..., 1 point to the kth ranked candidate, and 0 points to any non-ranked candidate.

Given the above preference orderings and the above-mentioned point assignment, the number of points awarded to candidates a, b, c, and d in the first counting round, are 114, 134, 110, and 72, respectively. Since the average number of points is 107.5 candidate d is deleted and a second counting round is conducted. The number of points awarded to candidates a, b, c in this round is 80, 93, and 85, respectively. As the average number of points in this round is 86, both candidate a and c are deleted so candidate b is elected. However, if all voters belonging to groups G2 and G6 (who are not very happy with the election of candidate b) decide not to rank (i.e., truncate) candidate b, then the number of points awarded to candidates a, b, c, and d, are 109, 85, 92, and 72, respectively. As the average number of points in this case is 89.5, candidates b, d are deleted so candidate c is elected. This result is of course preferred by the voters in groups G2 and G6 to the election of candidate b, thereby demonstrating the susceptibility of Nanson's procedure to the truncation paradox.

Example 3.5.16.3 demonstrates the vulnerability of Nanson's procedure to the no-show and twin paradoxes.

3.5.16.3 Example

This example is due to Hannu Nurmi (private communications, 25.5.2001 and 15.2.2010; this example appears also in section 10.5.2 in this volume). Suppose there are 19 voters whose preference orderings among four candidates, a, b, c, d, are as follows:

No. of voters	Preference ordering
5	$a \succ b \succ d \succ c$
5	$b \succ c \succ d \succ a$
6	$c \succ a \succ d \succ b$
1	$c \succ b \succ a \succ d$
2	$c \succ b \succ d \succ a$

Here the Borda scores of candidates a, b, c, d are 28, 31, 37, 18, respectively, and the average Borda score is 28.5. Therefore candidates a and d are eliminated, whereupon candidate b is elected under Nanson's procedure. But if, *ceteris paribus*, one of the two last voters abstains then candidate c – the abstainer's most preferred candidate – is elected under Nanson's procedure, thus demonstrating the vulnerability of this procedure to the no-show paradox.

We also have here an instance of the twin paradox: we have just seen that if there is only one voter with preference ordering $c \succ b \succ d \succ a$ then, *ceteris paribus*, candidate c will be elected under Nanson's procedure. But if he is joined by a twin with the same preference ordering then b will be elected under Nanson's procedure, thus demonstrating the vulnerability of this procedure to the twin paradox.

Example 3.5.16.4 demonstrates the vulnerability of Nanson's procedure to SCC.

3.5.16.4 Example

This example is due to Fishburn (1977, p. 486). Suppose there are 86 voters who must elect one out of four candidates, a, b, c, or d, under Nanson's procedure and whose preference orderings are as follows:

No. of voters	Preference ordering
20	$d \succ a \succ b \succ c$
20	$d \succ b \succ c \succ a$
12	$c \succ b \succ d \succ a$
28	$a \succ c \succ b \succ d$
3	$b \succ c \succ a \succ d$
3	$c \succ b \succ a \succ d$

Accordingly, the number of Borda points awarded to candidates a, b, c, and d are 130, 127, 127, and 132, respectively – so candidates b, c are deleted and in the second counting round candidate d gets more Borda points (52) than candidate a (34) and hence d is elected.

Now suppose that, *ceteris paribus*, candidate a drops out of the race. In this case the number of Borda points awarded to candidates b, c and d are 89, 89, and 80, respectively, so there is a tie (to be broken randomly) between b and c – in violation of SCC.

3.5.17 *Demonstrating Paradoxes Afflicting Schwartz's Procedure*

Except for being vulnerable to cyclical majorities and to strategic voting, Schwartz's procedure is vulnerable to the reinforcement, no-show, twin, truncation, and the pareto-dominated candidate paradoxes.

Example 3.5.17.1 demonstrates the vulnerability of Schwartz's procedure to the reinforcement paradox.

3.5.17.1 Example

This example is due to Fishburn (1977, p. 483). Suppose there are two districts, I and II. In district I there are five voters, three of whom have preference ordering $x \succ y \succ w \succ z$ and the remaining two voters have preference ordering $z \succ y \succ w \succ x$. Since x constitutes here the top preference of an absolute majority of the voters, x will be elected in district I according to Schwartz's procedure.

In district II there are four voters: one with preference ordering $y \succ x \succ z \succ w$, one with preference ordering $w \succ y \succ x \succ z$, one with preference ordering $z \succ w \succ y \succ x$, and one with preference ordering $x \succ z \succ w \succ y$. The social preference ordering here is cyclical [$z \succ w \succ y \succ x \succ z$] so all four candidates should be in the choice set in district II according to Schwartz's procedure.

It would therefore be reasonable to assume that if, *ceteris paribus*, the two districts are amalgamated into a single district of nine voters, then x should be in the choice set of the amalgamated district according to Schwartz's procedure. However, in the amalgamated district y becomes the Condorcet winner and hence is the only candidate in the choice set according to Schwartz's procedure – thus demonstrating its vulnerability to the Reinforcement paradox.

Example 3.5.17.2 demonstrates the vulnerability of Schwartz's procedure to the no-show and twin paradoxes. Unlike the demonstration of these paradoxes under other procedures, in order to demonstrate the vulnerability of Schwartz's procedure to these paradoxes one must assume whether the voters are risk-neutral, risk-averse, or risk-seeking. I shall assume that the voters are risk-neutral, i.e., when only the voters' ordinal (but not cardinal) preferences are known, I assume that a voter whose ordinal preferences between three candidates, a, b, c is $a \succ b \succ c$ will be indifferent between obtaining a tie between these three candidates which will be broken randomly and the election of candidate b with certainty. Similarly, I assume that if this voter's ordinal preferences among four candidates is $b \succ c \succ d \succ a$ he would prefer the election of candidate c with certainty than to obtain a tie among all four candidates which will be broken randomly. Using different examples it is of course possible to demonstrate these paradoxes also when one assumes that the voters are risk-averse or risk-seeking.

3.5.17.2 Example

This example is due to Hannu Nurmi (private communication, 1.3.2010; this example appears also in section 10.5.4 in this volume). Suppose there are 100 voters whose preference orderings among four candidates, a, b, c, d, are as follows:

No. of voters	Preference ordering
23	$a \succ b \succ d \succ c$
28	$b \succ c \succ d \succ a$
49	$c \succ d \succ a \succ b$

Here the social preference ordering is cyclical [$a \succ b \succ c \succ d \succ a$] and according to Schwartz's procedure all four candidates belong to the choice set.

Now suppose that, *ceteris paribus*, four of the 28 $b \succ c \succ d \succ a$ voters decide not to participate in the election. In this case c becomes the Condorcet winner – which the absentee voters certainly prefer over a tie among all candidates that will be broken randomly – thereby demonstrating the vulnerability of Schwartz's procedure to the No-Show paradox.

We have here also a demonstration of the twin paradox. We just saw that, *ceteris paribus*, if there are only 24 voters with preference ordering $b \succ c \succ d \succ a$ then candidate c is the Condorcet winner and is the only candidate belonging to the choice set according to Schwartz's procedure. But if, *ceteris paribus*, one adds another four twins with preference ordering $b \succ c \succ d \succ a$ then Schwartz's choice set includes all candidates – which is a less preferable outcome for these voters, thus demonstrating the vulnerability of Schwartz's procedure to the Twin paradox.

To demonstrate the vulnerability of Schwartz's procedure to the Truncation paradox we use again Example 3.5.13.1. In the first part of this example we obtained that candidate d is the Condorcet winner and hence is the sole candidate belonging to the Schwartz set. But, *ceteris paribus*, when the two voters whose preference ordering is $e \succ b \succ a \succ d \succ c$ decide not to reveal their last two preferences (thereby assuming that the probability that they prefer d to c is equal to the probability they prefer c to d), one obtains the following expected paired comparisons matrix:

	a	b	c	d	e
a	–	6	6	4	0
b	4	–	6	4	0
c	4	4	–	5	2
d	6	6	5	–	6
e	10	10	8	4	–

As can be seen from this matrix only candidates d, e belong to the Schwartz set (because each of these candidates either beats or ties with each of the other three candidates) – which is a preferred outcome for the above-mentioned two truncating voters over the certain election of candidate d – thereby demonstrating the vulnerability of Schwartz's procedure to the Truncation paradox.

This preference matrix can also be used to demonstrate the vulnerability of Schwartz's procedure to the SCC paradox. We have just seen that according to this preference matrix only candidates d, e belong to the Schwartz set. However, if *ceteris paribus*, candidate c is eliminated (by deleting the row c and column c from this matrix) then candidate d becomes the Condorcet winner and is elected by Schwartz's procedure – in violation of the SCC postulate.

Example 3.5.17.3 demonstrates the vulnerability of Schwartz's procedure to the pareto-dominated candidate paradox.

3.5.17.3 Example

This example is due to Fishburn (1973, p. 89; 1977, p. 478). Suppose there are three voters whose preference orderings among four candidates, a, b, c, d are as follows:

No. of voters	Preference ordering
1	$a \succ b \succ c \succ d$
1	$d \succ a \succ b \succ c$
1	$c \succ d \succ a \succ b$

Here the social preference ordering is cyclical ($a \succ b \succ c \succ d \succ a$) and according to Schwartz's procedure all four candidates belong to the choice set – this despite the fact that candidate b is dominated by a (because all voters prefer a to b) – thus demonstrating the vulnerability of this procedure to the pareto-dominated candidate paradox.

3.5.18 Demonstrating Paradoxes Afflicting Young's Procedure

Except for being vulnerable to cyclical majorities and to strategic voting, Young's procedure is vulnerable to the Condorcet loser, absolute loser, reinforcement, no-show, twin, truncation, preference inversion, and SCC paradoxes.

Example 3.5.11.1 can be used to demonstrate the vulnerability of Young's procedure to electing not only a Condorcet loser but also an Absolute Loser. In that example candidate d is an absolute loser (and hence also a Condorcet loser), but under Young's procedure d will be elected because for d to become a Condorcet winner only two voters must be removed from the 11-voter electorate (any two voters whose last preference is d), whereas for each of the other three candidates more than two voters must be removed in order for them to become a Condorcet winner.

Example 3.5.11.1 can also be used to demonstrate the vulnerability of Young's procedure to the preference inversion paradox because, as we have already seen, if all voters in Example 3.5.11.1 invert their preference orderings then d becomes an Absolute Winner and hence is elected under Young's procedure.

Example 3.5.12.3 can be used, *mutatis mutandis*, to demonstrate the vulnerability of Young's procedure to the reinforcement paradox. In that example candidate x is a Condorcet winner in district I and hence is elected in this district according to Young's procedure too. To become the Condorcet winner in district II only five voters must be removed (any five voters who prefer y to x), whereas for any of the other candidates to become a Condorcet winner in district II more than five voters must be removed. So according to Young's procedure candidate x is elected also in district II. But, as was demonstrated in Example 3.5.12.3, in the amalgamated district with 19 voters candidate y becomes the Condorcet winner and is therefore elected also according to Young's procedure – thereby demonstrating its vulnerability to the reinforcement paradox.

Example 3.5.18.1 demonstrates the vulnerability of Young's procedure to the no-show, twin, truncation, and SCC paradoxes.

3.5.18.1 Example

This example is due to Hannu Nurmi (private communication 22.2.2010). Suppose there are 39 voters whose preference orderings among five candidates, a, b, c, d, e, are as follows:

No. of voters	Preference ordering
11	$b \succ a \succ d \succ e \succ c$
10	$e \succ c \succ b \succ d \succ a$
10	$a \succ c \succ d \succ b \succ e$
2	$e \succ c \succ d \succ b \succ a$
2	$e \succ d \succ c \succ b \succ a$
2	$c \succ b \succ a \succ d \succ e$
1	$d \succ c \succ b \succ a \succ e$
1	$a \succ b \succ d \succ e \succ c$

These preference orderings can be depicted as the following paired comparisons matrix:

	a	b	c	d	e
a	–	11	22	24	25
b	28	–	12	24	25
c	17	27	–	24	13
d	15	15	15	–	25
e	14	14	26	14	–

The social preference ordering here is cyclical ($c \succ b \succ a \succ d \succ e \succ c$). The minimal number of voters one must remove in order for any of the five candidates to become a Condorcet winner is 12 (the 10 voters whose top preference is a and the two voters whose top preference is c) in order for e to become the Condorcet winner. So e is elected according to Young's procedure given this profile.

Now suppose that, *ceteris paribus*, 10 new voters whose preference ordering is $e \succ d \succ a \succ b \succ c$ join the electorate – thus presumably strengthening e's position. However, in this case we obtain the following paired comparisons matrix:

	a	*b*	*c*	*d*	*e*
a	–	21	32	24	25
b	28	–	22	24	25
c	17	27	–	24	13
d	25	25	25	–	25
e	24	24	36	24	–

which shows that candidate *d* is the Condorcet winner, hence the 10 added voters are better off abstaining – thus demonstrating the vulnerability of Young's procedure to the No-Show paradox.[19] Obviously twins are not always welcome here.

However, if the 10 added voters reveal only their top preference (*e*), then we obtain the following paired comparisons matrix:

	a	*b*	*c*	*d*	*e*
a	–	16	27	29	25
b	33	–	17	29	25
c	22	32	–	29	13
d	20	20	20	–	25
e	24	24	36	24	–

Here candidate *e*will be elected according to Young's procedure because for *e* to become the Condorcet winner in this case only two voters must be removed (any two voters whose bottom preference is *e*), whereas for any of the other candidates to become a Condorcet winner more than two voters must be removed – thus demonstrating that Young's procedure is vulnerable to the truncation paradox.

To demonstrate the vulnerability of Young's procedure to SCC let us look again at the paired comparison matrix of the 39 voters at the beginning of this example. We saw that given this matrix candidate *e* is elected under Young's procedure. Now suppose that, *ceteris paribus*, candidate *b* decides to withdraw from the race. But if, as a result, we cross out row *b* and column *b* in the paired comparison matrix, we see that candidate *a* becomes the Condorcet winner and hence elected by Young's procedure – thereby demonstrating its vulnerability to SCC.

Acknowledgements An earlier version of this chapter was presented in a symposium on *Electoral Reform and Its Problems*, London School of Economics and Political Science, 27 May 2010, and in a workshop on *Assessing Alternative Voting Procedures*, Chateau du Baffy, Normandy, France, 30 July – 2 August, 2010, sponsored by The Leverhulme Trust (Grant F/07–004 M).

[19] The added 10 voters also demonstrate that Young's procedure violates what Pérez (1995, p. 143) has called the *Monotonicity property in face of new voters*. This property requires that if candidate x is chosen in a given situation and then, *ceteris paribus*, a new voter is added whose top preference is x, then: (1) x must remain chosen for *Weak Monotonicity* to be satisfied, and (2) x must remain chosen and no one not chosen before should be chosen now in order for *Monotonicity* to be satisfied.

I wish to thank Vincent Chua, Rudolf Fara, Moshé Machover, and Maurice Salles for helpful comments. I am especially grateful to Hannu Nurmi and Nicolaus Tideman for privately sending me examples contained in this chapter.

References

Balinski, M., & Laraki, R. (2007a). A theory of measuring, electing and ranking. *Proceedings of the National Academy of Sciences of the United States of America (PNAS), 104*, 8720–8725.
Balinski, M., & Laraki, R. (2007b). Election by majority judgement: experimental evidence, (mimeograph). Paris: Ecole Polytechnique, Centre National De La Recherche Scientifique, Laboratoire D'Econommetrie, Cahier No. 2007–28. Downloadable from http://tinyurl.com/65q6cg
Balinski, M., & Laraki, R. (2011). *Majority judgment: measuring, ranking, and electing.* Cambridge: MIT.
Baujard, A., & Igersheim, H. (2009). Expérimentation du vote par note et du vote par approbation. *Revue Economique, 60*, 189–201.
Black, D. (1958). *The theory of committees and elections.* Cambridge: Cambridge University Press.
de Borda, J.C (1784). *Mémoire sur les élections au scrutin.* Histoire de l'Academie Royale des Sciences: Paris.
Brams, S.J. (1982). The AMS nominating system is vulnerable to truncation of preferences. *Notices of the American Mathematical Society, 29*, 136–138.
Brams, S.J., & Fishburn, P.C. (1978). Approval voting. *American Political Science Review, 72*, 831–847.
Brams, S.J., & Fishburn, P.C. (1983). *Approval voting*. Birkhäuser: Boston.
Brams, S.J., Kilgour, D.M., & Zwicker, W.S. (1998). The paradox of multiple elections. *Social Choice and Welfare, 15*, 211–236.
Campbell, D.E., & Kelly, J.S. (2002). Non-monotonicity does not imply the no-show paradox. *Social Choice and Welfare, 19*, 513–515.
Chernoff, H. (1954). Rational selection of decision functions. *Econometrica, 22*, 422–443.
Condorcet, M. (1785). *Essai sur l'application de l'analyse à la probabilité des décisions rendues à la pluralité des voix.* Paris: L'Imprimerie Royale.
Coombs, C.H. (1964). *A theory of data.* New York: Wiley.
Coombs, C.H., Cohen, J.L., Chamberlin, J.R. (1984). An empirical study of some election systems. *American Psychologist, 39*, 140–157.
Copeland, A.H. (1951). *A 'reasonable' social welfare function, mimeographed.* University of Michigan, Department of Mathematics, Seminar on Applications of Mathematics to the Social Sciences.
Doron, G. (1979). The Hare voting system is inconsistent. *Political Studies, 27*, 283–286.
Doron, G., & Kronick, R. (1977). Single transferable vote: an example of a perverse social choice function. *American Journal of Political Science, 21*, 303–311.
Farquharson, R. (1969). *Theory of voting*. New Haven: Yale University Press.
Felsenthal, D.S., & Machover, M. (1992). After two centuries should Condorcet's voting procedure be implemented? *Behavioral Science, 37*, 250–273.
Felsenthal, D.S., & Machover, M. (2008). The majority judgement voting procedure: a critical evaluation. *Homo Oeconomicus, 25*, 319–333.
Felsenthal, D.S., & Maoz, Z. (1988). A comparative analysis of sincere and sophisticated voting under the plurality and approval procedures. *Behavioral Science, 33*, 116–130.
Fishburn, P.C. (1973). *The theory of social choice*. Princeton: Princeton University Press.
Fishburn, P.C. (1974). Paradoxes of voting. *American Political Science Review, 68*, 537–546.
Fishburn, P.C. (1974a). Social choice functions. *SIAM Review, 16*, 63–90.

Fishburn, P.C. (1974b). On the sum-of-ranks winner when losers are removed. *Discrete Mathematics, 8*, 25–30.
Fishburn, P.C. (1974c). Subset choice conditions and the computation of social choice sets. *The Quarterly Journal of Economics, 88*, 320–329.
Fishburn, P.C. (1977). Condorcet social choice functions. SIAM. *Journal on Applied Mathematics, 33*, 469–489.
Fishburn, P.C. (1981). Inverted orders for monotone scoring rules. *Discrete Applied Mathematics, 3*, 27–36.
Fishburn, P.C. (1982). Monotonicity paradoxes in the theory of elections. *Discrete Applied Mathematics, 4*, 119–134.
Fishburn, P.C., & Brams, S.J. (1983). Paradoxes of preferential voting. *Mathematics Magazine, 56*, 207–214.
Gehrlein, W.V. (1983). Condorcet's paradox. *Theory and Decision, 15*, 161–197.
Gehrlein, W.V., & Lepelley, D. (2011). *Voting paradoxes and group coherence: The Condorcet efficiency of voting rules*. Berlin: Springer.
Gibbard, A. (1973). Manipulation of voting systems: a general result. *Econometrica, 41*, 587–601.
Hix, S., Johnston, R., McLean, I. (2010). *Choosing an electoral system: A research report prepared for the British Academy*. London: British Academy Policy Centre.
Hoag, C.G., & Hallett, G.H. (1926). *Proportional representation*. New York: Macmillan.
Holzman, R. (1988/89). To vote or not to vote: what is the quota? *Discrete Applied Mathematics, 22*, 133–141.
Kemeny, J. (1959). Mathematics without numbers. *Daedalus, 88*, 577–591.
Kemeny, J., & Snell, I. (1960). *Mathematical models in the social sciences*. Boston: Ginn.
Kramer, G.H. (1977). A dynamical model of political equilibrium. *Journal of Economic Theory, 16*, 310–333.
Lepelley, D., & Merlin, V. (2001). Scoring runoff paradoxes for variable electorates. *Economic Theory, 17*, 53–80.
Lepelley, D., Louichi, A., Smaoui, H. (2008). On Ehrhart polynomials and probability calculations in voting theory. *Social Choice and Welfare, 30*, 363–383.
McGarvey, D.C. (1953). A theorem on the construction of voting paradoxes. *Econometrica, 21*, 608–610.
McLean, I., & Urken, A.B. (Eds.) (1995). *Classics of social choice* (pp. 321–359). Ann Arbor: University of Michigan Press.
Merlin, V., & Saari, D.G. (1997). The Copeland method II: manipulation, monotonicity and paradoxes. *Journal of Economic Theory, 72*, 148–172.
Merlin, V., & Valognes, F. (2004). The impact of indifferent voters on the likelihood of some voting paradoxes. *Mathematical Social Sciences, 48*, 343–361.
Merlin, V., Tataru, M., Valognes, F. (2002). The likelihood of Condorcet's profiles. *Social Choice and Welfare, 19*, 193–206.
Moulin, H. (1988a). *Axioms of cooperative decision making*. Cambridge: Cambridge University Press.
Moulin, H. (1988b). Condorcet's principle implies the no-show paradox. *Journal of Economic Theory, 45*, 53–64.
Nanson, E.J. (1883). Methods of elections. *Transactions and Proceedings of the Royal Society of Victoria, 19*, 197–240.
Niemi, R.G., & Riker, W.H. (1976). The choice of voting systems. *Scientific American, 234*, 21–27.
Niou, E.M.S. (1987). A note on Nanson's rule. *Public Choice, 54*, 191–193.
Nurmi, H. (1998a). *Rational behavior and the design of institutions: concepts, theories and models*. Cheltenham: Edward Elgar.
Nurmi, H. (1998b). Voting paradoxes and referenda. *Social Choice and Welfare, 15*, 333–350.
Nurmi, H. (1999). *Voting paradoxes and how to deal with them*. Berlin: Springer.
Nurmi, H. (2004). A comparison of some distance-based choice rules in ranking environments. *Theory and Decision, 57*, 5–23.

Nurmi, H. (2007). Assessing Borda's rule and its modifications. In Emerson, P. (ed.) Designing an all-inclusive democracy: Consensual voting procedures for use in parliaments, councils and committees. Berlin, Heidelberg: Springer, ch. 6 (pp. 109–119).

Pérez, J. (1995). Incidence of no-show paradoxes in Condorcet choice functions. *Investigaciones Economicas, 19*, 139–153.

Plott, C.R. (1973). Path independence, rationality, and social choice. *Econometrica, 41*, 1075–1091.

Ray, D. (1986). On the practical possibility of a 'no-show paradox' under the single transferable vote. *Mathematical Social Sciences, 11*, 183–189.

Richelson, J.T. (1979). A comparative analysis of social choice functions I, II, III: a summary. *Behavioral Science, 24*, 355.

Riker, W.H. (1958). The paradox of voting and Congressional rules for voting on amendments. *American Political Science Review, 52*, 349–366.

Saari, D.G. (1984). The ultimate of chaos resulting from weighted voting systems. *Advances in Applied Mathematics, 5*, 286–308.

Saari, D.G. (1987). The source of some paradoxes from social choice and probability. *Journal of Economic Theory, 41*, 1–22.

Saari, D.G. (1989). A dictionary for voting paradoxes. *Journal of Economic Theory, 48*, 443–475.

Saari, D.G. (1994). *Geometry of voting*. New York: Springer Verlag.

Saari, D.G. (2000). Mathematical structure of voting paradoxes. *Economic Theory, 15*, 1–53, 55–102.

Saari, D.G. (2008). *Disposing dictators, demystifying voting paradoxes: social choice analysis.* New York: Cambridge University Press.

Satterthwaite, M. (1975). Strategy-proofness and Arrow's conditions: Existence and correspondence theorems for voting procedures and social choice functions. *Journal of Economic Theory, 10*, 187–217.

Scarsini, M. (1998). A strong paradox of multiple elections. *Social Choice and Welfare, 15*, 237–238.

Schultze, M. (2003). A new monotonic and clone-independent single-winner election method. *Voting Counts, 17*, 9–19. http://www.macdougall.org.uk/VM/ISSUE17/I17P3.PDF

Schwartz, T. (1972). Rationality and the myth of the maximum. *Noûs, 6*, 97–117.

Schwartz, T. (1986). *The logic of collective choice*. New York: Columbia University Press.

Simpson, P. (1969). On defining areas of voter choice. *Quarterly Journal of Economics, 83*, 478–490.

Smith, J.H. (1973). Aggregation of preferences with variable electorate. *Econometrica, 41*, 1027–1041.

Smith, W.D. (2000). *Range voting*. http://www.math.temple.edu/~wds/homepage/rangevote.pdf

Straffin, Ph.D. (1980). *Topics in the theory of voting*. Boston: Birkhäuser.

Tideman, T.N. (1987). Independence of clones as a criterion for voting rules. *Social Choice and Welfare, 4*, 185–206.

Tideman, N. (2006). *Collective decisions and voting: the potential for public choice*. Aldershot: Ashgate.

Young, H.P. (1974). An axiomatization of Borda's rule. *Journal of Economic Theory, 9*, 43–52.

Young, H.P. (1977). Extending Condorcet's rule. *Journal of Economic Theory, 16*, 335–353.

Young, P. (1995). Optimal voting rules. *Journal of Economic Perspectives, 9*, 51–63.

Young, H.P., & Levenglick, A. (1978). A consistent extension of Condorcet's principle. *SIAM Journal of Applied Mathematics, 35*, 283–300.

Chapter 4
Election Inversions by the U.S. Electoral College

Nicholas R. Miller

An *election inversion* occurs when the candidate (or party) that wins the most votes from an electorate fails to win the most electoral votes (or parliamentary seats) and therefore loses the election. Public commentary commonly uses terms such as "reversal of winners," "wrong winner," "divided verdict," and "misfire" to describe this phenomenon; the academic social choice literature adds such terms as "representative inconsistency," "compound majority paradox," "referendum paradox," and "majority deficit." Election inversions can occur under any two-tier electoral system, including the U.S. Electoral College. As is well known, the Electoral College actually produced a "wrong winner" in the 2000 Presidential election, and it has done so twice before.

In so far as this phenomenon may be "paradoxical," it is of a somewhat different character from most other paradoxes in the theory of voting and social choice, in that it may arise even if there are only two candidates, it is straightforward in nature, and its occurrence is readily apparent. However, the likelihood of inversions and the factors that produce them are less apparent, and there has been considerable confusion about the circumstances under which election inversions occur. For example, the susceptibility of the Electoral College to inversions is sometimes blamed on the small-state bias in the apportionment of electoral votes and/or the "non-proportional" or "winner-take-take-all" manner of casting state electoral votes, but inversions can occur in the absence of both factors.

With specific respect to the U.S. Electoral College, I first note the three historical manifestations of election inversions and identify and discuss one massive but "latent" inversion in more detail. I then use "uniform swing analysis" based on historical election data in order to estimate the frequency, magnitude, and direction of potential election inversions. Along the way, I identify three sources of election inversions – "rounding effects," "apportionment effects," and "distribution effects" –

N.R. Miller (✉)
Department of Political Science, University of Maryland Baltimore County, USA
e-mail: nmiller@umbc.edu

D.S. Felsenthal and M. Machover, *Electoral Systems*, Studies in Choice and Welfare,
DOI 10.1007/978-3-642-20441-8_4, © Springer-Verlag Berlin Heidelberg 2012

and examine their separate impacts on the likelihood of election inversions by the Electoral College.

4.1 The Problem of Election Inversions

The President of the United States is elected, not by a direct national popular vote, but by an indirect Electoral College system in which (in almost universal practice since the 1830s) separate state popular votes are aggregated by adding up state electoral votes awarded, on a winner-take-all basis, to the plurality winner in each state.[1] Therefore the U.S. Electoral College is a two-tier electoral system: individual voters cast votes in the first tier to choose between rival slates of "Presidential electors" pledged to one or other Presidential candidate, and the winning elector slates then cast blocs of electoral votes for the candidate to whom they are pledged in the second tier. Each state has electoral votes equal in number to its total representation in Congress and since 1964 the District of Columbia has three electoral votes. At the present time, there are 538 electoral votes, so 270 are required for election and a 269–269 electoral vote tie is possible.

To the best of my knowledge, the first theoretical work on election inversions was May (1948), who attempted to calculate the *a priori* probability of inversions based on a particular probability model of election outcomes. Several years earlier Schattschneider (1942) had noted in passing the "25%–75% rule" pertaining to election inversions that will be discussed later. Sterling (1981) provided an insightful geometric analysis of "Electoral College misrepresentation," a modified version of which will be fruitfully employed here. More recently, Nurmi (1999, 2001, 2002), Laffond and Laine (2000) and Feix et al. (2004) have addressed the general phenomenon of election inversions in social choice terms, and Chambers (2008) has demonstrated (in effect) that no neutral (between candidates or parties) two-tier electoral rule can satisfy "representative consistency," i.e., preclude election inversions. Merrill (1978) and Ball and Leuthold (1991) have provided empirically based estimates of the expected frequency of Electoral College election inversions, and Lahrach and Merlin (2012) have done related work with respect to French local government elections. The fact that the Electoral College can produce inversions is regularly cited by its critics (e.g., Peirce and Longley 1981; Abbott and Levine 1991; Longley and Peirce 1996; Edwards 2004), so its defenders (e.g., Best 1971; Diamond 1992; Ross 2004) must also address the question.

"Westminster" single-member-district parliamentary systems in the U.K., Canada, Australia, India, and New Zealand (prior to 1993) are likewise two-tier voting systems and have produced election inversions about as frequently as the U.S. Electoral College. Some examples are listed in Table 4.1. It can be seen that

[1] At present Maine (since 1972) and Nebraska (since 1992) use the "modified district system," under which electoral votes may be split. The 2008 election was the first actually to produce a split (in Nebraska).

Table 4.1 Election inversions in "Westminster" parliamentary systems

Country	Election	Leading parties	Pop. vote (%)	Seats
Britain	1929	Conservative	38.06	260
		Labour	37.12	287
Britain	1951	Labour	48.78	297
		Conservative	47.97	302
Britain	1974 (Feb.)	Conservative	37.90	297
		Labour	37.18	301
New Zealand	1978	Labour	40.44	40
		National	39.80	51
New Zealand	1981	Labour	39.01	43
		National	38.77	47
Canada	1979	Liberal	40.11	114
		Conservative	35.89	136

most of these election inversions were very close with respect to both votes and seats, but the case of Canada in 1979 shows that this is not always the case.

These parliamentary systems differ in two important respects from the U.S. Electoral College. First, "Westminster" systems have *uniform districts* – that is, the districts have equal weight (namely a single parliamentary seat), reflecting (approximately) equal populations and/or numbers of voters. In contrast, Electoral College "districts" (i.e., states) are unequal in both population and voters and likewise have unequal electoral votes.

Second, the popular vote percentages shown in Table 4.1, which can be seen to add up to substantially less than 100%, indicate that many of these parliamentary inversions occurred in elections in which third and perhaps additional minor parties received a substantial percent of the vote (and some seats). The presence of third parties can distort the relationship between votes and seats for the two leading parties. In contrast, Electoral College inversions, like most U.S. elections, have occurred in what were for all practical purposes two-candidate contests. Indeed, the following analysis deals entirely with two-party popular vote percentages and, with the exception of special consideration of the 1860 election, excludes Presidential elections in which third candidates carried one or more states and thereby won some electoral votes.

The U.S. Electoral College has produced the three manifest election inversions listed in Table 4.2.[2] All were very close with respect to popular votes, and two were very close with respect to electoral votes as well.

[2]The 1876 election was decided (just before inauguration day) by an Electoral Commission that, by a bare majority and straight party-line vote, awarded all of 20 disputed electoral votes to Hayes. The 1824 election is sometimes counted as an inversion, in that John Quincy Adams was elected President even though Andrew Jackson had received more popular votes (in the 18 out of 24 states in which presidential electors were popularly elected) than Adams. However, Jackson also won more electoral votes than Adams but not the required majority, so the election was decided by the House of Representatives, which elected Adams. In 1960, peculiarities with respect to Presidential

Table 4.2 The three historical election inversions by the U.S. Electoral College

Election	EC Winner [EV]	EC Loser [EV]	EC Loser's 2-P PV %*
1876	Hayes (R) [185]	Tilden (D) [184]	51.53%
1888	Harrison (R) [233]	Cleveland (D) [168]	50.41%
2000	Bush (R) [271]	Gore (D) [267**]	50.27%

* Two-party popular vote percent
** Gore lost one electoral vote to a "faithless elector"

Table 4.3 The 1860 election: a latent but massive inversion

Candidate	Party	Pop. vote	EV	Unified dem.	Unified opp.	EV
Lincoln	Republican	39.82%	180	39.82%	39.82%	169
Douglas	Northern Democrat	29.46%	12	} 47.55%	} 60.16%	134
Breckinridge	Southern Democrat	18.09%	72			
Bell	Constitutional Union	12.61%	39	12.61%		

In addition to these three historical instances, the Electoral College produced one massive but "latent" election inversion, which has been recognized as such by Sterling (1981) but by few others. Abraham Lincoln won an electoral vote majority 1860 on the basis of a plurality of less than 40% of the popular vote. The Democratic Party had split into Northern and Southern wings, each with its own Presidential candidate (Stephen Douglas and John Breckinridge, respectively) and a fourth candidate, John Bell, had been nominated by the remnants of the Southern Whig Party under the label of the Constitutional Union Party. The popular and electoral vote totals, as shown in Table 4.3, entail two manifest but inconsequential inversions – namely, Douglas won more popular votes but fewer electoral votes than either Breckinridge or Bell. Under a system of direct popular vote, the two Democratic candidates would have been "spoilers" against each other if we can suppose that, in the event of the withdrawal of one, the other would have inherited most of his support and would therefore have defeated Lincoln. However, under the Electoral College system, Douglas and Breckinridge were *not* spoilers against each other. Indeed, we can make the following strong counterfactual suppositions and still preserve a Lincoln electoral vote victory: (i) the Democrats successfully hold their Northern and Southern wings together and thereby win *all* the votes captured by each wing separately, (ii) the election is a typical "straight fight" and the Democrats also inherit *all* the votes of the Constitutional Union party; and, for good measure, (iii) the Democrats win *all* of New Jersey's electoral votes (which, for peculiar reasons, were split between Lincoln and Douglas). Even so, Lincoln still would have won the 1860 election on the basis of electoral votes. The final column of

ballot in Alabama make it unclear exactly how to determine the "popular vote" for President in that state, and thus also nationwide. One (somewhat implausible) reckoning of the Alabama popular vote makes Nixon the national popular vote winner, thereby making 1960 an election inversion. In any event, the 1960 election is excluded from this analysis because a third candidate won electoral votes from "unpledged electors" (see the Appendix).

Table 4.3 shows the results of this counterfactual 1860 election. The consequence of all these suppositions is that only 11 electoral votes (in California, Oregon, and New Jersey) would switch from the Republican to Democratic column. We will examine this counterfactual two-party variant of the 1860 election in more detail later.

In sum, a first cut at estimating the expected frequency of election inversions under the Electoral College – based on the historical record since 1828 (the first election in which almost every state selected presidential electors by popular vote) – is either $3/46 = 0.06$ or (counting the counterfactual 1860) $4/46 = 0.087$. However, with the exception of the counterfactual 1860 election, all inversions occurred in close elections and, considering only elections in which the popular vote winner's margin was no greater than about three percentage points, the expected frequency of inversions is considerably higher, namely $3/12 = 0.25$. Clearly an important determinant of the probability of an election inversion is the probability of a close division of the popular vote.

4.2 Popular Votes and Electoral Votes

We now turn to a more informative empirical analysis of election inversions that uses historical state-by-state popular vote percentages to construct the "Popular Vote-Electoral Vote" (PVEV) step-function for each.[3] The PVEV function is based on the kind of "uniform swing analysis" pioneered by Butler (1951), which has also been called "hypothetical (single-year) swing analysis" (Niemi and Fett 1986) and the "Bischoff method" (see Peirce and Longley 1981), and which has been employed by Nelson (1974), Garand and Parent (1991), and others in the context of assessing "partisan bias" in the Electoral College.

The PVEV function is a cumulative distribution function and is therefore (weakly) monotonic. It is a step-function because the "dependent variable" (EV) is discrete, assuming only whole number values and jumping up in discrete steps as the "independent variable" (PV) increases (essentially) continuously.

Let us consider the 1988 election as an example. We set up the template used in Fig. 4.1, showing the Democratic popular vote percent on the horizontal axis and the Democratic electoral vote on the vertical axis.[4] The Democratic nominee Michael Dukakis received 46.10% of the two-party national popular vote and won 112 electoral votes (one of which was lost to a "faithless elector"). This combination of Democratic popular and electoral votes is plotted in Fig. 4.1a.

---

[3]See the Appendix for details concerning this data.

[4]Remember that, here and elsewhere, popular vote percentages are put on a strictly two-party basis, excluding votes cast for third or other minor candidates, and (with the 1860 exception already noted) we consider only elections in which the two major candidates won all the electoral votes, thus putting everything on a strictly two-party basis. We therefore would reach exactly the same conclusions if we organized the figures in terms of Republican popular and electoral votes.

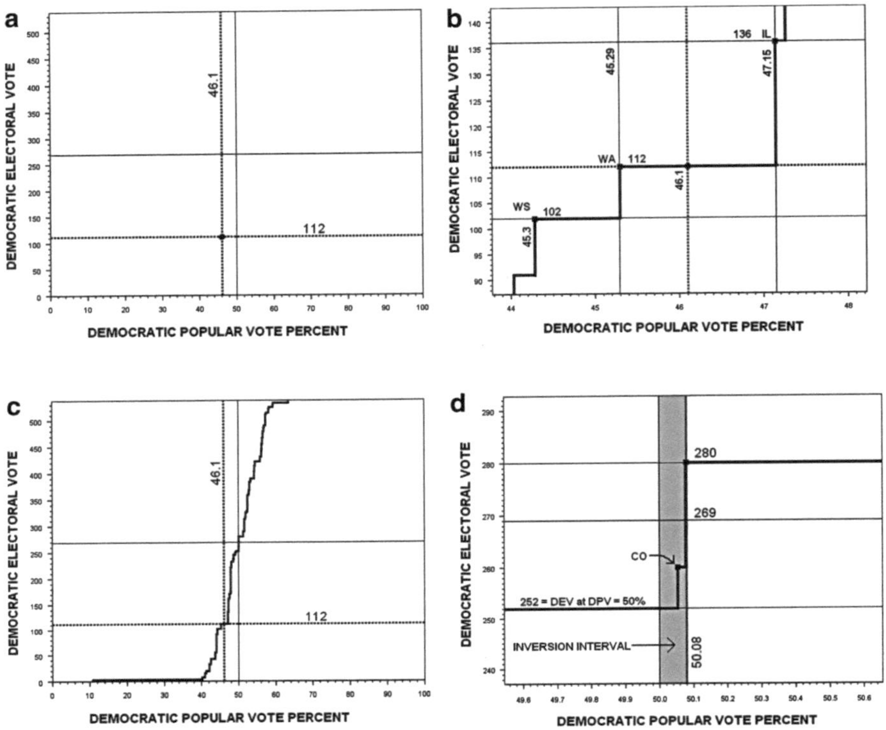

Fig. 4.1 The Democratic PVEV function in 1988 (**a**) General template (**b**) Zoom in on election outcome (**c**) The full PVEV function (**d**) Zoom in on inversion interval

Figure 4.1b zooms in on the neighborhood of this plotted point. Of all the states that Dukakis carried, he won Washington (with 10 electoral votes) by the smallest margin of 50.81%, so if the Democratic national popular vote of 46.10% were to decline by 0.81 percentage points (to 45.29%) *uniformly across all states*, Washington would tip into the Republican column and thereby reduce the Democratic electoral vote to 102, as shown in Fig. 4.1b. In like manner, of all the states that Dukakis failed to carry, he lost Illinois (with 24 electoral votes) by the smallest margin of 48.95%, so if the Democratic national popular vote of 46.10% were to increase by 1.05 percentage points (to 47.15%) uniformly across all states, Illinois would tip Democratic and thereby increase the Democratic electoral vote to 136, as also shown in Fig. 4.1b.

More generally, we can "swing" the Democratic vote downwards until the Democratic electoral vote falls to the logical minimum of zero and upwards until it increases to the logical maximum of 538, as is shown in Fig. 4.1c.[5] This chart

[5]Defining the uniform swing in terms of the absolute percent of the total popular vote means that highly lopsided state popular votes in conjunction with extreme swings can create hypothetical popular vote percentages that are less than 0% or greater than 100%. But this is of no practical

displays the PVEV function for 1988, over which Democratic popular support rises or falls uniformly across the states and translates into corresponding Democratic electoral vote totals.

While the full PVEV function in Fig.1c appears to go through the *two-way tie point* corresponding to PV $= 50\%$ and EV $= 269$, a moment's thought suggests that almost certainly it does not go *precisely* through this point. This becomes evident when we zoom in on the center of the chart in Fig. 4.1d. We see that (i) if Dukakis had won exactly 50% of the popular vote, he would have lost the election with only 252 electoral votes, and (ii) if he had won anything between 50.00% and 50.08% of the popular vote, he still would have lost the election with no more than 260 electoral votes. His popular vote percent required for an electoral vote majority is 50.08%. Thus there is an *inversion interval* that is $50.08\% - 50\% = 0.08$ percentage points wide, within which Dukakis would have won the popular vote but lost the election on the basis of electoral votes. Given the Democratic orientation of our analysis, the fact that the width of the interval is positive (e.g., $+0.08$) means that the Democratic candidate must win that more than 50% of the popular vote in order to win a majority of the electoral vote and reflects an anti-Democratic bias in the PVEV function. Conversely, a negative inversion interval means that the Democratic candidate can win a majority of the electoral vote with less than 50% of the popular vote and reflects a pro-Democratic bias in the PVEV function. In either event, the absolute width of the inversion interval is more consequential than the electoral vote split at the 50% popular vote mark, since the likelihood of an election inversion depends on the absolute width of the inversion interval, while the specific number of electoral votes that the "wrong winner" receives within this interval does not affect who is elected President. Moreover, while the Democratic electoral vote at the 50% popular vote mark determines whether the inversion interval lies below or above the 50% mark, the *magnitude* of the Democratic electoral vote deficit or surplus at the 50% mark is logically unrelated to the *width* of the inversion interval.

With an even number of electoral votes, there may also be a "tie interval," within which neither candidate has the required electoral vote majority. The historical PVEV functions for 1872, 1972, and 2008 exhibit tie intervals. If a "tie interval" were to span the 50% popular vote mark, there would be no inversion interval at all, but no historical PVEV exhibits such a "spanning" tie interval.

As is well known, the 2000 election produced an actual election inversion. We might think that this was because the PVEV function for 2000 was quite different from that in 1988 and, in particular, that it entailed a larger pro-Republican bias. Indeed, with exactly 50% of the popular vote, Gore would have won only 237 electoral votes, less than Dukakis's 252. But Gore would have won the election with 50.2664% of the popular vote, hardly more than that required for a Dukakis victory. Indeed, the two PVEV functions are very similar, as a comparison of Fig. 4.2a with Fig. 4.1d shows. The crucial difference between the two elections is the obvious

---

concern because our focus is on hypothetical elections that are close to the 50% mark with respect to the national popular vote.

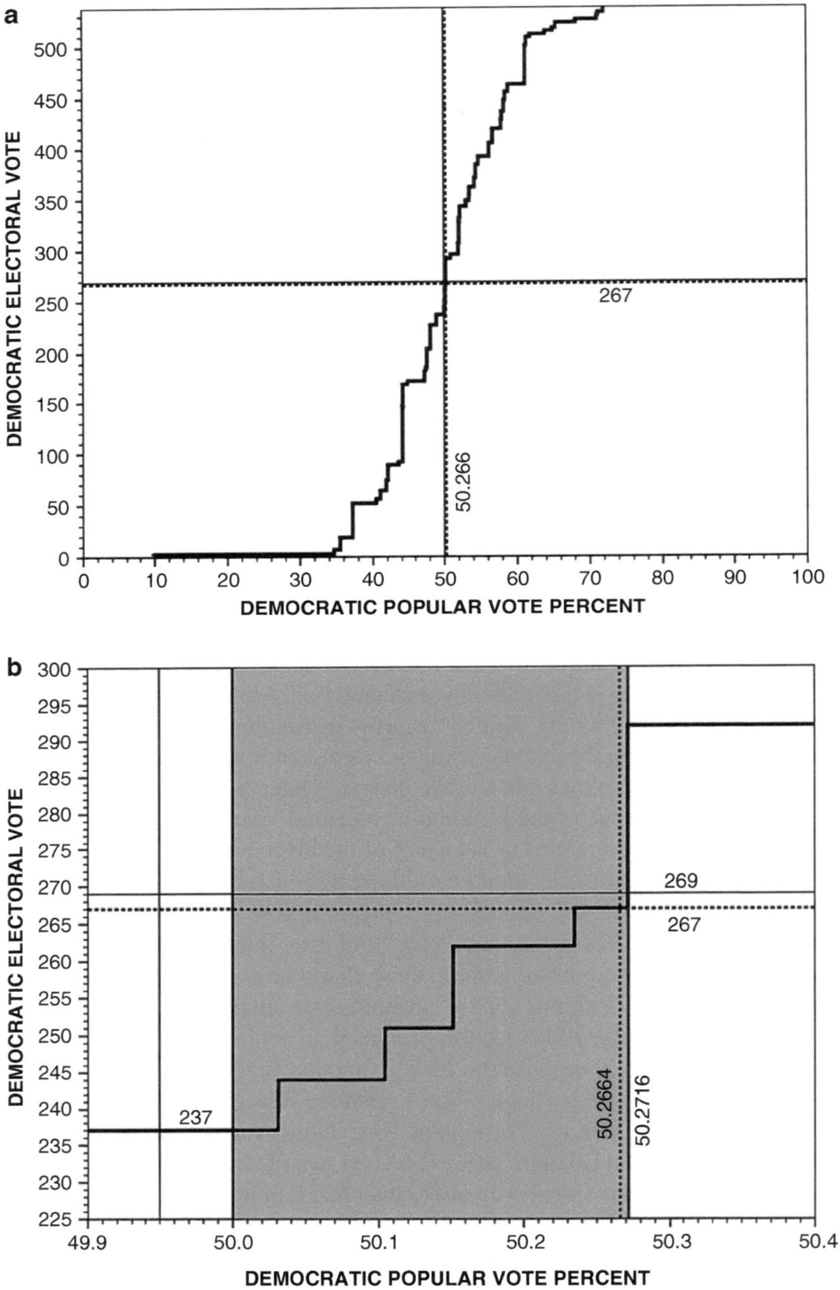

Fig. 4.2 The Democratic PVEV function in 2000 (**a**) The full PVEV (**b**) Zoom in on inversion interval

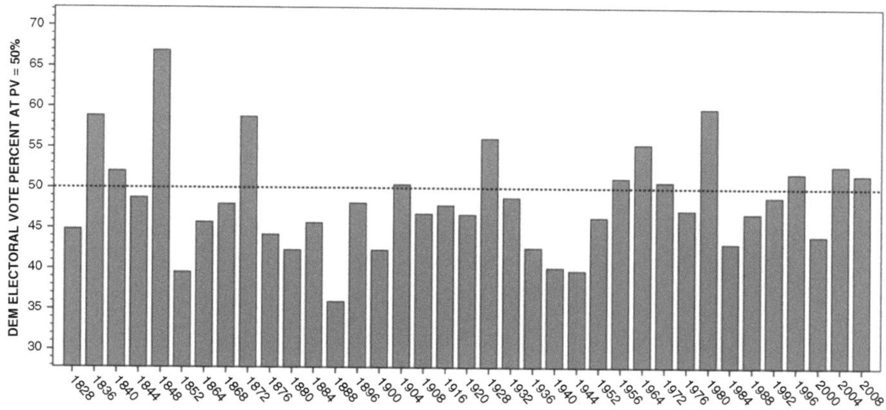

Fig. 4.3 Electoral votes at 50% of the popular vote: 1828–2008

fact that the actual 2000 election itself was *much* closer. The Democratic two-party popular vote percent was 50.2664%, putting it (just) within the 2000 inversion interval, as shown in Fig. 4.2b. (Even so, Gore would have won if the inversion interval in 2000 had been as small as in 1988.)

In contrast to both 1988 and 2000, in the counterfactual 1860 election a unified Democratic ticket would have won only 134 electoral votes out of 303; more astoundingly, the Democrats would have needed to win 61.26% of the popular vote for an electoral vote majority, producing an inversion interval 11.26 percentage points wide.

For every two-candidate Presidential election since 1828, Fig. 4.3 shows the Democratic percent of the electoral vote at the 50% popular vote mark. (Electoral vote percentages are given, since the number of electoral votes has changed over time.) Fig. 4.4a shows the Democratic popular vote percent required for an electoral vote majority (or tie). Figure 4.4b is derived from Fig. 4.4a and explicitly shows the width and direction (i.e., negative or pro-Democratic vs. positive or pro-Republican) of the inversion interval in each election.

Several observations can be drawn from these charts. First, as previously observed, if a bar that falls short of 50% in Fig. 4.3, the corresponding bar exceeds 50% in Fig. 4.4a and vice versa. However, the magnitudes of the two deviations are by no means strongly associated. For example, in 1888 the Democrat (Cleveland) would have won only about 36% percent of the electoral vote at the 50% popular vote mark, but he still would have won a majority of electoral votes with about 51% of the popular vote. (In one of the three historical election inversions, he lost the election with about 50.4% of the popular vote.)

Second, Fig. 4.3 shows that the Democratic (and therefore also Republican) percent of the electoral vote at the 50% popular vote mark often deviates strikingly from 50%. Moreover it is apparent that, over the whole time period, the Democratic electoral vote percent at the 50% mark has been far more likely to fall below 50% than to exceed it, indicating a historical anti-Democratic bias in the Electoral

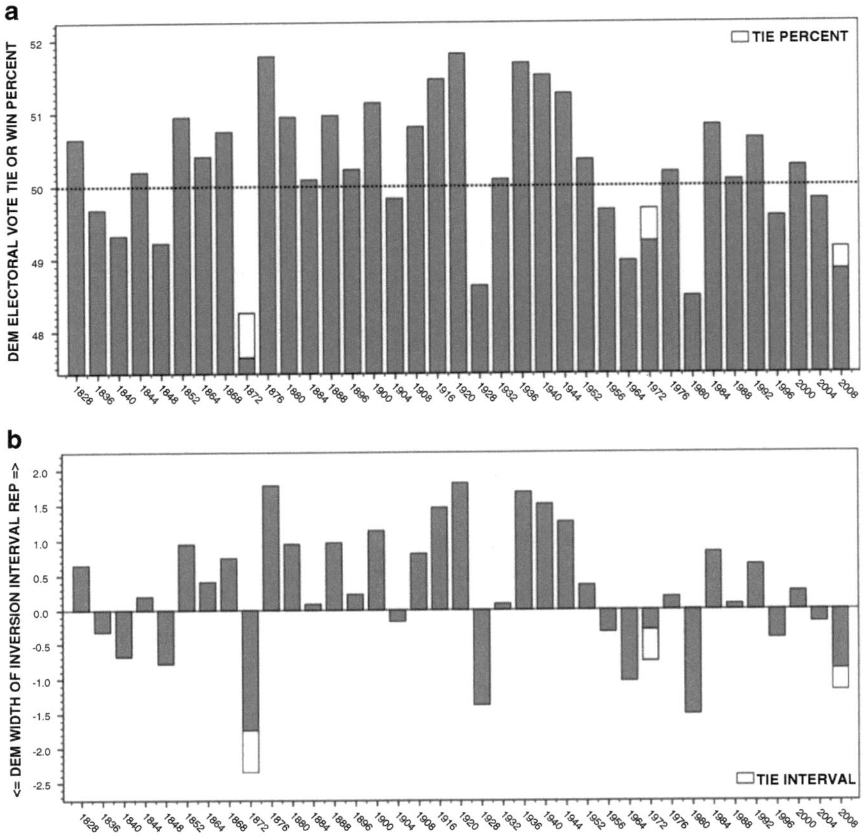

Fig. 4.4 Inversion intervals: 1828-2008 (**a**) PV required for EV majority (**b**) Width and direction of inversion intervals

College system. At the same time, the resulting inversion intervals shown in Fig. 4.4b are typically quite small, rarely exceeding two percentage points, and the mean of the absolute intervals (ignoring whether they reflect pro-Republican or pro-Democratic bias) is only 0.076 percentage points. Like the data in Fig. 4.3, they have exhibited an anti-Democratic bias more often than not. However, considering only elections from the mid-twentieth century on, the intervals have been smaller, rarely exceeding one percentage point and averaging about 0.5 percentage points, and exhibit no particular party bias.[6] Furthermore, the 1988 election turns out to have the smallest inversion interval on record.

---

[6]Garand and Parent (1991) employ a similar uniform swing analysis for Presidential elections from 1872 through 1984 but, instead of using each PVEV function directly, they use it to estimate the best fitting (with two parameters, "representational form" and "partisan bias") logistic S-curve to predict the electoral vote for the Republican candidate at the 50% popular vote mark. Using such

Based as they are on state-by-state data for all two-candidate Presidential elections, these results provide a more refined basis for estimating the likelihood election inversion by the Electoral College. Over all these elections, the Democratic two-party popular vote percent is approximately normally distributed with a standard deviation of about 6.2%. If we set its mean value at 50% (it is actually 49.17%), the mean absolute inversion interval of 0.76% implies a probability of an election inversion of approximately 0.048. If we consider only 1952 onwards, the popular vote SD increases to about 7.0% while the mean inversion interval falls to 0.47%, which implies a probability of an election inversion of only about 0.027. Considering only the six most recent elections, the average inversion interval falls further to 0.43%, but these elections have all been relatively close with a SD of 3.7%, which raises the probability of an inversion to about 0.046.

How the probability of inversion depends on the closeness of elections is more comprehensively displayed in Fig. 4.5a and b. The first figure separately stacks negative (pro-Democratic) and positive (pro-Republican) tie or inversion intervals on top of each other in order of their widths to give a sense of the how the frequency of inversions varies with closeness of the popular vote; this chart makes the historical Republican advantage very evident. Figure 4.5b stacks *absolute* (tie or) inversion intervals in the same manner; this chart can reasonably be interpreted as indicating the approximate probability of an election inversion as a function of the popular vote winner's margin above 50% of the two-party vote. If that margin is arbitrarily close to 50%, we can expect *a priori* that the probability of an inversion is about 0.5; if it is about 50.5%, Fig. 4.4b shows that the probability is about 0.25; if it is about 51%, the probability is about 0.125, and if it exceeds 52%, the probability is almost zero (in the absence of extreme sectional conflict like 1860). It is worth noting that Merrill (1978) and Ball and Leuthold (1991) produced quite similar estimates based on rather different methods.

4.3 Rounding Effects

The PVEV function for 1988 is almost symmetric. Figure 4.6 shows that, if we construct the Republican PVEV and superimpose it on the Democratic one, the two step functions, while distinct in detail, come very close to coinciding, not only in the vicinity of the 50% popular vote mark but throughout. If the two functions were to coincide precisely at the 50% mark, no inversion interval could exist. The

a curve produces quite different and usually much smaller Republican electoral vote percentages at the 50% popular vote mark. (Garand and Parent do not report inversion intervals, but their *S*-curves imply substantially wider intervals as well.) This is because the logistic curve estimated on the basis of the PVEV function is by assumption a symmetric S-shape and partisan bias shifts merely shifts the *S*-curve up or down the popular vote line. This mean that asymmetry anywhere in the PVEV data can shift the curve in the vicinity of the 50% popular, even if the PVEV function passes close to the perfect tie point. (Also see footnote 7.)

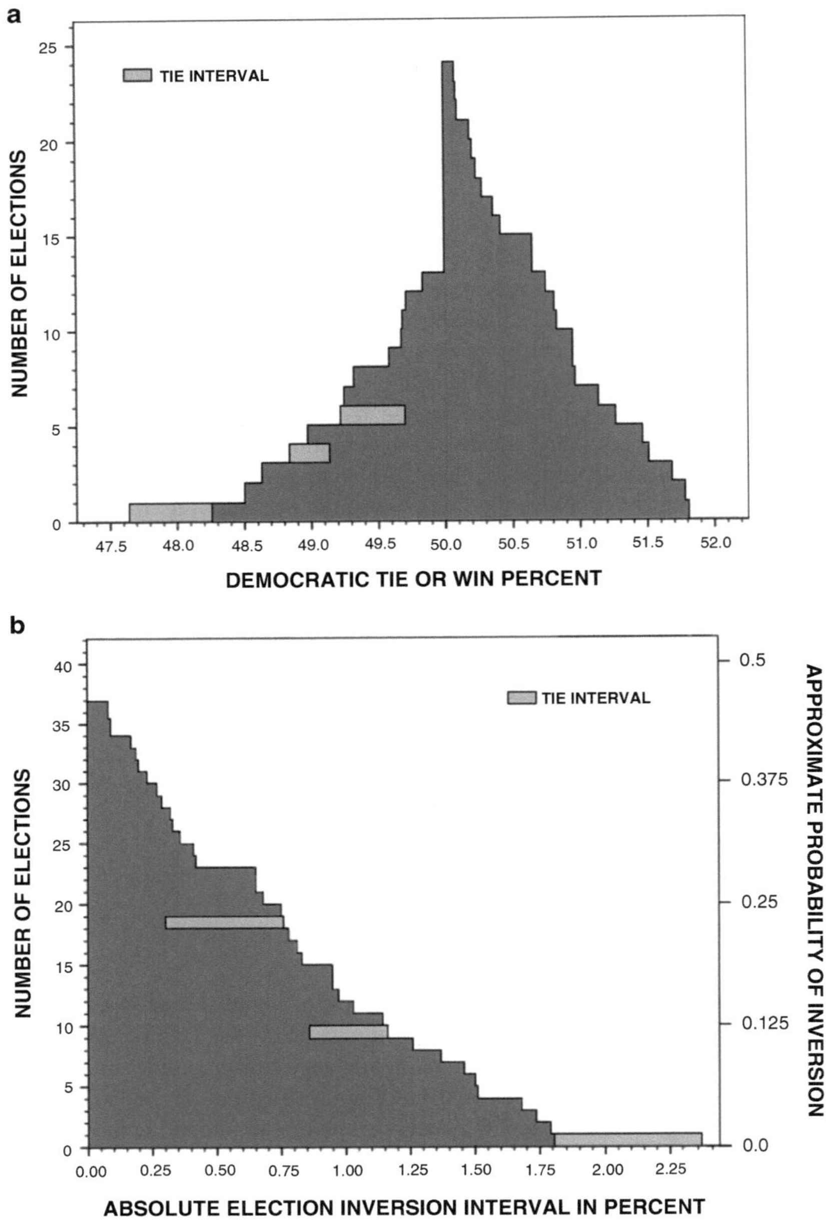

Fig. 4.5 Inversion intervals by closeness of popular vote: 1828–2008 (**a**) Intervals stacked by direction (**b**) Absolute intervals stacked

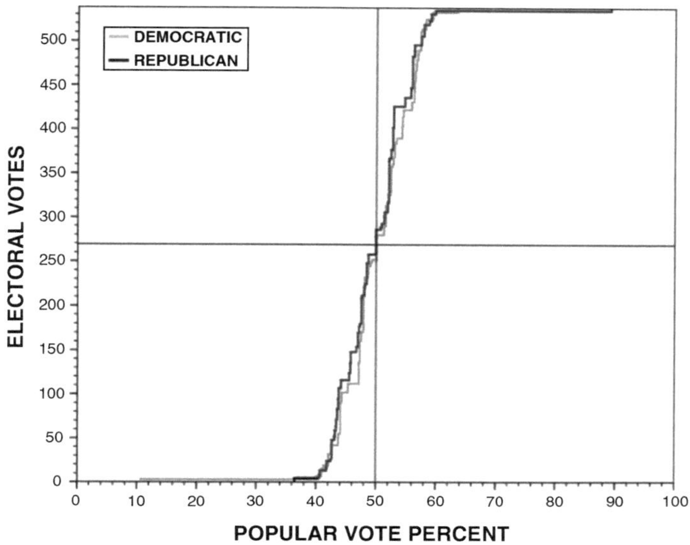

Fig. 4.6 Superimposed Democratic and Republican PVEV functions in 1988

small inversion interval that does exist results from what we might characterize as "rounding error" necessarily entailed by the fact that a PVEV function moves up or down in discrete steps as the popular vote swings up or down. For example, as the Democratic popular vote swings upwards, the pivotal state that gives the Democratic candidate 270 or more electoral votes almost certainly will not tip into the Democratic column *precisely* as the Democratic popular vote crosses the 50% mark but rather a little below or above the 50% mark, so an inversion interval of some magnitude essentially always exists. Clearly a specific PVEV function allows a "wrong winner" of one party only, depending on whether the inversion interval lies above or below the 50% Democratic popular vote mark.

However, suppose we plot other PVEV functions produced by small random perturbations in the actual 1988 state-by-state popular vote data. These PVEV functions will likely fall almost entirely within the "thickening" and "smoothing" of the actual PVEV function, as suggested in Fig. 4.7. The resulting "fuzzy" PVEV function passes through the two-way tie point even though almost certainly no specific "crisp" PVEV function does so. Figure 4.7 suggests that, if the 1988 election had been much closer and state-by-state votes had been slightly perturbed, Dukakis as well as Bush could have emerged as a "wrong winner."

In contrast, Fig. 4.8 shows the PVEV functions for both parties in 1940, which are clearly distinct almost everywhere. Moreover, even the "fuzzy" Democratic PVEV function clearly misses the two-way-tie point, as shown in Fig. 4.9. Figure 4.10 presents the most extreme case, namely the counterfactual version of 1860; clearly the "fuzzy" PVEV function would miss the two-way tie point by an even larger

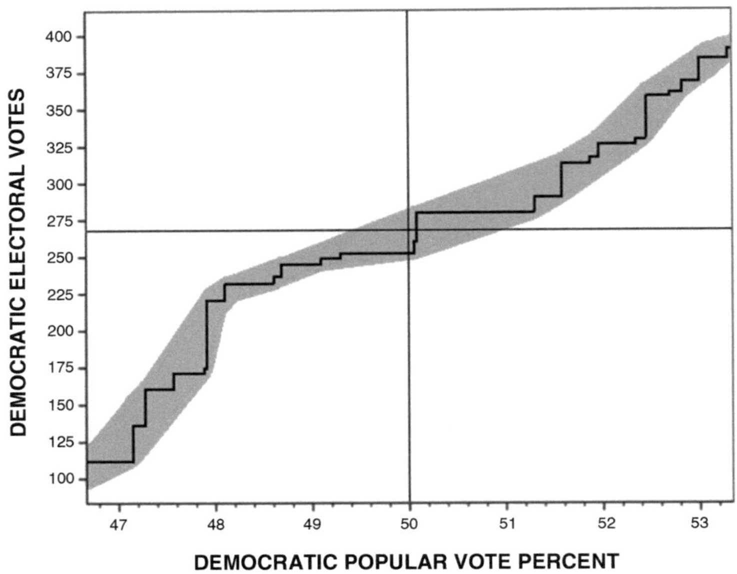

Fig. 4.7 The 'fuzzy' Democratic PVEV function in 1988

margin than in 1940. Given these PVEV functions in contrast to the 1988 one, inversions are much more likely to occur because they result, not from mere "rounding effects," but from a fundamental asymmetry in the general character of the PVEV function, particularly in the vicinity of the 50% mark. Moreover, even with fairly substantial perturbations of the state-by-state votes, "wrong winners" would almost always be Republicans, not Democrats.

The 1940 PVEV exemplifies in typical form – and the 1860 PVEV in exaggerated form – the substantial pro-Republican bias in historical PVEV functions in the vicinity of the 50% mark that results largely from the electoral peculiarities of the old "Solid South" throughout the first two-thirds of the twentieth century – in particular, its overwhelmingly Democratic popular vote percentages, combined with its strikingly low voting turnout. Though the *overall* bias might be deemed pro-Democratic, in the vicinity of the 50% popular vote mark the bias is pro-Republican. Consider the party PVEV functions for 1940 displayed in Fig. 4.8: the Democrats win more electoral votes than the Republicans do for almost all levels of popular vote support, but the Republicans win more in a narrow range in the vicinity of the 50% mark, which of course is precisely the range that matters.[7] The counterfactual 1860 case provides an even more extreme example.

[7]I believe that this consideration in part determines the Garand and Parent (1991) conclusion, based on smooth S-curves estimated on the basis of the entire PVEV, that the Electoral College has historically had a pro-Democratic bias.

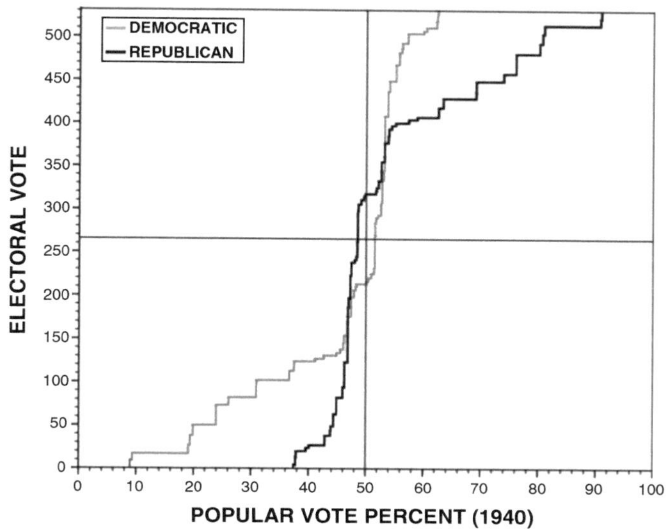

Fig. 4.8 Superimposed Democratic and Republican PVEV functions in 1940

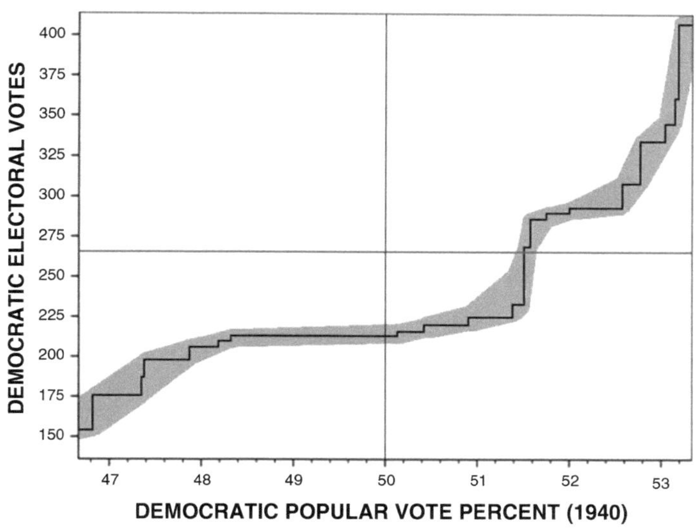

Fig. 4.9 The 'fuzzy' Democratic PVEV function in 1940

Bias in the PVEV function can result from either or both of two distinct phenomena: *apportionment effects* and *distribution effects*. The former refers to disproportionality between the popular votes cast within states and the electoral votes cast by states. As an example, the old "Solid South" had very low voting turnout (mostly reflecting the effective disenfranchise of potential black voters),

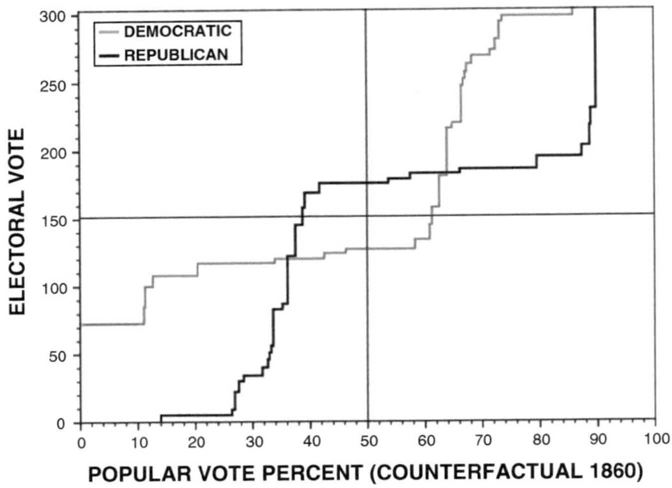

Fig. 4.10 The Democratic PVEV function in counterfactual 1860

with the result that Southern electoral votes were based on a much lower total popular vote than those of other regions. The latter reflects geographical patterns in the popular vote for the two candidates or parties that makes one candidate's distribution of popular votes more "efficient" in winning electoral votes than the other. As an example, the overwhelmingly Democratic popular votes in the "Solid South" did not win the Democrats any more electoral votes than more modest popular vote majorities would have, whereas the Republicans won most non-Southern states by more modest margins. Both apportionment and distribution effects by can produce election inversions by themselves and, in combination, they can either reinforce or counterbalance each other.

4.4 Apportionment Effects

In order to assess the magnitude and direction of apportionment effects, we start with the theoretical benchmark of a *perfectly apportioned* two-tier electoral system, in which apportionment effects are eliminated because electoral votes are apportioned among the states in a way that is precisely proportional to the total popular vote cast within each state (which requires that states be apportioned fractional electoral votes). In a perfectly apportioned system, a candidate who wins $X\%$ of the electoral vote carries states that collectively cast $X\%$ of the total popular vote.[8] This concept is introduced as an analytical tool; as a practical matter, an electoral system can be perfectly apportioned only retroactively – that is, after the popular votes in each state are cast and counted.

Apportionment effects encompass whatever may cause deviations from perfect apportionment. The U.S. Electoral College system is imperfectly apportioned, for at least six reasons.

1. House seats (and therefore electoral votes) must be apportioned in small *whole numbers*, and therefore cannot be *precisely* proportional to *anything*.
2. There are many *different methods* of apportioning whole numbers of seats or electoral votes on the basis of population, none of which is uniquely best (Balinski and Young 1982).
3. House (and therefore electoral vote) apportionments are anywhere from *two to ten years out-of-date* at the time of a Presidential election.
4. The apportionment of electoral votes is *skewed in favor of smaller states*, because all states are guaranteed a minimum of three electoral votes and (approximate) proportionality begins only after that.
5. The size of the House is not fixed by the Constitution and can be changed by law (as it frequently was until the early twentieth century), so the magnitude of the small-state bias can be reduced (or enhanced) by law, by increasing (or reducing) the size of the House.[9]
6. House seats (and therefore electoral votes) are apportioned to states on the basis of their *total population* and not on the basis of their (i) voting age population, or (ii) voting eligible population (excluding non-citizens, etc.), or (iii) number of registered voters, or (iv) number of actual voters in a given election.[10]

Similar apportionment imperfections apply (in greater or lesser degree) in all two-tier electoral systems.

While imperfect apportionment *may* create bias in a PVEV function, it *need not* do so. Overall bias depends on the extent to which states' advantages or disadvantages with respect to apportionment effects are correlated with their support for one or other candidate or party. The logically maximum bias that can arise from

[8]Note that this says nothing about the popular vote margin by which the candidate wins or loses states and, in particular, it does say or imply that the candidate wins $X\%$ of the national popular vote.

[9]See Neubauer and Zeitlin (2003) for an analysis of how changes in House size would have affected the 2000 Presidential election.

[10]In addition, until slavery was abolished by the Thirteen Amendment in 1865, House seats were apportioned on the basis of the total free population plus three fifths of "all other persons" (who certainly could not vote). While the Nineteenth Amendment, requiring all states to give women the right to vote on an equal basis with men, took effect in 1920, there was a preceding period in which some states allowed women to vote while others did not. This produced major apportionment effects. For example, in 1916 considerably more popular votes were cast in Illinois (which allowed women to vote) than in New York (which did not), even though Illinois had only 29 electoral votes compared with New York's 45.

imperfect apportionment can be determined by (i) ranking the states by their degree of advantage with respect to actual apportionment relative to perfect apportionment, (ii) cumulating both electoral votes and total popular vote shares over this ranking until an electoral vote majority is achieved, and (iii) noting the corresponding share of the national popular vote that has been accounted for. In recent elections this popular vote share has been about 45% but it was considerably smaller (about 32%) in the early twentieth century.

Let us examine the impact of apportionment effects in the counterfactual 1860 election, which was based on especially imperfect apportionment.

1. Southern states (in most cases for the last time[11]) benefitted from the three-fifth compromise giving them partial credit for their non-voting slave populations.
2. Southern states had on average smaller populations than northern states and therefore benefitted disproportionately from the small-state advantage in apportionment.
3. Even within the free population, suffrage was typically more restricted in the South than elsewhere (where close to universal adult male suffrage prevailed).
4. Turnout among eligible voters was generally lower in the South than the North.

But *all* of these apportionment effects favored the South and therefore the Democrats. Perfect apportionment would have *increased* the popular vote required for a Democratic electoral vote majority. Thus the massive pro-Republican election inversion was entirely due to distribution effects, and the inversion interval would have been even wider in the absence of the counterbalancing apportionment effects.

While perfect apportionment is presumably not feasible in practice, we can use it analytically. Figure 4.11 compares the 1988 Democratic PVEV functions and inversion intervals under actual and perfect apportionment. Clearly apportionment effects were very small in this election. Figures 4.12 and 4.13 make the same comparisons for 1940 and 1860. Here apportionment effects are very substantial at low Democratic popular vote percentages and remain quite substantial up to a bit over the 50% mark in 1940 and up to about the 60% mark in 1860. The 1940 chart reflects the typical "Solid South" effect that was displayed in a monolithic fashion in the 1860 election.

Note that the percent of the popular vote required for the Democrats to win a majority of electoral votes in 1860 is 61.26% under the actual apportionment and 62.51% under perfect apportionment. The difference between 61.26% and 62.51% of −1.25% indicates the impact of imperfect apportionment on the Democratic vote required for an electoral vote majority and, being negative, it indicates that the actual apportionment benefitted the Democrats. At the same time, the difference between 62.51% and 50% of +12.51% indicates the huge impact of distribution effects on the Democratic vote required for an electoral vote majority and, being positive, it indicates that the distribution effects harmed the Democrats.

[11]The few slave states that did not secede from the union retained this apportionment advantage in the 1864 election. By the time of the 1868 election, there were no "other persons."

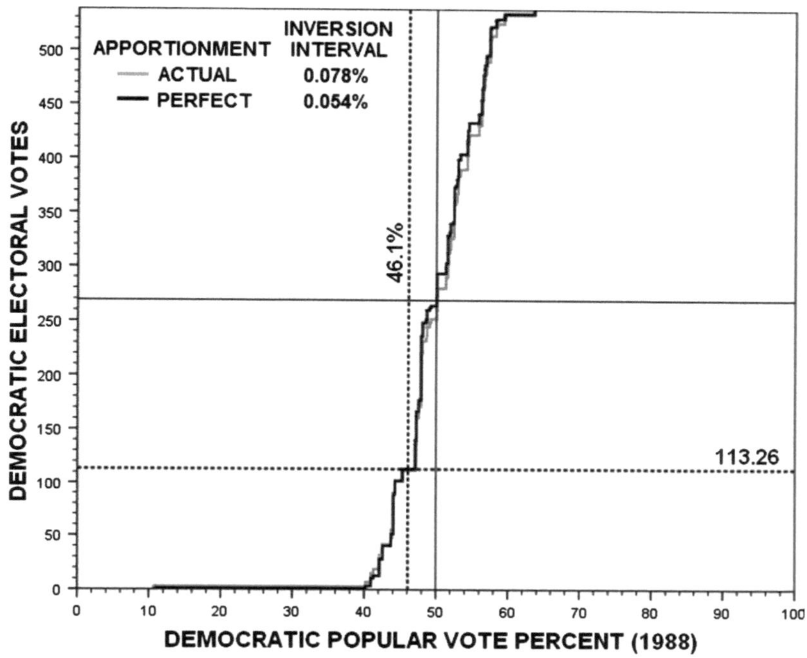

Fig. 4.11 The Democratic PVEV function in 1988 under perfect vs. actual apportionment

Figures 4.14 and 4.15 correspond to Figs. 4.3 and 4.4b by showing summary data for all elections under perfect apportionment.[12] One might expect that "perfecting" apportionment would typically reduce the width of overall inversion intervals and thereby reduce the frequency and of election inversions. Indeed, under perfect apportionment, Gore would have won the 2000 election with 274.92 electoral votes, and Tilden would have won the 1876 election with 182.174 electoral votes. However, under perfect apportionment Cleveland would have lost even more decisively to Harrison in 1888, winning only 135.76 electoral votes to Harrison's 265.24. Moreover, Wilson would have lost the 1916 election with only 238.57 electoral votes out 531, despite a modest majority of the popular vote. So with respect to actual election inversions, perfect apportionment would eliminate two but not all three instances and would create one new instance. Perhaps surprisingly, perfect apportionment actually increases the overall degree of Republican bias in the Electoral College system and, as a consequence this, considerably increases the average magnitude of absolute inversion intervals from 0.76% to 01.22%.

Finally, Fig. 4.16 decomposes each inversion interval into the contributions made by apportionment effects and distribution effects. We will look more closely at this figure in the concluding section.

[12]Since perfect apportionment requires fractional electoral votes, no electoral votes ties occur.

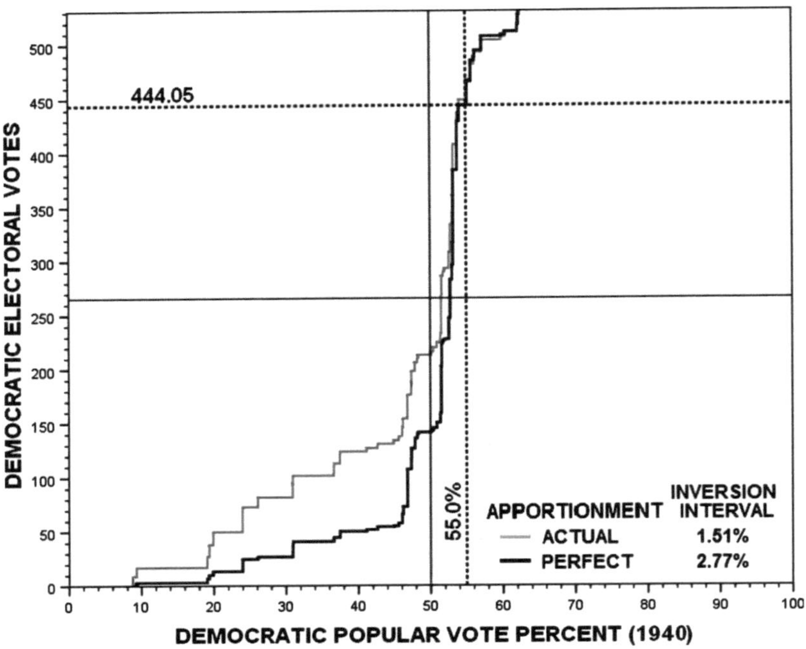

Fig. 4.12 The Democratic PVEV function in 1940 under perfect vs. actual apportionment

4.5 Distribution Effects

We can measure the impact of distribution effects on inversion intervals simply by calculating the difference between 50% and the percent of the vote received the Democrats at the 50% popular vote mark under perfect apportionment (as displayed in Fig. 4.16), but we can also examine distribution effects more directly.

Distribution effects in two-tier electoral systems result from the winner-take-all feature at the state (or district) level. Distribution effects can be powerful even with small uniform districts and/or perfect apportionment. If one candidate's (or party's) popular vote is more "efficiently" distributed over states (or districts) than the other's, an election inversion can occur even with perfect apportionment.

The simplest possible example of distribution effects producing an election inversion in a small, uniform, and perfectly apportioned district system is provided by nine voters in three districts. Suppose that the individual votes for candidates D and R in each district are as follows: (R,R,D) (R,R,D) (D,D,D). Thus the election outcome is as follows:

	Popular votes	Electoral votes
D	5	1
R	4	2

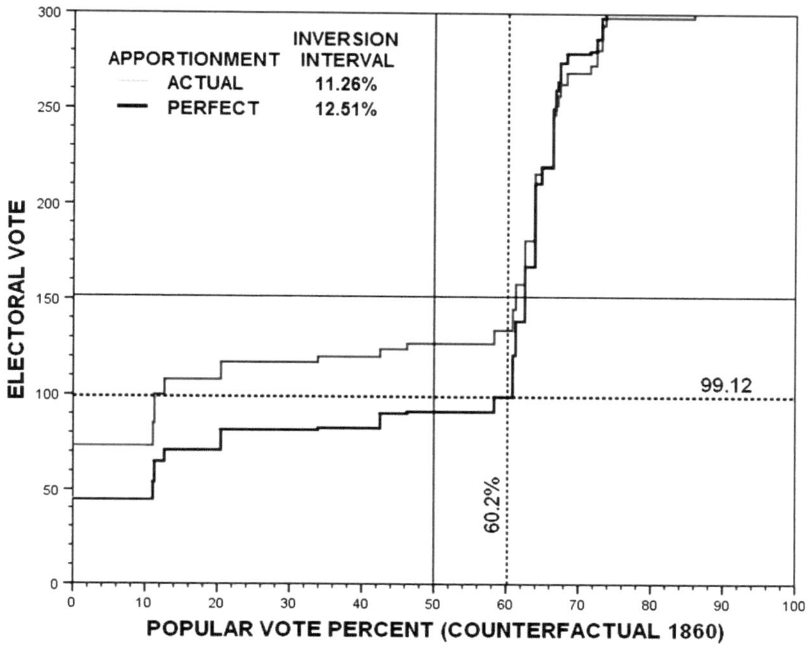

Fig. 4.13 The Democratic PVEV function in counterfactual 1860 under perfect vs. actual apportionment

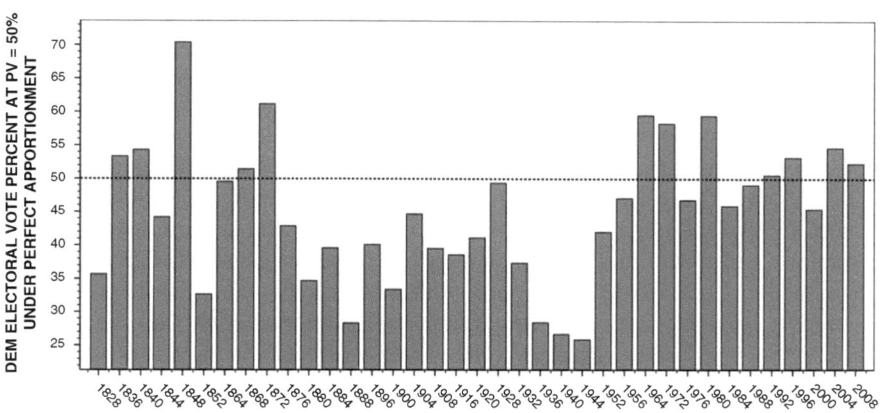

Fig. 4.14 Electoral votes at 50% of the popular vote with perfect apportionment: 1828–2008

Since R's votes are more "efficiently" distributed than D's (whose support is "wastefully" concentrated in the third district), R wins a majority of districts with a minority of popular votes.

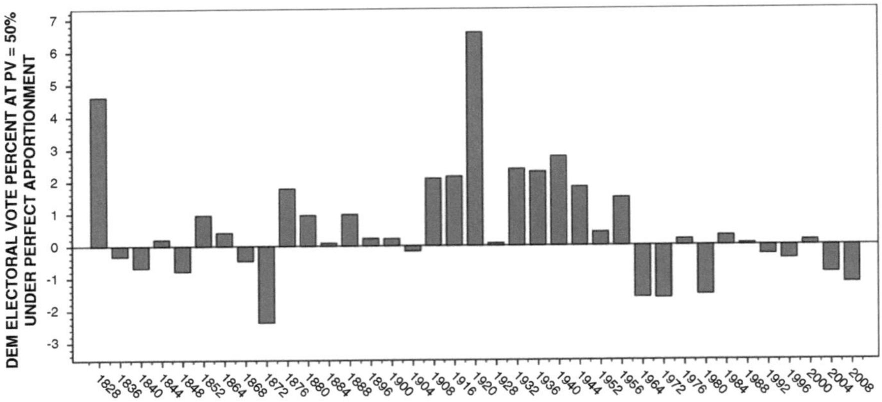

Fig. 4.15 Inversion intervals with perfect apportionment: 1828–2008

More generally, suppose there are k uniform districts each with n voters. To avoid the problem of ties, let us that assume both k and n are odd numbers. A candidate can win by carrying a bare majority of $(k+1)/2$ districts each with a bare majority of $(n+1)/2$ votes. Thus a candidate can win with as few as $[(k+1)/2] \times [(n+1)/2] = (n \times k + n + k + 1)/4$ efficiently distributed total votes. With $n = 3$ and $k = 3$, the last expression is $4/9 = 44.4\%$, but as n and/or k become large, the last expression approaches a limit of $(n \times k)/4$, i.e., 25% of the total popular vote.

Stated more intuitively, if the number of districts is fairly large and the number of voters is very large, the most extreme logically possible election inversion in a perfectly apportioned system results when one candidate or party wins just over 50% of the popular votes in just over 50% of the uniform districts or in non-uniform states that collectively have just over 50% of the electoral votes. These districts also have (just over) 50% of the popular vote (because apportionment is perfect). The winning candidate or party therefore wins just over 50% of the electoral votes with just over 25% of the popular vote. The other candidate, though winning almost 75% of the popular vote, loses the election, producing a massive election inversion. In the resulting PVEV, the inversion interval is just short of 25 percentage points wide. (If the candidate or party with the favorable vote distribution is also favored by imperfect apportionment, the inversion interval could be even greater.) This "25%–75%" rule pertaining to distribution effects was noted in passing by Schattschneider (1942, p. 70) and more formally by May (1948), Laffond and Laine (2000), and perhaps by others as well.

In the counterfactual 1860 Lincoln vs. anti-Lincoln scenario, the popular vote distribution over the states approached the logically extreme 25%–75% pattern more closely than in any actual Presidential election. In the counterfactual election, Lincoln carried all the northern (free) states except New Jersey, California, and Oregon, mostly by modest popular vote margins that rarely exceeded 60% and typically were closer to 50%. These states held somewhat more than half the

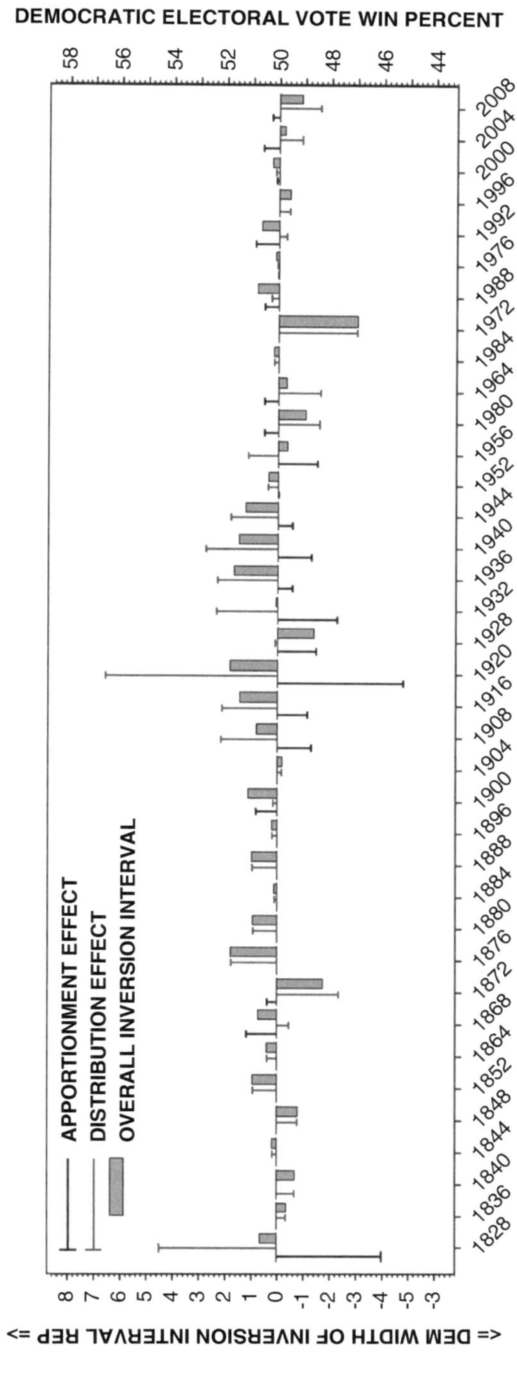

Fig. 4.16 Inversion intervals decomposed into apportionment and distribution effects: 1828–2008

electoral votes and a larger majority of the national popular vote. The anti-Lincoln opposition carried all the slave states by essentially 100% margins. (No Lincoln-pledged electors ran in any of the state that would subsequently secede from the Union.) The opposition also carried California and Oregon by substantial margins and New Jersey by a narrow margin. All together these states held somewhat less than half of the electoral votes and substantially less than half of the national popular vote.

Sterling (1981) has devised an insightful geometric construction to visualize "Electoral College misrepresentation. " A "Sterling diagram" is a histogram that displays the popular vote split between the two candidates in each state, where states are ranked in order from the strongest to weakest for the winning party, with the width of each state "bar" proportional to its total popular vote.[13]

Figure 4.17a shows the Sterling diagram for the 1988 election. Selected "bars" are explicitly drawn in and labeled by state. Running from the most Republican state of Utah to the least Republican "state" of the District of Columbia, it is Michigan, beating out Colorado by about 0.03%, that tips the Republican electoral vote over the 270 mark. Once Michigan is in their column, the Republicans are carrying states with 49.43% of the national popular vote, as indicated by the vertical dashed line. The fact that this falls below the 50% mark reflects the (very small and previously noted) apportionment effect favoring the Republicans in 1988.[14] The area of the whole rectangle making up the Sterling diagram represents all 100% total national (two-party) popular vote. The shaded area below the tops of the bars represents the 53.9% of the popular vote won by Republican Bush and the unshaded area above the top of the bars represents the 46.1% of the popular votes won by Democrat Dukakis.

Figure 4.17b demarcates different portions of the total Republican popular vote in 1988. The dark shaded portion represents the portion of the total popular vote essential for 270 electoral votes. This is essentially the 25% given by the "25%–75% rule" (except that, because apportionment effects work slightly in favor of the Republicans, it is actually slightly less than 25%). The lightly shaded portion

[13]I follow Sterling by orienting these charts to the party that actually won the election, rather than to the Democratic party.

[14]Note that the interval between 49.43% and 50% is not directly related to the inversion interval. The inversion interval is the difference between 50% and *the smallest national popular vote percent* for a candidate that produces an electoral vote majority. This interval is the difference between 50% and *the share of the total popular vote cast by the smallest set of states* (ranked by party strength) *that produces an electoral vote majority*. In the absence of apportionment effects, this interval would be zero. If states were instead ranked from in order from strongest to weakest for the Democratic party, Michigan would again be the pivotal state and, once the Democrats win Michigan, they would be carrying states with 54.58% of the national popular vote; note that Michigan (which cast 4.01% of the national popular vote) counts in both totals. Such percentages (and the corresponding electoral vote splits) can deviate substantially from a 50–50 split, because pivotal states are typically big states, and tiny shifts the national popular vote split between the two candidates can shift a pivotal state one way or another and thus have a big impact on the percent of the national popular (and electoral) vote cast by states carried by one or other candidate.

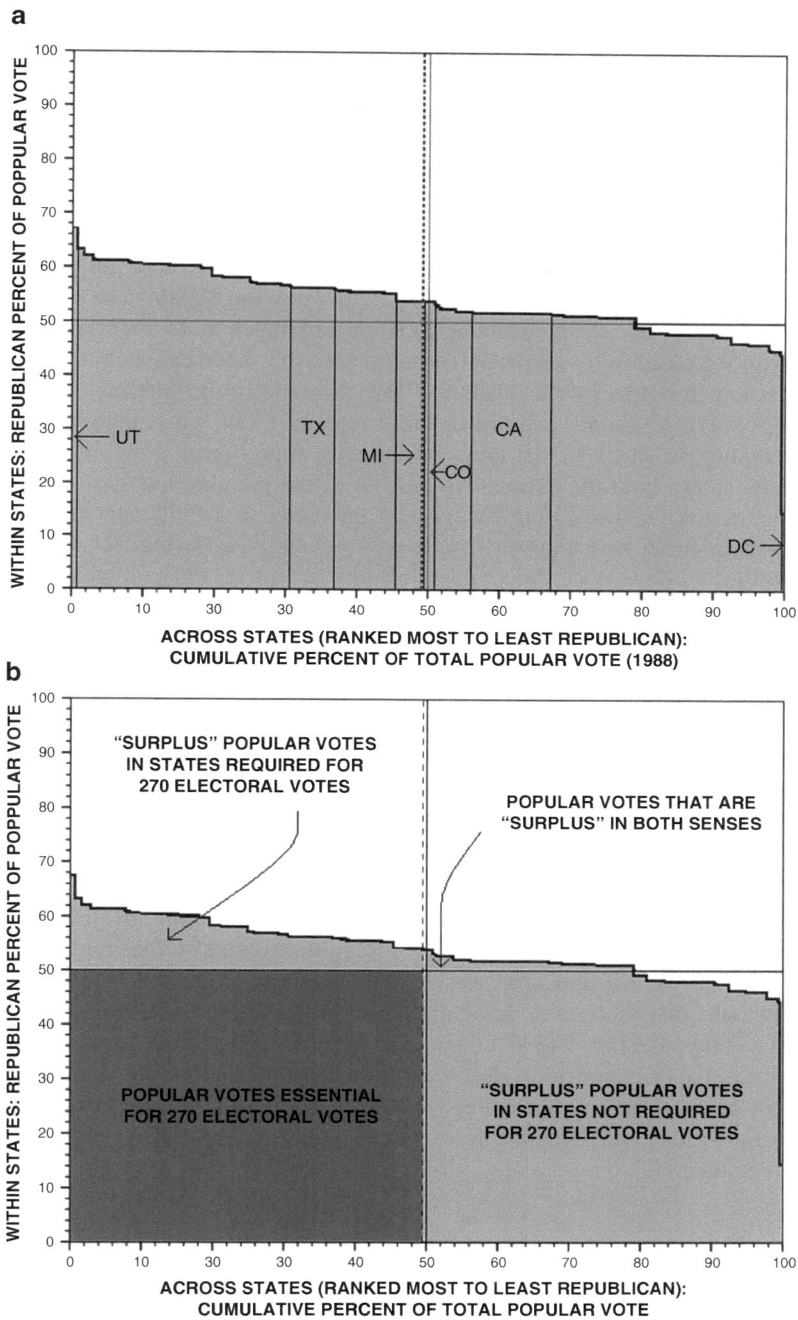

Fig. 4.17 The Sterling diagram for 1988 (**a**) The basic Sterling diagram (**b**) The Sterling diagram showing "surplus" votes

represents "surplus" (or "wasted") Republican popular votes, which fall into three different quadrants of the diagram: (i) "surplus" Republican votes (in excess of 50%) in the states essential for 270 electoral votes, (ii) all Republican votes up to 50% in "surplus" states (in excess of 270 electoral votes), and (iii) all Republican votes that are "surplus" in both respects.

We can get a direct measure of overall distribution effects in this or any other election by comparing the percent of all votes that are "surplus" for each party and thereby determine which party has the most efficient distribution of votes. In Fig. 4.17b, it appears that the Republicans "wasted" slightly more "surplus" votes than the Democrats. This comes about in part, because the Republicans have a (very small) advantage due to apportionment effects and, much more important, because they won the election by a substantial margin (so the shaded portion of Fig. 4.17b extends into the upper-right quadrant). We can modify the Sterling diagram by making two adjustments to remove these factors. First, we reallocate electoral votes among the states so that they are perfectly apportioned. Now the horizontal axis now shows both the cumulative percent of the popular vote cast in, and the cumulative percent of electoral votes cast by, the states, so we no longer must specify where an electoral vote majority is achieved. Second, we "swing" the Republican vote uniformly downward (by simply shifting the tops of the state bars uniformly downward) until the election is a *perfect tie*, in the specific sense that the popular vote is tied in the pivotal state (Michigan) that produces 270 electoral votes, and the median (state) bar has a height of 50% and no Republican votes appear in the upper-right quadrant.[15]

Figure 4.18 shows the Sterling diagram for 1988 with these two adjustments. In the adjusted diagram, 50% of the total popular votes (precisely those in the upper-left and lower-right quadrants) are "surplus" to one other party. The adjusted diagram shows how this fixed proportion of surplus votes is divided between the two parties. In the absence of distribution effects (and in a perfect tie election with no apportionment effects), surplus votes would be equally divided between the two parties (25% for each). We see that in 1988 surplus votes are almost equally divided at the perfect tie point: 25.24% for Republicans and 24.76% for Democrats. This fact that the Republicans "wasted" slightly more votes than the Democrats demonstrates that there is no logical connection between the *overall* distribution effects displayed in Fig. 4.18 and the impact of distribution effects on the width and direction of the inversion interval. If in 1988 the Republicans had won their strongest states by more modest margins, they would have had fewer "surplus" votes than the Democrats, rather than slightly more, but this would have had no impact on the inversion interval.

[15]In this sense, the 2000 election was only 537 votes away from a perfect tie. Note that a perfect tie in this sense is almost certainly not a "two-way tie," since almost certainly the national popular vote is not tied. While the popular vote is likely to be very close, the only logical constraint remains that given by the 25%-75% rule.

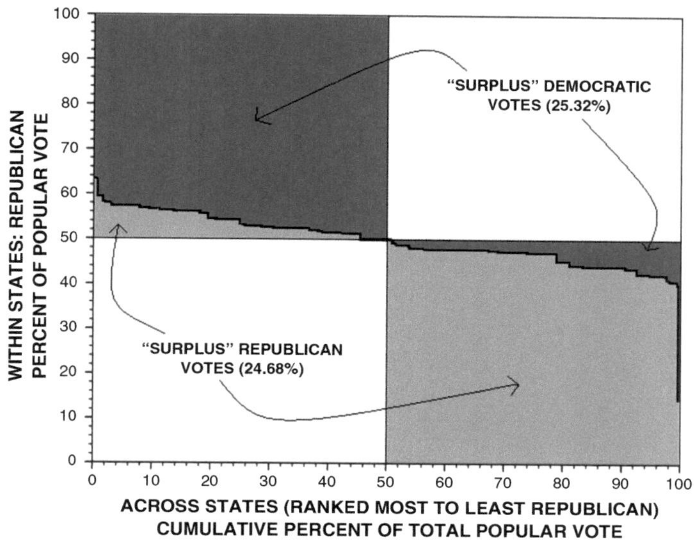

Fig. 4.18 The adjusted Sterling diagram for 1988

The 1988 election provides an example of an election with very small distribution effects. Once again the counterfactual 1860 election provides by far the most spectacular example of huge distribution effects (that totally overwhelm somewhat more modestly pro-Democratic apportionment effects). Figure 4.19 shows the standard Sterling diagram for 1860. Due the Democratic apportionment advantage, the Republicans had to carry states casting 60% of the popular to win, and they actually carried states casting about 67% of the popular vote. Figure 4.20 shows the Sterling diagram adjusted to show perfect apportionment and a uniform swing against the Republicans just sufficient to bring about a perfect tie election. It thereby isolates distribution effects and shows the massive Republican advantage in this respect: of the 50% of all votes that are by definition surplus, 38.9% were "wasted" by the Democrats and only 11.1% by the Republicans.[16]

Figure 4.21 shows the Democratic advantage or deficit in each election with respect wasted votes as the difference between 25% and the actual percent of Democratic popular votes that are surplus, along with the net impact of distribution effects on the inversion interval previously shown in Fig. 4.16. It can be seen that these two quantities track each other quite closely but, as we would expect given the considerations previously mentioned, they are less than perfectly related.

[16]This Republican advantage with respect to surplus vote could be reckoned as even greater, since the 11.1% takes no account of the fact that the swing required to create a tie election makes the Republicans popular vote negative in the Southern states in which they actually won (literally) zero votes.

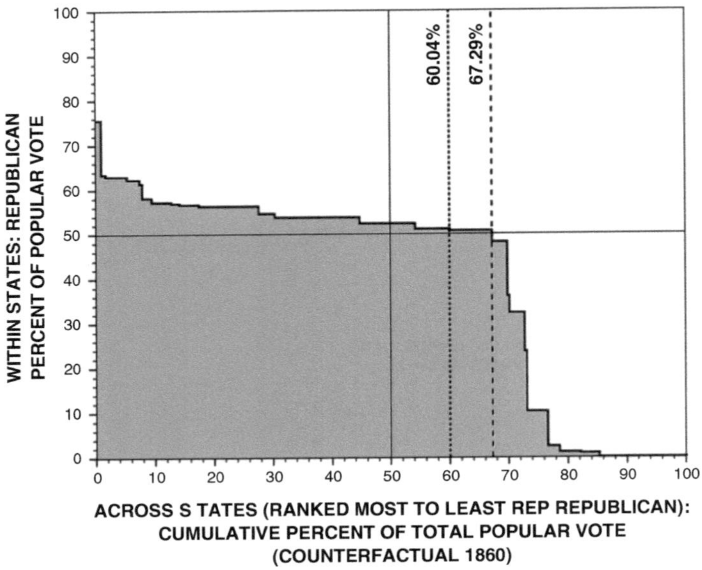

Fig. 4.19 The Sterling diagram for counterfactual 1860

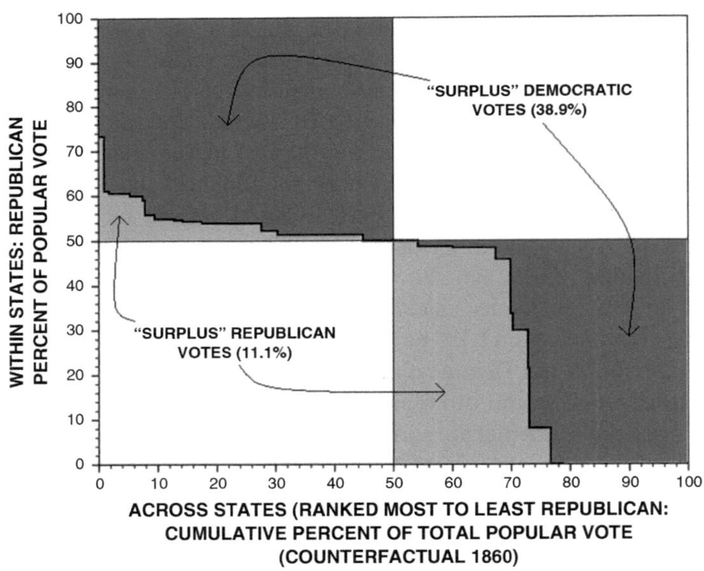

Fig. 4.20 The adjusted Sterling diagram for counterfactual 1860

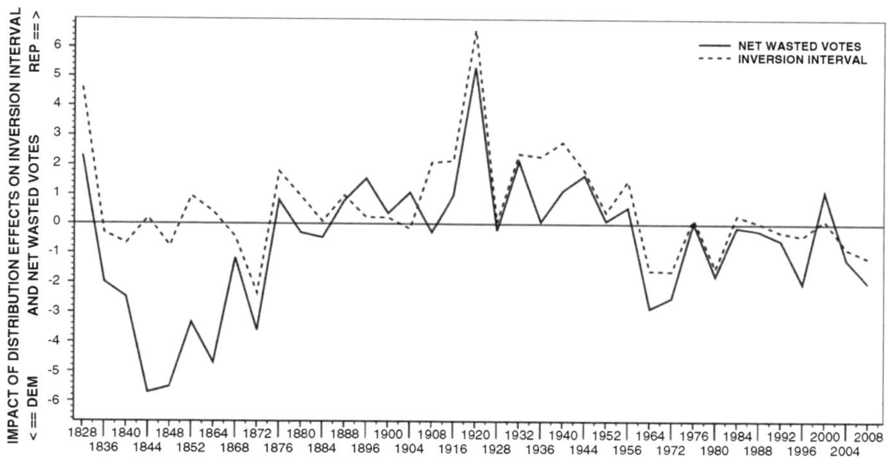

Fig. 4.21 Surplus votes and distribution effects: 1828–2008

4.6 Conclusions

Let us more closely examine Fig. 4.16, which shows the magnitude and direction of inversion intervals and their decomposition into apportionment and distribution effects for every election. A number of observations are in order.

First, it may seem surprising that the impact of apportionment effects on inversion intervals is precisely zero in as many as fourteen elections, given the highly imperfect apportionment underlying the Electoral College system. However, we must recall that even highly imperfect apportionment can have little impact on an inversion interval if party support over states is not correlated with their apportionment advantages. Beyond this, zero impact on an inversion interval does not mean that the PVEV functions under perfect and actual apportionment are everywhere identical, only that they coincide at the 50% popular vote mark – that is, when we cumulate both actual and perfectly apportioned electoral votes, the same state is pivotal under both apportionments. The fact that pivotal states are likely to be large states with many electoral votes (under both apportionments) reinforces this observation.

Second, perhaps the most striking fact conveyed by Fig. 4.16 is that, out of the 22 elections in which both apportionment and distribution effects have an impact on the inversion interval, these effects work in opposite rather than reinforcing directions in 19 of these cases, and they thereby tend to produce relatively small overall inversion intervals.

Third, the fact the apportionment and distributions typically work in opposite directions is largely an artifact of another overall pattern Fig. 4.16, which is that in general apportionment effects have favored Democrats while distribution effects have favored Republicans. Since the latter effects have generally been somewhat stronger than the former, the Electoral College has historically exhibited a small but significant pro-Republican bias. (In this respect, the counterfactual 1860 election merely exaggerates the overall pattern.) The pro-Republican bias is further enhanced by the fact that, in all four elections in which apportionment and distribution effects reinforced one another, they did so in the pro-Republican direction.

Such patterns become more understandable when we take account of the chronological order of these cases. It then become evident the "overall pattern" noted above really is the product of a particular historical era extending across the first two-thirds of the twentieth century. Until the early twentieth century, no consistent pattern is evident, perhaps reflecting relatively loose party ties in the early party system followed by the disruptive events leading to and during the Civil War and subsequent Reconstruction. During most of this period – though with the conspicuous exceptions of 1856 (excluded from this analysis) and 1860 (included in its counterfactual variant) – politics was largely non-sectional, with both parties typically carrying states and winning electoral votes in all regions of the country.

Electoral maps that show which party carried each state may suggest that the "solid South" emerged in immediately in the first election following the end of Reconstruction in 1877, for such maps show that the Democrats won the electoral votes of every Southern state in 1880 and for decades thereafter. However, such maps do not show that until the early twentieth century, Republicans consistently won a substantial minority of votes in Southern states, based in large part on the support of not yet disenfranchised black voters. But beginning in 1890, Southern states began to establish "Jim Crow" regimes that entailed (in addition to racial segregation) suppression of both the black and Republican vote (and a good deal of the white vote as well). "Jim Crow" becomes fully evident in the 1908 election, which begins a string of ten elections (plus three excluded from this analysis) through 1956 in which apportionment effects consistently favored the Democrats (as a result of vote suppression and low turnout in the South) and distribution effects favored the Republicans (as a result of "wastefully" large Democratic popular vote majorities in the South), with the latter outweighing the former in their impact on inversion intervals and producing an overall pro-Republican bias in the Electoral College system.[17]

This string of elections ends with collapse of the "Jim Crow" system in the late 1950s and early 1960s under the pressure from the civil rights movement and federal intervention. Beginning in 1964, after passage of the Civil Rights Act sponsored by Democratic President Johnson and opposed by the Republican nominee Barry Goldwater, the old white Democratic South began to switch its party allegiance from the Democratic to the Republican side, so the partisan impact of apportionment and distribution effects was reversed. Thereafter, as the federal Voting Rights Act took full effect, turnout increased to normal levels in the South and heavy black support provided the basis for a substantial (but rarely winning) Democratic popular vote in Southern states. In the modern era, apportionment and distribution effects (and

[17] Key (1949) provides the definitive treatment of politics in the old "solid South."

inversion intervals) are relatively small and, in so far as they exist, typically reverse of the earlier pattern by favoring Republicans (who typically win by large margins in the small states of the Great Plains and inter-mountain West that are favored by the apportionment of electoral votes) and Democrats respectively.

Further research along these lines can proceed in a number of directions, including the following.

The elections with more than two candidates that have been excluded from this analysis can be included in various ways. First, in the manner of the treatment of the 1860 election here, scenarios may be created by combining the votes for candidates in various ways to produce counterfactual two-party contests. Second, the popular vote for a third candidate can be frozen, while major party votes are allowed to "uniformly swing" against each other. Typically this will show an Electoral College "deadlock interval" (of which a tie interval is a special case) in addition to or, more commonly, instead of an inversion interval.[18] Finally, multi-candidate or multi-party elections (such as are increasingly found in "Westminster" parliamentary systems) may be analyzed in their full complexity by considering hypothetical uniform swing between all pairs of parties. However, this will require complicated analytical methods.

A number of "reforms" of the U.S. Electoral College (apart from its total abolition and replacement by a one-tier direct election system) have been proposed over the years, including several "district" and "proportional" plans. In addition, the method of apportioning electoral votes among the states might be revised in various ways – in particular, to reduce or remove the small-state apportionment advantage. It remains to be determined whether such reforms would make election inversions more or less likely.

The notion of perturbing of a "crisp" PVEV function to create a "fuzzy" one, which was treated informally here, can be treated more formally by simulating elections on the basis of given PVEV function with random fluctuations (i.e., *non-uniform* swings) at the state and/or regional level.

Finally, a theoretically productive approach would be to estimate the probability of election inversions in random or "Bernoulli" elections, in which voters decide how to vote by independently flipping fair coins. Such elections can be easily simulated, and this is the same probability model that provides a practical interpretation of the *absolute Banzhaf voting power measure* – namely, that voting power is the probability of casting a decisive vote in such an election. Voting power analysis has well-known applications to two-tier electoral system such as the Electoral College (Felsenthal and Machover 1998). Indeed, Feix et al. (2004) have already estimated, by means of simulations, the probability of election inversions in uniform and perfectly apportioned two-tier electoral systems. This probability quickly approaches a limit of about 0.205 as the number of districts increases.

---

[18]Elections with substantial third-candidate popular votes, such as 1980, 1992, and 1996) may also be fruitfully analyzed in this way, even if the third candidate won no electoral votes in the actual election.

My own preliminary work along the same lines indicates that the probability of inversions is somewhat greater than this in Electoral College simulations, but the extent to which this is due to non-uniform districts or to imperfect apportionment is as yet unclear. One advantage of the random election approach is that systematic distribution effects are removed and estimates of inversion probabilities therefore reflect only of the properties of the electoral institutions themselves (i.e., with respect to the Electoral College, the manner of apportioning and casting electoral votes) and not more contingent features pertaining to the geographical basis of party support in any particular historical period.

In the meantime, we can conclude that the probability of an election inversion by the existing U.S. Electoral College is quite small and largely dependent on the closeness of the popular vote. This probability is now smaller than in some times past and, unlike those earlier times, is more or less equally likely to make the Democratic or Republican candidate the "wrong winner."

Appendix: Presidential Election Data

The 1828–2004 Presidential election data used here comes from Congressional Quarterly's *Guide to U.S. Elections* (2005), which is based on the Interuniversity Consortium for Political and Social Research (ICPSR) Historical Election Returns file. See the p. xvi in the *Guide* for further details. The 2008 data comes from David Leip's *Atlas of U.S. Elections* at **http://uselectionatlas.org/**, which is based on information from state election agencies. For present analytic purposes, it was necessary or expedient to make the following adjustments in the data.

1. All state and national popular vote percentages are based on the major two-party vote only, excluding popular votes cast for third-party and other minor Presidential candidates.
2. Apart from 1860 (for which we consider the "Republican vs. anti-Republican" counterfactual two-party variant), the following elections are set aside because third-party candidates won electoral votes by carrying at least one state:

1832	Wirt (Anti-Masonic Party) won 8 electoral votes;
1856	Fillmore (American Party) won 8 electoral votes;
1860	Brekinridge (Southern Democrat) won 72 electoral votes and Bell (Constitutional Union Party) won 39 electoral votes;
1892	Weaver (Populist Party) won 22 electoral votes;
1912	T. Roosevelt (Progressive Party) won 88 electoral votes;
1924	LaFollette (Progressive Party) won 13 electoral votes;
1948	Thurmond (Southern Democrat) won 38 electoral votes;
1960	Byrd (Southern Democrat) won 14 electoral votes (cast by "unpledged" electors);
1968	Wallace (American Independent Party) won 45 electoral votes.

3. Despite significant third-candidate popular votes in 1980 (Anderson), 1992 (Perot), and 1996 (Perot), these elections are not excluded because Anderson and Perot carried no states and therefore won no electoral votes. The popular votes for Anderson and Perot are excluded from popular vote totals (like popular votes for minor candidates in all elections).

4. Because of the general-ticket system for electing party-pledged electors, each state's electoral vote is normally undivided. However, divisions in state electoral votes occur in three circumstances:

   (a) when a "faithless" elector violates his or her pledge and casts a "protest" electoral vote for another candidate, which occurred in 1948, 1956, 1960, 1968, 1972, 1976, 1988, 2000, and 2004);
   (b) when electors are elected from districts rather than statewide, and each major-party candidate carries at least one district, as happened in Michigan in 1892 and Nebraska in 2008; and
   (c) when electors are elected at-large but individually rather than on a general ticket, as happened with some frequency in the 19th century, in several states in 1912, and in Alabama in 1960.

   Consistent with the almost universal practice and present analytical purposes, all calculations assume that states cast undivided electoral votes for the state popular vote winner. (Thus, McCain in 2008 is credited with all five electoral votes from Nebraska and Gore in 2000 is credited with 267 electoral votes). When electors are elected at-large but not on a general ticket system, standard records of the Presidential vote by state (including those relied on here) credit each Presidential candidate with the popular vote for his party's leading elector.

5. The South Carolina legislature appointed presidential electors through 1860. These electors were always Democrats, but in 1832 and 1836 they cast their electoral votes for an "Independent Democrat" rather than the national Democratic party nominee. The Delaware legislature appointed electors in 1828 (pledged to National Republican J.Q. Adams), the Florida legislature appointed electors in 1868 (pledged to Democrat Horatio Seymour), and the Colorado legislature appointed electors in 1876 (pledged to Republican Rutherford Hayes).

   In the calculations pertaining to the actual Electoral College, South Carolina is counted as voting 100% Democratic but casting 0% of the national popular vote, and Delaware, Florida and Colorado are treated in a parallel manner. For purposes of making perfect apportionment calculations for Delaware, Florida, and Colorado, I use the total popular vote for governor in the same year (or in 1829 in the case of Delaware) to take the place of the (non-existent) popular vote for Presidential electors. This data came from Tables 7.8, 7.9 and 7.6, pp. 264–275, of Walter Dean Burnham's *Voting in American Elections* (2010). However, the South Carolina legislature appointed the governor as well as Presidential electors through 1860. Therefore I use the total vote for U.S. House candidates to take the place of the Presidential vote. This was calculated using Tables 2.2–2.4 (Potential Electorate Estimates), pp. 115–119, together with Table 8.3b (Estimated House Turnout), pp. 401–410, in Burnham's book.

6. In 1836, Whig presidential electors were pledged to different candidates in different states. The popular and electoral votes for the three Whig candidates are simply added up to get a national Whig popular vote percent and electoral vote total, so the calculations treat 1836 as a normal two-party election.
7. In 1860, Democratic "fusion" (i.e., anti-Lincoln) elector slates that included both prospective electors pledged to Douglas and others to Breckinridge (and, in at least one state, several pledged to Bell) were run in a number of Northern states, sometimes in competition with "pure" Douglas slates (see Fite 1911, p. 223). None of the "fusion" slates won, but they make apportioning popular vote support between Douglas and Breckinridge a somewhat arbitrary matter. But since we use only the counterfactual version of 1860 in which Lincoln runs against a unified opposition, we can sidestep these complexities.
8. In 1872, the Democratic (and "Liberal Republican") candidate Horace Greeley died after the Presidential election but before the casting of electoral votes. Three Democratic electors in Georgia cast electoral votes for their deceased nominee, while the other Democratic electors scattered their votes among four living candidates. Congress refused to count the three Greeley electoral votes from Georgia, and it also refused to count electoral votes (cast for Republican Ulysses Grant) from Arkansas and Louisiana, due to disruptive conditions in those states. The scattered Democratic electoral votes (including the three rejected votes for Greeley) are counted toward the Democratic total and the Arkansas and Louisiana popular and rejected electoral votes are counted toward the Republican total, so the calculations treat 1872 as a normal national election (apart from the absence of Mississippi, Texas, and Virginia, which had not yet been readmitted to the union).

References

Abbott, D.W., & Levine, J.P. (1991). *Wrong winner: the coming debacle in the electoral college*. New York: Praeger.

Balinski, M.L., and Young, H.P. (1982). *Fair representation: meeting the ideal of one man, one vote*. New Haven: Yale University Press.

Ball, W.J., & Leuthold, D.A. (1991). Estimating the likelihood of an unpopular verdict in the electoral college. *Public Choice, 70*, 215–224.

Best, J. (1971). *The case against direct election of the president*. Ithica: Cornell University Press.

Burnham, W.D. (2010). *Voting in American elections*. Bethesda: Academica Press.

Butler, D.E. (1951). Appendix: An examination of the results. In H. G. Nichols (Ed.), *The British general election of 1950*. London: Macmillan.

Congressional Quarterly (2005). *Guide to U.S. elections*. CQ Press.

Chambers, C.P. (2008). Consistent representative democracy. *Games and Economic Behavior, 62*, 348–363.

Diamond, M. (1992). The electoral college and the American idea of democracy. In W. Berns (Ed.), *After the people vote: a guide to the Electoral College*. Washington, DC: AEI.
Edwards, G.C., III (2004). *Why the Electoral College is bad for America*. New Haven: Yale University Press.
Felsenthal, D.S., & Machover, M. (1998). *The measurement of voting power: theory and practice, problems and paradoxes*. Cheltenham: Edward Elgar.
Feix, M.R., Lepelly, D., Merlin, V.R., & Rouet, J.L. (2004). The probability of conflicts in a U.S. Presidential type Elections. *Economic Theory, 23*, 227–257.
Fite, E.D. (1911). *The Presidential campaign of 1860*. New York: Macmillan.
Garand, J.C., & Parent, T.W. (1991). Representation, swing, and bias in U.S. Presidential elections, 1872–1988. *American Journal of Political Science, 35*, 1011–1031.
Key, V.O., Jr (1949). *Southern politics in state and nation*. Random House: New York.
Laffond, G., & Laine, J. (2000). Representation in majority tournaments. *Mathematical Social Sciences, 39*, 35–53.
Lahrach, R., & Merlin, V.R. (2012). Assessing the probability of the referendum paradox: the French local election case. Chapter 5 in this volume.
Longley, L.D., & Peirce, N.R. (1996). *The Electoral College primer*. New Haven: Yale University Press.
May, K. (1948). Probabilities of certain election results. *The American Mathematical Monthly, 55*, 203–209.
Merrill, S., III (1978). Empirical estimates for the likelihood of a divided verdict in a Presidential election. *Public Choice, 33*, 127–133.
Nelson, M. (1974). Partisan bias in the Electoral College. *Journal of Politics, 36*, 1033–1048.
Niemi, R.G., & Fett, P. (1986). The swing ratio: an explanation and an assessment. *Legislative Studies Quarterly, 11*, 75–90.
Neubauer, M.G., & Zeitlin, J. (2003). Outcomes of Presidential elections and House size. *PS: Politics and Political Science, 36*, 221–224.
Nurmi, H. (1999). *Election paradoxes and how to deal with them*. Berlin: Springer.
Nurmi, H. (2001). Referendum paradox in the U.S. Presidential elections. Prepared for presentation at the First General Conference of ECPR, Canterbury.
Nurmi, H. (2002). *Voting procedures under uncertainty*. Berlin: Springer.
Peirce, N.R., & Longley, L.D. (1981). *The people's President: The Electoral College in American history and the direct vote alternative, rev. ed.* New Haven: Yale University Press.
Ross, T. (2004). *Enlightened democracy: the case for the Electoral College*. Dallas: Colonial Press.
Schattschneider, E.E. (1942). *Party government*. New York: Holt, Rinehart and Winston.
Sterling, C.W. (1981). Electoral College misrepresentation: a geometric analysis. *Polity, 8*, 425–449.

Chapter 5
Which Voting Rule Minimizes the Probability of the Referendum Paradox? Lessons from French Data

Rahhal Lahrach and Vincent Merlin

5.1 Introduction

A major goal of democracy is to achieve equal representation of the citizens. Though equal representation can be easily achieved when all the voters directly select a president or decide on a policy through a referendum, the issue is not that simple for indirect democracy. A crucial question thus relates to the choice of the "best" two-tier voting rules. More precisely, how many mandates should be allocated to each jurisdiction (examples being electoral constituencies, local jurisdictions, regions, states, countries) in this type of system? In this chapter,we will suggest a solution to this problem by using the electoral data of French local elections.

The most natural manner is to allot the seats proportionally to the population, in order to give "equal right" to each citizen. In the United States, the number of representatives allotted to each State is directly proportional to its population, and is recomputed after each census. The election of the president by the American electoral college follows the same principle: a State i with population n_i obtains $S + R$ electors where S is the senator number (2 per State) and R is the State representative number in the house, which is proportional to its population. The candidate who obtains a majority of votes in the State i obtains all its mandates[1] and

[1] The only exceptions are the States of Maine and Nebraska.

R. Lahrach
Mohammed First University-EST Oujda, Université Mohammed Premier, BP:473, 60000 Oujda, Maroc
e-mail: rahhal.lahrach@est.ump.ma

V. Merlin (✉)
CREM, Université de Caen Basse Normandie and CNRS, 19 rue Claude Bloch, 14032 Caen Cedex, France
e-mail: vincent.merlin@unicaen.fr

D.S. Felsenthal and M. Machover, *Electoral Systems*, Studies in Choice and Welfare, DOI 10.1007/978-3-642-20441-8_5, © Springer-Verlag Berlin Heidelberg 2012

the candidate who obtains a majority of the votes in the electoral college becomes the new president. The properties of the different apportionment methods, and their use in the US electoral college, are well documented by Balinski and Young (1982).

In the EU, the mandate number allotted to each country by the treaty of Nice were vaguely proportional to a law of the type $n_i^{1/2}$, but with enormous fluctuations and a stair curve type. Each new enlargement of the EU caused a negotiation between the former EU Members, and no specific rule ever guided the number of mandates per country. The project of the European constitution was an attempt to define a more rigorous method, and was proposed to reconcile the "one Man-one vote" principle with the "one State-one vote" by the means of double-keys of vote: to be approved, a proposal should be supported by 55% of the countries of which their populations constituted at least 65% of the EU population. At last, this decision rule was enforced by the Lisbon Treaty.

Thus, when we analyze the American and European cases, we find systems which swing between the pure federal system "one State-one vote" (the second key in the European constitution, and the premium +2 per State in the electoral college) and the more democratic proportional representation. Clearly, for all these systems, the results were the culmination of a political negotiation between small States and big States. For these historical cases, there was also no reference to specific normative criterion which could be used in order to specify what should be the good voting process for a federal decision process. This confusing situation is also due to the fact that the implementation of "equity" principle between all the citizens is easy in a direct election, but not in federal systems.

In this paper, we will use a simple normative criterion to evaluate the different two-tier voting methods: *a method is said to be majority efficient if it minimizes the probability that a decision is taken with a majority of mandates at the federal level though it is supported by a minority of voters over the whole federation.* It is equivalent to the concept of *Condorcet efficiency* that has been developed in Social Choice Theory in order to discriminate among the voting rules on their capacity to pick out the Condorcet winner whenever it exists.[2] Our criterion is also equivalent to the minimization of the likelihood of the *referendum paradox* as defined by Nurmi (1999): A referendum paradox occurs whenever a decision taken by representatives elected in local jurisdictions conflicts with the decision that would have been adopted if the voters had directly given their opinion through a referendum.

The first theoretical results on the likelihood of the paradox for two-tier voting systems can be found in a paper due to May (1948), that has received little publicity until recently.[3] The study of Feix et al. (2004) also evaluated a priori the likelihood of the referendum paradox under some probabilistic assumptions for equal size jurisdictions. Feix et al. (2010) extended the analysis to jurisdictions

[2]For more on this literature, see Gehrlein (2006). Notice that the majority winner is always well defined in a two candidate election.

[3]We thank Hannu Nurmi and John Roemer who mentioned to us the existence of this reference.

with different population sizes. In these papers, the authors generated the results of many elections with urn models to assess the frequency of the paradox, but did not use real electoral data. These different models suggest that the paradoxes could be encountered in 12–28% of the cases when the elections between two parties become close.

However, the referendum paradox is not just a theoretical object; such strange political situations happen, a well known case being the election of George W. Bush against Al. Gore in 2000. Lahrach and Merlin (2010) tried to assess the empirical frequency of this paradox, by using data from French local elections. Indeed, France is divided into 102 "départements", each one governed by a local assembly called the "conseil général". Every département is divided into cantons (on average 39 cantons per département) that designate each a councilor by the mean of a plurality run-off system: either a candidate must arrive first, second or receive the support of at least 10% of the registered voters to qualify for the second stage, where the plurality rule is used to select the final winner. Thus, cantonal elections allow the population to choose their representatives to the district council of their département. Though it is not easy to reconstruct the results of a two party system from the French electoral data, by using the results of seven elections in 93 départements, and by focusing on the seats called at the second stage only, Lahrach and Merlin suggested that a referendum paradox could be identified in 9.68% of the cases. For more details about the methodology, see Lahrach (2009), and Lahrach and Merlin (2010).

In this chapter, we will use the same data set in a different way. Historically, the cantons were created after the French revolution on a geographical basis. Typically, a canton would gather all the population within a few kilometers from a central village or city. As time passes, when the population of a canton had grown, it was possible to split it in two (or more) new cantons. For example, the city of Caen is now divided into ten cantons, while the boundaries of many surrounding cantons have not changed since two centuries. However, the splitting process has never been done on a rigorous basis. As a consequence, the populations of the cantons are quite different, and we can find in some départements a population ratio between the smallest and the biggest cantons as huge as 20! On this basis, it is not surprising that Lahrach and Merlin (2010) found many occurrences of the paradox, as mal-apportionment is a common feature in French cantonal elections. We will not here discuss directly the sources of the referendum paradoxes, whether due to mal-apportionment or gerrymandering.

The main objective of this paper is to seek for the allocation of the mandates in a two-tier voting system that minimizes the probability of the referendum paradox. The US and European cases tell us that there are many possible rules we can use. We will here focus on the class of α-rules, where the number of mandates received by each jurisdiction is proportional to n_i^α, with n_i the population of jurisdiction i, and $\alpha \in [0, 1]$. Although this assumption is restrictive, it covers the pure federalism case ($\alpha = 0$ and each jurisdiction is represented by one vote mandate), the square-root rule case ($\alpha = 1/2$), and the pure proportionality case ($\alpha = 1$). More precisely, we will use the same set of data (cantonal elections) to identify which parameter $\alpha^\star$

minimizes the probability of the paradox if we change the weight of a representative, by allocating a_i mandate for canton i according to the law $a_i = n_i^\alpha$.

In Sect. 5.2, we discuss the literature on the normative criteria that we can use in order to evaluate the merits of the different two-tier voting rules. Section 5.3 presents in details the French cantonal elections, some of the results from Lahrach and Merlin (2010) and, finally, our estimation about the best α-rule. We discover that the likelihood of the referendum paradox is minimized for low values of α, with $\alpha^\star = 0.4$. In Sect. 5.4, we try to explain this result, by expressing the difference of votes between the Left and the Right, as a function of the population size of the cantons. We conclude our paper by a discussion in Sect. 5.5.

5.2 Normative Criteria for Two-Tier Voting Rules

In order to evaluate the qualities and flaws of the different two-tier voting systems, various normative criteria were proposed in the literature. The contributions can be roughly divided into two categories. Historically, the first ones were related to the literature on power indices, and the objective was to give to each voter the same influence on the decision-making process. New criteria based on total utility maximization were recently proposed.

5.2.1 Equalizing Power

The main objective of the literature on voting power is to define rigorously the concepts of power and influence, and to propose indicators for its objective measure. The starting point is the notion of a *swing* or *decisive* voter. A voter is said to be decisive whenever his or her vote can change the result of the election. The power of a voter is then his a priori probability of being decisive.

Although it seems simple to allocate mandates proportionally to the sizes of populations, this process does not take into consideration the combinatorial properties of mandates, which often implies a gap between voting weight and voting power. Moreover, for two-tier voting systems, the direct proportionality disregards the relationship between the influence of an individual in his jurisdiction and the power of that same jurisdiction at the higher level.

The most famous solution to this problem, which also constitutes the first normative criterion concerning the two-tier voting evaluation, is due to Penrose (1946, 1952) and Banzhaf (1965). We will briefly present their assumptions and reasonings. A more detailed description of tools and concepts about the power indices can be found in Felsenthal and Machover (1998).

Consider a two-tier voting system with m jurisdictions. There are n_i voters in jurisdiction i, and the representative of jurisdiction i at the higher level controls

a_i mandates.[4] For the sake of simplicity, assume that two parties A and B, are in competition in all the jurisdictions, and that the majority rule is used both at the local and federal level. The key assumption made by Penrose and Banzhaf is to assume that each voter will choose A or B independently with equal probability. This assumption is called the independence assumption (Straffin 1977). Under this assumption, Penrose (1946, 1952) noticed first that the probability to be decisive for an individual j residing in jurisdiction i is in inverse proportion to the square root of the population in this jurisdiction ($1/\sqrt{n_i}$). This probability has to be multiplied by the probability that the representative of jurisdiction i, equipped with a_i mandates, is decisive in the federal assembly. In parliaments where the majority rule is enforced to make decision, Penrose noted that the relative voting power measurement of two representatives tends asymptotically to their relative voting weight. This result, known as Penrose theorem, is only an approximation and can even become false for some peculiar games. But this property is held (roughly) in the majority of real cases.[5] To sum up, the probability of being decisive for voter j in the jurisdiction i is in inverse proportion to the square root of the population in his jurisdiction ($1/\sqrt{n_i}$) and roughly proportional to the the number of mandates a_i of his representative. Thus, equal treatment in terms of voting power is carried out when each jurisdiction obtains a number of mandates proportional to the square root of its population. This result, also discovered independently by Banzhaf (1965) is known as the Penrose square root rule. It is today a classical reference for many studies on federal unions and two-tier voting rules. The book and the papers by Felsenthal and Machover (1998, 2001, 2004) are perfect examples of this tradition. Maaser and Napel (2007) also prove that the square root principle applies with a continuum of alternatives on the line.

Nevertheless, this approach can be contested. Indeed, in reality, the voter seldom makes a choice between the proposals independently from the others (here between the two parties) before expressing its vote. By analyzing the last 50 years of electoral data, Gelman et al. (2004) proved that the Straffin independence assumption had to be rejected for the elections of the senators, the representatives and the American Electoral College. Similar conclusions are drawn from the study of the electoral data that were gathered all over Europe. Thus, if we want to make recommendations on the choice of a two-tier voting rule, more realistic behaviorial models of the electors are needed, coupled with deeper data analysis.

Moreover, all this advanced literature argues that citizens of different jurisdictions should have the same power, i.e. the same probability to be decisive. But a traditional argument against this approach remarks that the voters have a vague notion of their ability to influence the results, and that a voter's influence is for any event extremely low within federal systems like the United States or the UE, of a

[4]The number of mandates may not be an integer in this study.

[5]For more on the Penrose approximation, the reader can check the recent works by Lindner and Machover (2004), Chang et al. (2006), Lindner and Owen (2007), Feix et al. (2007), Maaser and Napel (2007), and Slomiczyński and Życzkowski (2007).

magnitude of the size $a_i/\sqrt{n_i}$! Thus, it is not very likely that the citizens would fight for a representation just based on this faint decisiveness notion!

Other concepts such as equal opportunity of success are surely more bitting. Rae (1969) was the first author who proposed a clear definition of success: by employing the assumption of independence he defined his index as the probability of being on the victorious side. However, it is well-known that the Rae index is related to that of Penrose-Banzhaf. It then gives an equivalent normative recommendation (see Felsenthal and Machover (1998), Laruelle and Valenciano (2005)). Hence, the same critique applies: Laruelle and Valenciano (2008) show that for many voting rules, the differences in terms of success between most of the rules are so negligible that this concept may not convince the citizens either.

In fact, the recent development of new criteria is due to a shift to the aggregated level, with measures based upon the total utilities of the members of the society. One does not seek any more to equalize probabilities of being decisive or probabilities of success for each citizen, but to maximize the welfare of the society defined as the sum of the individual utilities.

5.2.2 Maximizing Total Utility

The idea that the total utility of the citizens is the good measure for the welfare of the society is often linked to the works of Bentham (1789). As we will see, this idea can be implemented in many ways for the analysis of a two-tier voting rule.

Felsenthal and Machover (1999) suggest that, for a federal union, the average difference between the size of the majority camp among all the citizens and the number of citizens who agree with the decision made by the majority of the representatives in the States should be minimized. This criteria can be considered as an utilitarian one, in the sense that a satisfied (resp. dissatisfied) voter gets a $+1$ (resp -1) utility level. It gives an estimation for the loss of utility of the society when the decision is not supported by a majority of voters. They prove that, under the independence assumption, the Penrose square root rule still applies as a solution to the problem of the choice of the best two-tier voting rule.

Barberà and Jackson (2006) generalize the same idea. They assume that in a two party election, candidate A's partisans obtain a utility $u_j = 1$ if she is elected (and 0 in the other case), whereas the partisans of B obtain a utility $u_j = v$, $v \in [0, +\infty[$ if their preferred candidate is elected (and 0 in the other case). They also assume, that, at the federal level, a motion passes if it is supported by $q\%$ mandates, with q possibly different from 1/2. Then, the optimal voting rule for two-tier election systems is the one that maximizes the expected total utility of voters. Barberà and Jackson called their criteria the *efficient utility* principle. Their first results are very general in the sense that they do not depend on a particular model of probability. Beisbart et al. (2005) compared seven various possible decision rules for the European Union on their capacity to choose motions which will have a positive total utility for its citizens, while rejecting the bad policies. But the probabilistic foundations of their model are different, in the sense that each country is modeled by a unique representative agent.

Depending on the assumption that they use to model the behavior of the electors, these authors will either recommend that the number of mandates should be directly proportional to the population size (Barberà and Jackson 2006) or proportional to the square root of the population (Felsenthal and Machover 1999). But all these criteria suffer from the same critique: How to measure the utilities? Is it normatively appropriate to add up individual's utilities?

5.2.3 *The Principle of Majority Efficiency: Minimization of the Referendum Paradox Frequency*

In social choice literature, a *Condorcet winner* is a candidate who is able to defeat any other opponent in pairwise comparisons. The majority efficiency is defined as the probability for a given rule to select the Condorcet winner whenever it exists (see Gehrlein (2006)). As said in the introduction, for two candidate elections, maximizing the majority efficiency is tantamount to reducing the probability of the referendum paradox. This criterion can be compared with the utilitarian principle presented in the previous section. The main difference between the utilitarian criterion discussed above is that the former ignores the importance of the paradox. It only attempts to estimate the number of situations where a majority of voters are frustrated, but do not evaluate the magnitude of the paradox, either by counting unhappy voters, as in Felsenthal and Machover (1999) or by summing-up the utilities, as in Barberà and Jackson (2006) and Beisbart et al. (2005). Moreover, since the referendum paradox has been popularized by the media after the U.S. elections of 2000, this criterion could be accepted by public opinion more easily than any other criteria. So it is not just a theoretical recommendation, but also a practical tool that can be implemented. And using sufficiently rich electoral data, we can evaluate different two-tier voting rules as to their propensity to elect the Condorcet winner.

The first application of this criterion was presented in a paper due to Kenneth May (1948), that has been recently rediscovered. May considers a two party competition in a federal union with m states of equal population n. Then May reduces the problem to an urn model: in each state, there are $n + 1$ balls, each one marked with a number from 0 to n. They correspond to all the possible results for party A. Then, an election consists of drawing in each state one of these balls independently. Candidate A may win a majority of votes in a majority of states, but the sum of the numbers may be inferior to $(nm)/2$, meaning that candidate B got more votes on average. This is a referendum paradox in modern terms, and May, using several central limit theorems, proves that its probability tends to 1/6 as n and m tend to infinity.

Unaware of this first result, the same problem was studied by Feix et al. (2004), using not only May's assumption, but also the independence assumption. In a recent article, Feix et al. (2010) generalized the analysis to unequal population size among the jurisdictions, both for May's model and Straffin's independence assumption. Since it is difficult to study all methods of seats distribution, Feix et al. (2010) focused on the family of α-rules. In other words, it is assumed that the mandate vectors $\tilde{a} = (a_1, \ldots, a_m)$, is fully characterized by the parameter α, $\alpha \in [0, \infty]$ such that $a_i = n_i^{\alpha}$ $\forall i = 1, \ldots, m$. Although this assumption is restrictive, it covers the pure federalism case ($\alpha = 0$ and each state has a mandate), the square-root rule case ($\alpha = 1/2$), the pure proportionality case ($\alpha = 1$) and even the dictatorship of the largest state ($\alpha = \infty$).

It is important to recall that the main objectives of the study of Feix et al. (2010) were twofold. First, they wanted to give the first evaluations on the majority efficiency of two-tier voting rules for the class of α-rules, when the population of the jurisdictions are unequal. Secondly, they wanted to find the optimal α-rule, both under Straffin's independence assumption and May's assumption. To perform their task, they used computer simulations to generate a vast number of elections. Under May's assumption, $\alpha = 1$, that is the proportional rule, is clearly the best voting rule. The result is less clear under the independence assumption: for a small number of jurisdictions, the optimal α seems to be close to 0.45 and it progressively tends toward 0.5 as the number of jurisdictions increases.

5.2.4 What Have We Learnt?

Indeed, the conclusion of the theoretical models is that whatever the criterion adopted (the equalization of power, the utility maximization, the minimization of the referendum paradox), the choice of the optimal α-rule for the apportionment of the mandates seems to be driven by the underlying probability assumption we are using to randomly generate elections. The independence assumption seems to always lead to the square root rule, while all the other models would recommend to apportion the mandates in exact proportion with the populations. These conclusions thus call for applied works. Gelman et al. (2004) have already proven that in the US case, the square root rule principle should be abandoned. More precisely, they show that the margin in percentage points between the Democratic and the Republican parties was stable, whatever the population of the state is. This contradicts a prediction of the independence assumption, that is, that the margin between the two major American parties should shrink in percentage points as the states become more populated. Gelman et al. (2004) found that the margin is proportional with $n_i^{0.9}$. But they were reluctant to make recommendations about the 0.9-rule since it does not have the platonic attraction of the proportional rule. As a consequence, they endorse the actual system for the Electoral College in their conclusion.

The objective of this paper will be somehow similar: using the electoral data from French local elections, can we suggest that one α rule is better?

5.3 The Optimal Rule in the Cantonal Elections

5.3.1 Data

In France, the cantonal elections allow the population to choose their representatives to district councils of their département. Indeed, every département is divided into cantons (on average 39 cantons per département) that designate a councilor by the means of a plurality run-off system. A councilor is elected for 6 years, but the renewal of the district council is done partly every 3 years. In other words, every 3 years, half of the seats are subject to an election. To be elected in the first round, the candidate must get the absolute majority of the votes cast, as well as a number of votes at least equal to a quarter of the registered voters. To be qualified to the second round, it is necessary to arrive, first, second, or to get more than 10% of the registered voters at the first stage. During the second round, the candidate who obtains a plurality of votes is elected.

The data in the present study are extracted from the electoral database carried out by the Quetelet center (based upon Home Department data) and the LASMAS (based upon French Statistical institute: INSEE). They are the data of cantonal elections of Metropolitan France, which took place between 1985 and 2004. We excluded from the database Paris, Corsica, and Overseas territories which either have their own specific voting rules, or present a much diverse spectrum of political forces. Thus, the cantonal elections present an important series of data, with 93 départements voting every 3 years. Thus, we have 651 (93×7) cases to study. For more detail, we have approximately 1,915 cantons per election year, that is 13,405 election data over a period of 20 years. Moreover, over the period, no political camps constantly dominated the scene: The results were in favor of the Right camp in 1985, 1988, 1992 and 1994, while the Left won in 1998, 2001 and 2004. In 1992, 1998, and 2004, the extreme right was able to gather a bit more then 10% of the electorate in the first round, but just got a handful of candidates elected.

5.3.2 Methodology

To evaluate the frequency of the paradox, it is necessary for us to build a pertinent method to measure this occurrence while taking into account the specificities of the cantonal ballot in comparison with the idealized structure where only two candidates are in competition in each jurisdiction. For the cantonal elections, one must overcome the following three obstacles in order to establish an effective measure:

1. The presence of more than two major candidates in many cantons,
2. The existence of two rounds instead of one,
3. The frequent presence of a third party in the second round.

To circumvent these problems, we will consider the seats called at the second round in each département, and make our analysis on this basis only. We thus consider an artificial assembly which is a subset of the real one. By doing so, we eliminate the cases where a very popular candidate manages to win in the first round, and concentrate on the competitive seats. Due to the fact that only the top two candidates plus the few ones that reach the 10% threshold can reach the second stage, our analysis is also simplified. We are sure to have a Left candidate and a Right candidate in most of the cases, with sometime the presence of an Extreme Right candidate or an Independent as a third runner. Thus the aggregation of the data is simplified: we add up the number of votes and the number of seats obtained by each camp (Left, Right, Extreme Right) in each département for the seats called at the second stage, and check whether the camp which got the majority in terms of seats also had a plurality of votes.

Though this way to define a referendum paradox may be considered as crude, Lahrach and Merlin (2010) tested three other ways to aggregate the data. Whatever the solution proposed, they observed a paradox in about 10% of the elections. For the measure defined above, Table 5.1 lists the 63 cases out of 651 where a paradoxical situation occurs. For each year, we list the names of the problematic départements. The next columns give the number of votes cast for the second round as well as the number of seats called. Next, we present the percent of expressed votes and the number of seats won by each political camp. Table 2 presents the 50 cases where the results are deadlocked in terms of seats. They can be viewed as a weaker form of the paradox. The French electoral law states that in this case, the elder candidate is elected as president of assembly. As the president is granted with a tie breaking vote, in about 50% of these cases, the winner in terms of votes will not be able to control the council.

5.3.3 *What is the Best α-Rule?*

In this section, we will exploit the fact that the French cantons are very different in terms of population size, in order to search the "optimal" α-rule when we wish to minimize the likelihood of the referendum paradox. Contrarily to May (1948) or Feix et al. (2004,2010), we are no more dealing with abstract urn models to draw our conclusions: we are working with a set of 651 different local assembly elections. 7,108 seats out of 13,405 were called at the second round.

While in reality each councilor has one vote in the local assembly, we will assume that he controls $a_i = n_i^\alpha$ mandates. As usual, the camp with a majority of mandates will control the assembly. We will test 11 values of α, from 0 to 1, with an increment of 0.1. For $\alpha = 0$, we report two numbers. Indeed, as soon as $\alpha > 0$, the number of mandates is no longer an even number of seats and automatically, a large fraction of these cases will become paradoxical. In order to avoid an extreme discontinuity at the point $\alpha = 0$, we have to consider that some of these elections as paradoxical. First, we consider that all the deadlocked elections are paradoxical

Table 5.1 Départements where the paradox occurs for seat called at the second round

Year	Département	Ballots	C	Left		Right		Ext. Right	
				% vote	C	% vote	C	% vote	C
2004	AVEYRON	54,367	14	50.07	6	49.93	8	0.00	0
	CANTAL	10,378	3	45.81	2	54.19	1	0.00	0
	LOIRET	107,869	15	41.92	5	41.35	10	8.82	0
	MAINE ET LOIRE	108,908	13	47.15	8	52.85	5	0.00	0
	VOSGES	62,019	8	41.99	5	48.92	3	9.09	0
	YONNE	69,623	18	47.58	7	44.51	11	7.60	0
2001	ARDENNES	48,123	14	54.33	6	45.67	8	0.00	0
	BOUCHES DU RHONE	245,176	21	51.94	10	45.21	11	2.84	0
	CHARENTE	59,614	12	51.95	5	48.05	7	0.00	0
	CORREZE	48,717	13	53.64	6	46.36	7	0.00	0
	GARD	140,061	20	49.26	9	44.19	11	6.55	0
	LOIR ET CHER	40,818	8	41.28	4	48.11	0	2.61	0
	MAINE ET LOIRE	56,865	9	48.83	5	51.17	4	0.00	0
	MARNE	40,212	8	42.53	5	55.24	3	2.23	0
	NORD	290,989	28	50.34	13	45.01	15	4.65	0
	OISE	125,507	19	44.39	11	53.37	8	2.24	0
	SEINE MARITIME	140,163	20	52.49	9	47.51	11	0.00	0
	TARN	59,543	13	51.71	4	48.29	9	0.00	0
1998	CHER	58,701	17	52.36	7	43.12	10	0.00	0
	EURE	69,778	14	43.45	8	44.45	6	10.12	0
	FINISTERE	140,788	20	48.31	14	51.69	6	0.00	0
	ILLE ET VILAINE	120,552	20	47.37	13	50.38	7	0.00	0
	INDRE	30,574	7	53.23	3	46.77	4	0.00	0
	MAINE ET LOIRE	76,606	12	46.32	7	53.68	5	0.00	0
	MEURTHE ET MOSELLE	110,430	23	51.96	10	45.92	13	1.61	0

(continued)

Table 5.1 (continued)

Year	Département	Ballots	C	Left		Right		Ext. Right	
				% vote	C	% vote	C	% vote	C
	SEINE MARITIME	159,181	25	56.26	12	41.36	13	2.37	0
	SEINE ET MARNE	173,953	22	46.15	10	44.50	12	9.35	0
	TARN ET GARONNE	36,864	11	52.51	4	45.32	7	2.17	0
	VOSGES	73,148	13	47.27	7	50.78	6	1.95	0
	YONNE	42,700	14	45.52	4	42.11	9	8.15	0
1994	AISNE	67,574	12	50.44	4	49.56	8	0.00	0
	ALLIER	59,860	13	48.38	8	51.62	5	0.00	0
	ALPES DE HAUTE PROVENCE	19,956	9	55.86	4	44.14	5	0.00	0
	ARDECHE	52,323	12	49.31	7	50.69	5	0.00	0
	CHARENTE MARITIME	77,736	18	47.43	10	52.57	8	0.00	0
	DOUBS	73,976	13	49.17	7	50.83	6	0.00	0
	JURA	56,537	15	40.77	8	51.50	6	0.00	0
	MEURTHE ET MOSELLE	129,529	20	44.20	12	55.80	8	0.00	0
	OISE	122,963	16	45.47	9	47.69	7	6.84	0
	VIENNE	47,840	10	49.27	8	50.73	2	0.00	0
1992	HERAULT	176,548	24	43.37	14	45.34	10	11.29	0
	LOIR ET CHER	37,857	8	49.11	5	50.89	3	0.00	0
	HAUTE SAONE	40,334	11	48.16	4	47.96	7	1.27	0
	SEINE MARITIME	173,340	25	50.45	11	47.19	14	1.21	0
	VAL DE MARNE	137,299	20	41.03	13	51.37	7	7.59	0

Table 5.1 (continued)

Year	Département	Ballots	C	Left		Right		Ext. Right	
				% vote	C	% vote	C	% vote	C
1988	ARDENNES	35,897	13	53.45	6	46.55	7	0.00	0
	AVEYRON	11,995	4	49.24	3	50.76	1	0.00	0
	GERS	26,767	8	45.65	5	54.35	3	0.00	0
	LOIRE	91,604	15	50.71	7	49.29	8	0.00	0
	MANCHE	47,220	11	48.48	4	45.82	7	0.00	0
	MEURTHE ET MOSELLE	92,137	17	49.41	9	50.59	8	0.00	0
	MEUSE	20,146	7	49.43	2	47.28	5	0.00	0
	NIEVRE	27,090	10	59.93	4	40.07	6	0.00	0
	SAONE ET LOIRE	67,849	15	49.50	10	50.50	5	0.00	0
	ESSONNE	123,418	18	52.84	6	47.16	12	0.00	0
1985	BOUCHES DU RHONE	277,353	21	42.90	11	44.11	9	12.99	1
	CORREZE	28,863	7	49.01	4	50.99	3	0.00	0
	EURE ET LOIR	65,738	10	48.87	6	51.13	4	0.00	0
	HERAULT	167,100	21	47.90	12	49.37	9	2.73	0
	SEINE MARITIME	149,386	21	53.20	10	46.80	11	0.00	0
	TERRITOIRE DE BELFORT	30,315	8	49.09	5	50.91	3	0.00	0
	SEINE SAINT-DENIS	189,631	20	49.90	11	50.07	9	0.03	0
	VAL DE MARNE	144,917	18	49.40	11	50.60	7	0.00	0

C Councillors

Table 5.2 Départements where the result is deadlocked for seats called at the second round

Year	Département	Vote	C	Left		Right		Ext. Right	
				% vote	C	% vote	C	% vote	C
2004	CREUSE	10,471	4	52.78	2	47.22	2	0.00	0
	INDRE	34,500	6	52.29	3	36.46	3	0.00	0
	MARNE	84,044	14	51.47	7	42.97	7	5.56	0
	OISE	124,863	16	43.02	8	41.95	8	15.04	0
2001	CANTAL	14,302	4	53.63	2	46.37	2	0.00	0
	CHARENTE MARITIME	73,102	16	46.77	8	51.51	8	0.00	0
	LOIRE	114,252	16	48.14	8	51.86	8	0.00	0
	LOZERE	14,950	6	46.05	3	53.95	3	0.00	0
	MORBIHAN	95,515	14	46.02	7	53.98	7	0.00	0
	PAS DE CALAIS	143,674	22	53.81	11	45.23	11	0.00	0
	SEINE ET MARNE	135,215	16	47.89	8	52.11	8	0.00	0
	SOMME	70,528	14	44.00	6	40.27	6	0.00	0
	VAL DE MARNE	131,071	22	48.92	11	51.08	11	0.00	0
1998	CHARENTE	65,831	16	47.73	8	52.27	8	0.00	0
	CORREZE	41,696	12	47.37	6	52.63	6	0.00	0
	LOIR ET CHER	48,985	12	41.95	6	55.81	6	2.24	0
	HAUTE LOIRE	32,429	10	43.57	5	56.43	5	0.00	0
	RHONE	185,184	22	46.70	11	34.93	11	18.09	0
	HAUTE SAONE	33,477	10	52.79	5	43.32	5	3.89	0
1994	AVEYRON	32,728	8	50.48	4	49.52	4	0.00	0
	CHER	54,706	14	44.68	7	55.32	7	0.00	0
	EURE	75,428	14	45.38	7	49.33	7	5.30	0
	ILLE ET VILAINE	128,755	20	47.21	10	50.76	10	0.00	0
	INDRE ET LOIRE	62,465	11	47.74	5	48.87	5	0.00	0
	SARTHE	73,324	14	49.76	7	50.24	7	0.00	0

Table 5.2 continued

Year	Département	Vote	C	Left		Right		Ext. Right	
				% vote	C	% vote	C	% vote	C
	SOMME	85,748	16	48.08	8	50.99	8	0.93	0
	TARN	59,141	12	49.20	6	50.80	6	0.00	0
1992	AUDE	53399	10	58.79	5	41.21	5	0.00	0
	LANDES	50,612	8	52.35	4	47.65	4	0.00	0
	LOT	39,415	12	48.09	6	51.91	6	0.00	0
	LOZERE	4,727	4	53.86	2	46.14	2	0.00	0
	VAUCLUSE	96,238	12	38.49	6	40.86	6	19.48	0
	SEINE SAINT-DENIS	178,218	20	35.39	10	40.96	10	23.66	0
1988	AIN	47,490	12	45.97	6	54.03	6	0.00	0
	AISNE	54,245	10	56.95	5	43.05	5	0.00	0
	ALLIER	58,755	14	52.98	7	47.02	7	0.00	0
	ALPES DE HAUTE PROVENCE	15,049	8	45.37	4	54.63	4	0.00	0
	ARDECHE	45,375	10	47.31	5	52.69	5	0.00	0
	CHARENTE	33,255	8	53.31	4	46.69	4	0.00	0
	LOIR ET CHER	25,792	6	50.55	3	49.45	3	0.00	0
	LOIRE ATLANTIQUE	69,539	12	50.58	6	49.42	6	0.00	0
	MORBIHAN	70,421	10	47.99	5	52.01	5	0.00	0
	HAUTES PYRENEES	33,333	10	60.12	5	39.88	5	0.00	0
	SEINE MARITIME	144,576	24	55.23	12	44.77	12	0.00	0
	DEUX SEVRES	32,795	8	34.87	4	65.13	4	0.00	0
	VENDEE	14,300	2	51.09	1	48.91	1	0.00	0
	VIENNE	49,032	12	51.86	6	48.14	6	0.00	0
1985	CREUSE	19,690	6	49.14	3	50.86	3	0.00	0
	GIRONDE	108,765	14	48.39	7	51.61	7	0.00	0
	LOIRET	56,409	10	44.10	5	55.90	5	0.00	0

Table 5.3 The number of paradoxical cases as a function of alpha

Year	0	0.1	0.2	0.3	0.4	0.5	0.6	0.7	0.8	0.9	1
1985	11–9.5	8	7	7	7	7	7	8	7	7	8
1988	24–17	15	15	15	14	14	16	16	17	17	18
1992	11–7	6	6	6	6	6	6	6	7	7	8
1994	18–15	17	17	17	17	18	20	20	20	22	21
1998	18–15	15	14	15	14	16	16	17	18	19	19
2001	21–16.5	15	15	14	13	11	12	12	13	13	14
2004	10–8	6	6	5	5	8	8	9	8	8	9
Total	113–88	82	80	79	78	80	85	88	90	93	97
Likelihood	0.173–0.135	0.126	0.122	0.121	0.119	0.122	0.130	0.135	0.138	0.142	0.149

cases. This is an extreme interpretation, but we can argue that the paradox occurs as soon as the winner in terms of votes does not have a majority in terms of seats. The second number proposes another interpretation. We consider that half of the tied outcomes are paradoxical, by considering a tie breaking rule that selects the minority camp as a winner with probability 0.5. These results are summarized in Table 5.3.

The conclusion from this exercise are quite surprising. In total (or average), the minimum number of these paradoxical situations is given by $\alpha^\star = 0.4$, but all the values between 0.2 and 0.5 lead to 78, 79 or 80 cases among 651. When we make analysis per year, $\alpha = 0.4$ is always minimal, except in 2001 where the minimum is given by $\alpha = 0.5$. The results obtained for $\alpha = 0$ is either high (with a strict interpretation of the paradox) or just slightly above the optimal value. The number of paradoxical cases increase regularly from $\alpha = 0.4$ to $\alpha = 1$. Thus, in contradiction with traditional wisdom, giving to each canton its real weight is far from being the optimal solution for French cantonal elections if we wish to minimize the probability of the referendum paradox! On the contrary, a proposition based upon the square root rule would not be so far from the ideal situation.

5.3.4 Comparison with Theoretical Models

We briefly mentioned in the previous sections the theoretical models that have been used to assess a priori the probability of the referendum paradox. Straffin's independence assumption assumes that each voter will pick randomly and independently any of the candidates with probability 0.5. May's model proposes to draw the results of a jurisdiction from an urn containing one ball for each of the possible result. Feix et al. (2010) used both models to estimate a priori the probability of the referendum paradox. They first draw randomly the populations of the different jurisdictions for 1,000 different federations. Next, they use Monte Carlo simulations to generate 1,000,000 random elections. Finally, they made α vary. Their conclusion are reproduced in Fig. 5.1 for the independence assumption and in Fig. 5.2 for May's

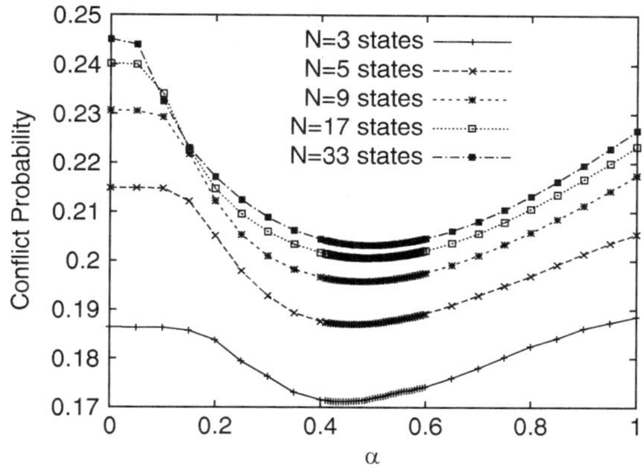

Fig. 5.1 Comparing different voting rules on their ability to avoid the referendum paradox under Straffin's independence (from Feix et al. (2010))

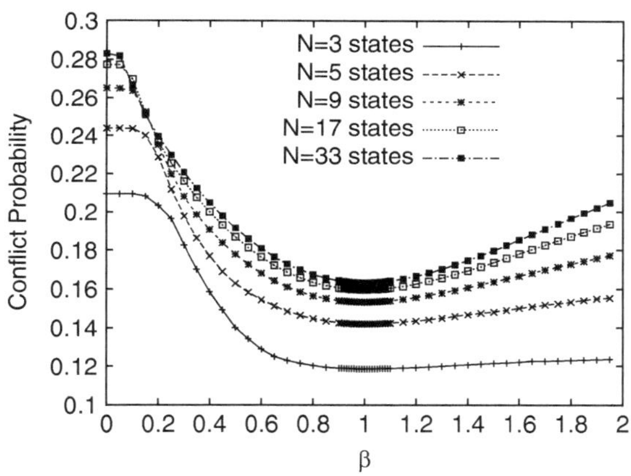

Fig. 5.2 Comparing different voting rules on their ability to avoid the referendum paradox under May's assumption (from Feix et al. (2010))

model.[6] One can immediately observe that the optimal value of α is slightly inferior to 0.5 in the first case, and exactly equal to 1 in the second case. In Fig. 5.3, we present our results. Notice that in contrast to Feix et al. (2010), we merged the data for federations of different size.

By comparing these figures, we first notice that the values obtained from French electoral data are of the same magnitude, but lower, than the ones proposed by the theoretical models. This is both frustrating and encouraging. Frustrating, as clearly

[6] In Feix et al. (2010), the authors consider that $a_i = n_i^{\beta}$ when they use May's model.

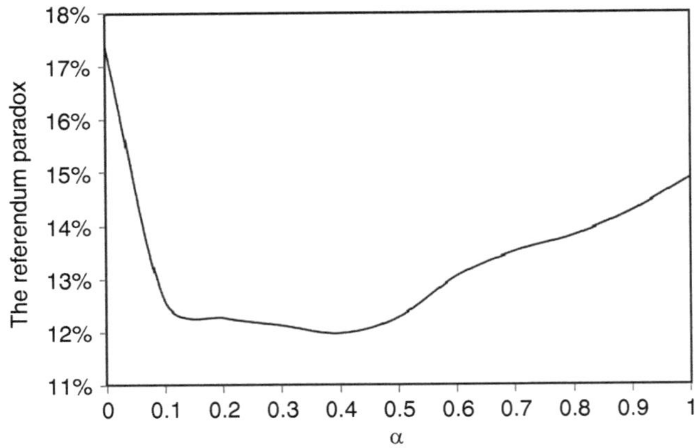

Fig. 5.3 Average probability of the referendum paradox in function of α

the models do not predict accurately the reality. But also encouraging, as the gap between reality and theory is not that huge. Similarly, all the curves present the same U-shape, but disagree on the optimal value for α.

Data analysis with more details shows that there are paradoxical situations that disappear and others appear according as α varies, leading to this U-shape. A possible explanation is that these evolutions are related to the primary sources of the referendum paradox. The notion of mal-apportionment directly refers to the discrepancies in term of populations size among the jurisdictions. The purpose of gerrymandering is to either concentrate opposition votes into a few districts to gain more seats for the other camp in surrounding districts (called packing), or to diffuse its strength across many districts (called dilution). When the paradox is mainly related to the mal-apportionment, it should disappear by applying the α-rules law as α increases. On the contrary, if gerrymandering is the main source of the paradox, paradoxical situations are reinforced via the winner takes all principle when α increases. Thus, the rise of α allows us to correct the distortion due to the effects of mal-apportionment but not that of gerrymandering. A third reason will be explored further in the next section, that is the relationship between the Left and the Right margins, and the population size.

5.4 A Rough Analysis of the Relationships Between Victory Margins and Populations Size

In this section, we study the link between the difference in votes between Left and Right (measured in absolute terms) and the sizes of districts. In fact, if the large districts are more competitive than the smaller ones, then they have, in relative terms, less influence than they should on the determination of the winner in terms of votes. Then, it may not be wise to attribute to them a weight exactly proportional to their population.

To test this relationship we propose a very simple theoretical model :

$$e_i = k n_i^{\beta} \tag{5.1}$$

where n_i is the size of the canton i, e_i is the vote difference between the Left and Right parties, measure in absolute terms, k is a constant and β is a parameter. Equation (5.2) can be written:

$$\ln(e_i) = \ln(k) + \beta \ln(n_i) \tag{5.2}$$

In our data set, we will only consider the cases where the seats were called in the second round, as in the previous sections. Our sample now contains 7,108 observations. Thus, we will use for our estimation the method of ordinary least squares (OLS) and we will try:

$$Lecart_i = c + \beta Lpop_i + \varepsilon_i \tag{5.3}$$

with ε_i *idd*, $\ln(e_i) = Lecart_i$, $c = \ln(k)$, $ln(n_i) = Lpop_i$.

The model's estimation is presented in Table 5.4. Statistical tests (−0.848456) show that the coefficient of the constant is not significant. So we will specify the model without the constant. Table 5.5 present new results. We obtain

$$Lecart = 0,668402 L pop$$

This shows that the link between the difference in votes and population size is significant and smaller than one, of magnitude 0.67. The elections in larger cantons are more competitive when measured in percentage points.

Table 5.4 The result of the estimate of OLS with constant: the variation according to the population size

Variable	Coefficient	Std. error	t-statistic	Prob.
C	−0.137016	0.161487	−0.848465	0.3962
LPOP	0.682938	0.017197	39.71361	0.0000
R-squared	0.181635	Mean dependent var		6.252329
Adjusted R-squared	0.181520	S.D. dependent var		1.297556
S.E. of regression	1.173897	Akaike info criterion		3.158817
Sum squared resid	9,792.314	Schwarz criterion		3.160749
Log likelihood	−11,224.43	F-statistic		1,577.171
Durbin-Watson stat	1.877386	Prob(F-statistic)		0.000000

Dependent variable: LECART
Method: Least squares
Sample: 7,108
Included observations: 7,108

Table 5.5 The result of the estimate of the OLS without constant: the variation according to the population size

Variable	Coefficient	Std. error	t-statistic	Prob.
LPOP	0.668402	0.001483	450.8013	0.0000
R-squared	0.181552	Mean dependent var		6.252329
Adjusted R-squared	0.181552	S.D. dependent var		1.297556
S.E. of regression	1.173874	Akaike info criterion		3.158637
Sum squared resid	9,793.306	Schwarz criterion		3.159603
Log likelihood	−11,224.79	Durbin-Watson stat		1.876711

Dependent variable: LECART
Method: Least squares
Sample: 7,108
Included observations: 7,108

Table 5.6 The result of the estimate of the OLS without constant: the variation according to the cast votes

Variable	Coefficient	Std. error	t-statistic	Prob.
LEXPRIM	0.729638	0.001610	453.1180	0.0000
R-squared	0.189621	Mean dependent var		6.252329
Adjusted R-squared	0.189621	S.D. dependent var		1.297556
S.E. of regression	1.168073	Akaike info criterion		3.148729
Sum squared resid	9,696.760	Schwarz criterion		3.149696
Log likelihood	−11,189.58	Durbin-Watson stat		1.640379

Dependent variable: LECART
Method: Least squares
Sample: 7,108
Included observations: 7,108

We repeat the same exercise by replacing the variable $Lpop_i$ by $Lexprim_i$, where $Lexprim_i$ is the logarithm of the sum of cast ballots for the two major parties in the canton i. The estimation results (see Table 5.6) according to the OLS model show that there is a strong correlation between these two variables, and that, again, the β coefficient is significatively lower than 1 (0.729638).

The fact that the difference in votes between the right and the left does not grow linearly with the size of the population may explain partially that we observed the optimal α around 0.4–0.5. Indeed, a victory in a large district will relatively "weigh" less than a victory in a district of small size.

5.5 Conclusion

The objective of this study was to use a relatively rich data base to study the likelihood of the referendum paradox from an empirical point of view. The French cantonal elections have this feature, as we have in our data set the results for 651 elections for local parliaments. The negative side is that, with at least ten parties able to get more than a few percents, French politics can hardly be modeled by an ideal two party competition. We had thus to impose some restrictions on the elections and seats we could include in our study, to mimic as closely as possible what would have been a simple two party competition between Left and Right, by focusing on seats called at the second stage only. On this basis, we estimate the likelihood of paradoxical situations at about 10%.

Another feature of the French voting system is the inequality in terms of populations among the different cantons. We exploited this flaw to determine the method of apportionment among the α-rules family, which minimizes the frequency of the referendum paradox. The result of our empirical work suggests the square root rule is close to be the optimal solution for French local elections. The intuitive explanation for this phenomenon lies in the slow growth in the differences of votes between Left and Right compared to the population growth. The elections become closer when the canton is larger. As the smaller cantons only have a few thousands inhabitants, one can conjecture that the communication costs are lower in the rural areas, and the electors perceive more easily the qualities and faults of the candidates. In the larger cantons of urban areas, the debates are more likely to be driven by national issues than by local ones, leading to a higher degree of competition.

In contrast, the square root rule is not adequate in the American system. The study of Gelman et al. (2004), examining mainly larger jurisdictions with hundreds of thousands or millions of inhabitants, found that the relation between the electoral margins and the populations was almost linear. It would be extremely interesting to replicate this study with more data sets, to understand whether we can conjecture that the square root rule principle would dominate for local assemblies, while the proportionality principle could be more adapted for national or supra national two-tier voting rules.

Acknowledgements Some parts of the research reported in this paper were presented at the Leverhulme Trust sponsored 2010 Voting Power in Practice workshop held at Chateau du Baffy, Normandy, from 30 July to 2 August 2010. We are grateful to the participants of this workshop for comments and suggestions. We also thank an anonymous referee for his remarks. Special thanks are due to Dan Felsenthal for is careful reading of the last version of the paper. This paper is part of the SOLITER project selected in the "Gouverner, administrer" program by the French Agence Nationale pour la Recherche (ANR) and has benefited from the ANR-08-GOUV-054 grant.

References

Balinski, M. L. & Young, P. (1982). *Fair representation: Meeting the ideal of one man, one vote*. London: New Haven

Banzhaf, J. F. (1965). Weighted voting doesn't work: A mathematical analysis. *Rutgers Law Review, 19*, 317–343.

Barberà, S. & Jackson, M. O. (2006). On the weights of nations: Assigning voting weights in a heterogeneous union. *Journal of Political Economy, 114*, 317–339.

Beisbart, C., Bovens, I., & Hartmann, S. (2005). A utilitarian assessment of alternative decision rules in the council of ministers. *European Union Politics, 6*(4), 395–418.

Bentham, J. (1789). *An introduction to the principles of morals and legislation*. Oxford: Clarendon Press

Chang, P. L., Chua, V. C. H., & Machover, M. (2006). L.S. Penrose's limit theorem: Tests by simulation. *Mathematical Social Sciences, 51*, 90–106.
Feix, M., Lepelley, D., Merlin, V., & Rouet, J. (2004). The probability of conflicts in a U.S. presidential type election. *Economic Theory, 23*(2), 227–257.
Feix, M., Lepelley, D., Merlin, V., & Rouet, J. (2007). On the voting power of an alliance and the subsequent power of its members. *Social Choice and Welfare, 28*(2), 181–207.
Feix, M., Lepelley, D., Merlin, V., & Rouet, J. (2010). Majority efficient representation of the citizens in a federal union. Working Paper.
Felsenthal, D. S. & Machover, M. (1998). *The measurement of voting power: Theory and practice, problems and paradoxes*. Northampton, MA: Edwar Elgar, Cheltenham edition
Felsenthal, D. S. & Machover, M. (1999). Minimizing the mean majority deficit: The second square root rule. *Mathematical Social Sciences, 37*, 25–37.
Felsenthal, D. S. & Machover, M. (2001). The treaty of Nice and qualified majority voting. *Social Choice and Welfare, 18*, 431–464.
Felsenthal, D. S. & Machover, M. (2004). Analysis of QM rules in the draft constitution for europe proposed by the european convention, 2003. *Social Choice and Welfare, 23*, 1–20.
Gelman, A., Katz, J. N., & Bafumi, J. (2004). Standard voting power indexes don't work: An empirical analysis. *British Journal of Political Sciences, 34*, 657–674.
Gerhlein, W. V. (2006). *Condorcet paradox*. Berlin: Springer.
Lahrach, R. (2009). *Essais en théorie du choix social: Pouvoir, paradoxes et règles de vote*. PhD thesis, Université de Caen Basse Normandie.
Lahrach, R. & Merlin, V. (2010). Assessing the probability of the referendum paradox: The French local election case. Working Paper.
Laruelle, A. & Valenciano, V. (2005). Assesing success and decisiveness in voting situations. *Social Choice and Welfare, 24*, 171–197.
Laruelle, A., & Valenciano, V. (2008). *Voting and collective decision making: Bargaining and power*. London: Cambridge University Press.
Lindner, I. & Machover, M. (2004). L.S. Penrose's limit theorem: Proof of some special cases. *Mathematical Social Sciences, 47*, 37–49.
Lindner, I. & Owen, G. (2007). Cases where the Penrose limit theorem does not hold. *Mathematical Social Sciences, 53*, 232–238.
Maaser, N. & Napel, S. (2007). Equal representation in two-tier voting systems. *Social Choice and Welfare, 28*, 401–420.
May, K. (1948). Probabilities of certain election results. *The American Mathematical Monthly, 55*, 203–209.
Nurmi, H. (1999). *Voting paradoxes and how to deal with them*. Berlin: Springer.
Penrose, L. S. (1946). The elementary statistics of majority voting. *Journal of the Royal Statistical Society, 109*, 53–57.
Penrose, L. S. (1952). *On the objective study of crowd behaviour*. London: H. K. Lewis & Co.
Rae, D. W. (1969). Decision-rules and individual values in constitutional choice. *American Political Science Review, 63*, 40–63.
Slomiczyński, W. & Życzkowski, K. (2007). From a toy model to the double square-root voting system. *Homo Oeconomicus, 24*, 381–400.
Straffin, P. D. (1977). Homogencitv, independence and power indices. *Public Choice, 30*, 107–118.

Chapter 6
A Gentle Combination of Plurality Vote and Proportional Representation for *Bundestag* Elections

Olga Birkmeier, Kai-Friederike Oelbermann, Friedrich Pukelsheim, and Matthias Rossi

6.1 Introduction

In the decision of 3 July 2008 the German Federal Constitutional Court called upon the *Bundestag* to amend the Federal Election Act (*Bundeswahlgesetz*) in order to avoid negative voting weights, and set a deadline of 30 June 2011 (Schreiber 2009; Meyer 1994). Though the issue did not stimulate much public discussion, several problem analyses and solutions were proposed in the literature (Arnim von 2008; Behnke 2010a; Holste 2009; Isensee 2010; König 2009; Meyer 2009; Nohlen 2009; Pukelsheim 2008; Pukelsheim and Rossi 2010, Prittwitz von http://www.volkervonprittwitz.de/anders_waehlen.htm, Roth 2008). Some of the proposals redesign the *Bundestag* system in quite a drastic way.

The present paper assembles an overview of those options that stay close to the existing system. These options (and others) have been developed under the assumption that the Court's decision demands an amendment definitely eliminating negative voting weights (Sect. 6.2). However, if the Court's decision is interpreted in a direction to eliminate negative voting weights in practice, to the greatest possible extent, notwithstanding their existence in theory, then the *Bundestag* enjoys a greater margin of discretion to amend the electoral system (Sect. 6.3). With a view towards this more liberal interpretation of the Court's decision, we propose another option which may be a viable way to amend the electoral act. The new option is called the "gentle combination" of plurality vote for the election of persons and of a

O. Birkmeier · K.-F. Oelbermann · F. Pukelsheim (✉)
Institute for Mathematics, University Augsburg, 86135 Augsburg, Germany
e-mail: olga.birkmeier@math.uni-augsburg.de; oelbermann@math.uni-augsburg.de; pukelsheim@math.uni-augsburg.de

M. Rossi
Chair for Constitutional and Administrative Law, European Law and Legislative Doctrines, University Augsburg, 86135 Augsburg, Germany
e-mail: rossi@jura.uni-augsburg.de

D.S. Felsenthal and M. Machover, *Electoral Systems*, Studies in Choice and Welfare, 151
DOI 10.1007/978-3-642-20441-8_6, © Springer-Verlag Berlin Heidelberg 2012

proportional representation list system (Sect. 6.4), and appears to be attractive from various points of view (Sect. 6.5).

To begin with we briefly outline the electoral system for the German *Bundestag*. The system combines the election of persons, proportional representation, and federalism. The voters' constituency votes serve to elect (by plurality) a constituency candidate, within each of 299 single-member constituencies. The voters' list votes are cast for a party's State list. The allocation of seats then is carried out in three stages. First, 598 *Bundestag* seats are apportioned in proportion to the parties' nationwide count of list votes. Next, each party's national seat contingent is allocated among the States where the party presented a candidate list. Finally, the seats of a party in a State are filled by the party candidates who were the plurality winner in their single-member constituency, and then the remaining seats are filled from the candidate list which the party presented in that State.

In cases when a party's constituency winners in a State exceed the party's share of seats in the proportional representation allocation, the constituency winners retain their seats and the *Bundestag* is expanded beyond the original house size of 598. For example, the 2009 election resulted in 21 overhang seats for the CDU and 3 overhang seats for the CSU. The creation of overhang seats, beyond the originally intended house size of 598 seats, is closely related to the occurrence of negative voting weights. Overhang seats may originate from an over-proportional share of constituency seats. During many decades this was the predominant view, and was tolerated because the support of the voters was considered sufficient justification for the creation of overhang seats even though they reach beyond the proportional representation share of seats. But the diverging success of constituency votes and of list votes can also be exaggerated by voters withholding their list vote from the party of their choice, thus exercising a negative voting weight (Meyer 1994).

Such a situation was manifest during the 2005 *Bundestag* election. The death of a candidate in the Dresden I constituency necessitated a by-election in which too many votes in favor of the CDU would have cost that party a *Bundestag* seat. Due to press coverage and Internet activities, about ten thousand CDU voters withheld their list vote from the CDU. This incident led to the 3 July 2008 decision of the Court and its call upon Parliament to amend the Federal Election Act in order to avoid any negative voting weights.

6.2 Four Options Close to the Current System

Negative voting weights may arise because the current Federal Election Act combines a proportional representation system with the election of persons and, additionally, takes into account the federal structure of Germany with its 16 States. Depending on how the three components (proportional representation, election of persons, and federal structure) are combined, the resulting electoral systems may differ.

We present an overview of four options that maintain the current ballot design with its two votes and thus stay particularly close to the present electoral system. Option F intends to honor the federal structure of Germany and treats each State as a separate electoral district (2.1). Thus constituency votes and list votes acquire quite a different character and entail serious constitutional problems.

As an alternative system the Court decision mentions the "trench system" (*Grabenwahlsystem*), here labeled Option G (2.2). Option G disassociates the two components of proportional representation and the election of persons, by filling one half of the *Bundestag* seats through plurality votes in single-seat constituencies, and the other half through proportional representation. As a result, representation of voters of smaller parties is practically cut in half which, presumably, is hard to communicate to the public.

In Option P the election of persons dominates over the proportional representation component (2.3). In contrast, Option V focuses on the proportional representation dominance (2.4). Both options reproduce the current electoral system in "regular instances" where overhang seats do not occur. These options take some corrective actions only in exceptional situations where the electoral results lead to overhang seats. Option P leaves the results of the election of persons untouched and adjusts the results from the proportional representation calculations, Option V proceeds inversely.

Reaching beyond options that stay close to the current system, there is a greater leeway for a re-arrangement of the system components to avoid negative voting weights. For example, Behnke favors the implementation of two-seat constituencies (Behnke 2010b). Meyer opts for a one-vote ballot which then is doubly evaluated, once as a constituency vote and once as a list vote (Meyer 2009). Prittwitz proposes a system with a one-vote ballot, where a federal list of candidates is formed *ex post* on the basis of how many constituency votes each candidate drew (Prittwitz http://www.volkervonprittwitz.de/anders_waehlen.htm).

6.2.1 *Option F: Separate per-State Apportionments*

Option F proposes to disentangle the three components of the *Bundestag* electoral system by carrying out the proportional representation calculations separately for each of the 16 States. This would presuppose that, sometime during the legislative period, the Constituency Commission allocates the 598 *Bundestag* seats to the 16 States. On election day, the seat contingent of a State is apportioned among parties proportionally to the number of list votes they draw. Hence the current system of a single federal election would be re-arranged into 16 separate, but simultaneously conducted, State elections.

Option F runs into conflict with the unitary character of the *Bundestag*. According to Article 38 I of the Basic Law (*Grundgesetz*) the members of the *Bundestag* shall be representatives of the whole people (*Vertreter des ganzen Volkes*), rather than representatives of 16 separate "State people". Moreover, if in a State a party features more constituency winners than are proportionally warranted (that is, the party wins "overhang seats"), the list votes for that party in that State would become totally ineffective and irrelevant. In the 2009 election, one fifth of all list votes would have been invalidated. This is in strong contrast to the imprint on the ballot sheets informing citizens that it is the list vote that is decisive (*maßgeblich*) to determine the composition of the *Bundestag*. Option F runs the risk of violating the principle of electoral equality that is guaranteed in the Basic Law.

6.2.2 Option G: Trench System

Another possibility is the "trench system" wherein half of the *Bundestag* seats are allocated through plurality, while the allocation of the other half is based on proportional representation. At present, constituency sizes vary considerably. The variation is judged constitutionally acceptable because the decisive votes are the list votes, not the constituency votes. If, as in the trench system, constituency votes also become decisive, constituency sizes would need to be equilibrated in a better way. However, no initiative has been taken to do so, and the time until 30 June 2011 seems to be running out.

Option G is not just adopting minimal corrections to the current system in order to avoid negative voting weights, but is calling for radical changes. The effect will be that representation of voters of smaller parties is cut in half of their current representation. The seats thus freed are re-allocated to the larger parties.

6.2.3 Option P: Plurality Vote Dominant

This option treats constituency votes as untouchable. In each constituency, the candidate with the most constituency votes receives the constituency seat. It is common jargon to call such a seat a "direct seat" (*Direktmandat*). The proportional representation calculation is then carried out conditional on the direct seats already allocated. The adjustments take a slightly different form whether they apply to the super-apportionment (among parties on the federal level), or to the sub-apportionments (within parties, across the 16 States).

On the level of the super-apportionment calculation, the *Bundestag* house size will be increased beyond the initial 598 seats until all direct seats are accommodated by the proportional representation calculations. For example, in the 2009 elections the CSU won 45 direct seats, while its list votes justified only 42 seats. In such a case Option P raises the size of the *Bundestag* to 641 seats such that the proportional share of the CSU increases to 45 seats and hence carries all direct seats. This enlargement strategy is possible because the proportional representation calculations are evaluated using the divisor method with standard rounding (Webster/Sainte-Laguë). Up to 2008 the Federal Election Act prescribed the Hare quota method with residual fit by largest remainders (Hamilton) where the enlargement strategy is hampered by paradoxical effects (Alabama paradox).

In the sub-apportionment calculations, where for each party its nationwide seats are apportioned among its State lists, Option P uses the direct-seat restricted divisor method with standard rounding. This method guarantees that, for each party's State list, the seat allocation meets or exceeds the number of direct seats of that party in that state, while the remaining seats are allocated according to the proportionality principle. Option P combines proportional representation and the election of persons so that the more advantageous of the two components prevails.

Option P yields results identical with the current electoral system as long as no overhang seats intervene. In exceptional cases when overhang seats occur, the adjustment mechanisms of Option P take effect and produce slight deviations from the current system.

6.2.4 Option V: Proportional Representation Dominat

Option V discards the rule that the plurality winner in a constituency is guaranteed a seat. To begin with, Option V executes the super-apportionment and the per-party sub-apportionments, strictly following the proportional representation principle. These numbers determine the maximum number of seats allocated to the parties' State lists.

Hence the cases that need to be taken care of are those where the number of proportionality seats of a party in a State falls short of its number of plurality winners. In such a case some of the plurality winners will not get a seat. Therefore the law must provide an abstract and *ex ante* rule who among the plurality winners will get a seat, and who will have to stay back. To this end Option V relies on the constituency vote counts and excludes those plurality winners who drew the fewest constituency votes. Of course, this would be fair only if constituency sizes are approximately equal.

Option V, too, reproduces the results from the current electoral system in the regular instances when the electoral results do not lead to overhang seats. Otherwise the plurality winners with the fewest constituency votes remain without a seat unless the number of direct seats is supported by the number of proportionality seats. In any case, the *Bundestag* house size of 598 seats is met exactly. Option V is not new, but was used in the State of Bavaria for the 1954 election of the Bavarian State Parliament. The law was challenged to the Bavarian Constitutional Court who judged the regulation to be constitutional. The logic and the argumentation of the judgement would seem to carry over to present day conditions (Table 6.1).

6.3 Contents and Meaning of the Law-Making Mandate

The possibility to gently combine personal and proportional representation (Option S below) reduces both, the number of overhang seats and the likelihood that negative voting weights occur, although it does not avoid either of these outcomes

Table 6.1 Overview of options

Option	Constituency seats	Proportionality seats	Level of federalism	House size	Negative voting weight
F Per-state apportionment	Retained	Retained	16 states	598+x	Impossible
G Trench system	Retained	Retained	Germany	598	Impossible
P Plurality vote dominant	Retained	Adjusted	Germany	598+y	Unlikely
V Proportional vote dominant	Adjusted	Retained	Germany	598	Impossible
S Gentle combination	Adjusted	Adjusted	Germany	598+z	Unlikely

entirely. The question arises whether the implementation of the gentle combination is nevertheless compatible with the law-making mandate of the Court's 3 July 2008 judgement. An answer to this question requires a short analysis of the general significance of law-making mandates of the Constitutional Court (3.1), and then a closer examination of the specific content and scale of the law-making mandate of the 3 July 2008 decision (3.2).

6.3.1 *Law-Making Mandates of the Constitutional Court*

These law-making mandates considerably tackle the antagonism between the principles of democracy and the rule of law, both of which shape, according to Articles 20 and 28 of the Basic Law, the constitutional order of the Federation as well as the 16 States.

While the principle of democracy guarantees the legislator a considerable degree of legislative leeway, the principle of the rule of law commands the adherence to the constitutional principles in Articles 20 III, 1 and 1 III Basic Law. Due to the separation of powers the Court is the only authority to overrule political decisions transposed into legislation by the legislator, who is democratically legitimized through direct elections, in case the limits mentioned were crossed.

There is a gradual difference between the competence to overrule acts of parliament on the one hand and the task to enact new regulations within a certain period of time on the other. While the annulment of rules affects the question 'how' the legislator makes use of its leeway, the law-making mandate concerns the question 'whether' the legislator makes use of its leeway at all. Nevertheless the parliamentary legislator, being confronted with law-making mandates of the Court, has never openly revolted against them, for good reasons. First of all, in a modern and open constitutional state the legislator has no unlimited margin of discretion anymore. International and supranational guidelines as well as the constitution itself oblige and confine the legislator. There are various rudimental and generally formulated passages in the constitution – like Article 38 III Basic Law – that therefore have to be observed by the legislator.

Furthermore, the Court issues concrete law-making mandates only if the law in question is of such importance that its immediate repeal would entail even more unconstitutional consequences than its retention. Hence, law-making mandates are not a more severe, but rather a milder consequence of a violation of the constitution. The rule in question does not immediately become inapplicable but shall remain in force until the legislator enacts a law that is in line with the constitutional principles. Thus the Court restricts the legislator's margin of discretion in order to ensure its legislative leeway. Therefore, law-making mandates by the Constitutional Court are – subject to their concrete contents – basically not only permissible but also necessary to demonstrate the separation of powers between the Court which preserves the constitution, and the legislator who acts in political terms. Law-making mandates can eventually be regarded as an expression of a delayed temporary unconstitutionality, as the affected rule does not become void immediately but only if the law-making mandate is not realized within the set time limit.

Regarding the law-making mandate of 3 July 2008, the Constitutional Court's encroachment upon the legislator's margin of discretion turns out to be even milder, also because it actually merely repeats and specifies the legislative mandate embodied in the constitution in Article 38 III of the Basic Law. Calling upon the law-maker to adopt a rule in conformity with the constitution within a certain time period is in accord with the constitutional separation of powers. It is more the expression of the Constitutional Court's responsibility enshrined in the constitution and does not imply that it is exceeding its authority on executive powers.

6.3.2 Contents and Scale of the Law-Making Mandate

Answering the question about content and scale of the concrete law-making mandate is more difficult. In headnote no. 2 of its judgement the Court obliges the legislator to adopt a rule that is in accordance with the constitution until 30 June 2011. This *per se* vaguely formulated mandate refers to headnote no. 1 in which the Court holds the Federal Election Act violated the principles of equal and direct suffrage "insofar as it is thus made possible that an increase in list votes would lead to the loss of seats of a party's State list, or that the loss of list votes would lead to an increase of seats of a party's State list." According to the wording, the mere possibility of the occurrence of votes with negative weight would immediately trigger the violation of the constitutional voting principles, and be unconstitutional. This would indicate that the amendment by the legislator must rule out even the slightest theoretical possibility of the occurrence of these effects regardless of how unlikely they may be. At second glance, however, at least four reasons make us doubt that the central reasoning of the Court must be understood in such an uncompromising manner.

First of all, the legislative leeway with regard to the necessity to specify the Federal Election Act is not only determined by a single issue but by a variety of guidelines. The principles set up in Article 38 I Basic Law, which protect the voters' decisions and the impact of their votes, have to be respected just like the principle of equal opportunities for political parties in Articles 21 I and 3 I Basic Law and the MPs' equal status rights in Article 38 I. In addition an appropriate reference to the federal system may be included. This does not directly result from Article 38 I Basic Law, but is the consequence of the federal character of the States' constitutional order. To orient the whole electoral system exclusively towards preventing even the slightest possibility of the occurrence of votes with negative weight would unduly confine the already restricted legislative leeway and would prevent the legislator from focusing on other aspects.

Secondly, one cannot deduce from the Court's concrete remarks the strict prohibition of votes with negative weight. While the law-making mandate, briefly summarized in headnotes no. 1 and 2 seems to be absolute, the Court's reasoning shows that it knows about the difficulties of the various requirements of the electoral system. The Court refers to it as a "complex system of regulations" and acknowledges that "even a minor change to the Federal Election Act could, in the present constellation, possibly lead to far-reaching structural changes." The Court would seem to conclude from this only the long deadline for the legislator to amend the electoral law. At the same time, however, it knowingly accepts the occurrence of votes with negative weights in the (then) upcoming 2009 *Bundestag* election. Votes with negative weight are thus not to be avoided at any expense but have to be balanced with other interests.

Thirdly, the Court has, in earlier decisions, continuously approved of the constitutional restriction of the legislative leeway, which can almost be called magical because this restriction is not consistent and does not always have to be strictly adhered to. The issue of the votes with negative weight was – albeit named differently – already brought before the Court in a 1997 case dealing with a lawsuit filed by the State of Lower Saxony. Yet the Court did not treat this problem, let alone declare it unconstitutional. Therefore many academic commentators have pointed out the contradiction between the present statement of unconstitutionality, and the Courts former reasoning concerning overhang seats. Of course, the Court may – not least because of the changing in the personnel – from time to time cautiously adjust its own judicature. Also, it may be that in 1997 the Court concentrated less on the votes with negative weight and more on the overhang seats. However, considering the concrete circumstances of the 2008 decision, the crucial aspect seems to be that the votes with negative weight could be used "tactically" in the constellation of by-elections, and that the ballot box failed to be a black box for all voters, but seemed transparent to some. In view of this incidence, declaring the votes with negative weight illegal may appear to be uncompromising, but it should rather be seen as an effort to prevent tactical voting. Leaving aside the question of how such a "tactical" voting decision could be distinguished from a "real" voting decision, we would stress that such a voting behavior is peculiar to the very special constellation of a by-election. If no by-election takes place, the effect of votes with negative weight cannot be operationalized.

Fourthly, we feel that it is necessary to relativize the prohibition of votes with negative weight because the electoral law as a whole should not be determined by the pursuit of certain abstract ideals but by its concrete usage. More than in any other field of law, the Court requests the legislator to consider electoral law as *law in action* and to adjust the *law in books* if necessary. In view of this, the Court's interest is not to prevent even the slightest possibility of the occurrence of votes with negative weight in order to maintain an uncorrupted electoral law but primarily, to prevent this effect from being abused in concrete cases. Therefore a rule to reduce the occurrence of votes with negative weight and in addition, to reduce the necessity of by-elections by adequately regulating the move-up procedure of substitute candidates (*Nachrückregelung*), would fully meet the requirements of the Court's law-making mandate.

By-elections are always a massive encroachment on the principle of equal suffrage. For this reason it is desirable to prevent by-elections if possible. This effort is only marginally related to the issue of votes with negative weights. As the gentle combination, to be detailed below, regulates the relation between personal and proportional representation and thus arithmetically binds the two components to each other, we feel encouraged to also use it to prevent by-elections.

We feel that these considerations are sufficient to justify the proposal of a gentle combination between personal and proportional representation (Sect. 6.4) and to discuss its qualitative features (Sect. 6.5). Yet, if the constitutional principles demand the definite elimination of negative voting weights, or if application of the gentle combination will prove not to decrease negative voting weights to an extend we had hoped for, there is a tighter variant, Option W, which definitely does away with negative voting weights. However, the system-stabilizing and self-healing powers of Option W are inferior to those of Option S.

6.4 Option S: The Gentle Combination

Our proposal is based on two considerations. Firstly, the Court's call for an amendment should be satisfied by staying as close as possible to the current electoral system. At present, neither the political actors nor the interested public indicate that they want a novel electoral design. Therefore Option S leaves all regular situations intact and touches only upon the exceptional cases that may complicate the current system.

Secondly, we follow the goal formulated in Section 1 I of the Federal Election Act, according to which the members of the *Bundestag* are elected in accordance with the principles of proportional representation combined with the election of persons (*nach den Grundsätzen einer mit der Personenwahl verbundenen Verhältniswahl*).

Thus we concentrate on the parts of the electoral system that constitute the elements combining proportional representation and the election of persons. We regard the "combination" of proportional representation and of the election of persons as a fundamental goal that calls for a deliberate and careful system design. Option S appears to be a compromise solution mediating between Options P and V where one of the two system components dominates over the other.

Option S respects the obligation that every constituency is represented in parliament by (at least) one of the constituency candidates. Deviating from the current law, however, we do not imply that the plurality winner is the sole person who can represent the constituency. In exceptional cases one of the competing candidates may represent the constituency.

Exceptional cases may arise in order to reduce the occurrence of negative voting weights. A constituency where the exceptional rule is applied is called an *exceptional constituency*. Exceptional constituencies may be represented in Parliament by candidates who have been voted into Parliament over their State list, rather than attracting a plurality of constituency votes.

Extending the privilege of parliamentary representation from the plurality winner to other constituency candidates admittedly mean a certain change of paradigm. However, the change of paradigm focuses on a judicial view rather than on a practical view. Factually, only a third of the *Bundestag* members are the sole representatives of their constituency. Two-thirds of the constituencies are represented by two or more *Bundestag* members. This shows that, when proportional representation is combined with the election of persons, the quote of "my constituency" does not have the same meaning as in a pure plurality system. The quote only indicates that the representative is at home in this constituency, and that this is where he or she has the center of political activities outside Parliament. Even this criterion may occasionally fail since many candidates are not residents of "their" constituency, but just stand as candidates in addition to acquiring a position on the party's State list.

The exceptional case, where a constituency is represented by some constituency candidate other than the plurality winner, becomes relevant only when there are more plurality winners of a party in a State than there are proportionality seats for that party in that State. According to the current law such situations generate overhang seats (*Überhangmandate*), which in the law are referred to as a "discrepancy number" (*Unterschiedszahl*). Such a discrepancy may occur in the super-apportionment when the 598 *Bundestag* seats are allocated among parties in proportion to their nationwide list votes. On the other hand a discrepancy number may also emerge in one of the sub-apportionments when the nationwide seats of a party are distributed among that party's State lists.

Confusing cross-references in the current law inhibit a transparent description of the rules. Moreover, they appear inappropriate in that they equate unequal procedures. The super-apportionment procedure concerns the composition of the *Bundestag* with respect to the political spectrum. Quite in contrast, the sub-apportionments reflect the federal structure and adjoin a geographical dimension. Political representation and federal representation are two distinct issues.

For this reason Option S proposes different amendments for the (nationwide) super-apportionment, and for the (per-party) sub-apportionments. In the super-apportionment among parties, a modest increase of the house size avoids overhang seats (4.1). During the within-party sub-apportionments, "overhang seats of a new kind" may occur (4.2). In either case we illustrate Option S by means of the 2009 *Bundestag* election.

6.4.1 Gentle Super-Apportionment

Striving for a modest amendment, the gentle super-apportionment does not come to bear in regular cases, but only in exceptional cases. Option S proceeds as follows. It starts out by apportioning all 598 *Bundestag* seats mentioned in Section 1 I of the Federal Election Act.

6.4.1.1 Regular Cases

A regular case comprises those instances when a party is apportioned at least as many proportionality seats as are needed nationwide to accommodate its plurality winners. In case all parties constitute a regular case, the super-apportionment terminates.

6.4.1.2 Exceptional Cases

An exceptional case occurs when the plurality winners of a party outnumber its nationwide proportionality seats. In 2009, the CSU featured 45 plurality winners, but was allocated just 42 proportionality seats. In fact, this the only instance in the post-war history of the Federal Republic of Germany where the nationwide allocation of seats fell short of the total number of plurality winners.

In exceptional cases Option S proceeds as follows. All plurality winners of a given party in a given State are ranked by decreasing constituency vote counts. The plurality winners with the largest counts receive a seat until the number of proportionality seats is exhausted. Hence Option S does not allocate more seats than are made available through the initial house size of 598 seats.

There remain a few exceptional constituencies where the constituency vote count of the plurality winner is too low to be allocated a seat. For these exceptional constituencies Option S makes special arrangements. It checks in the constituency of the plurality winner whether there are competing candidates who get their seat via their State list. If so, the constituency is considered to be represented in the *Bundestag*, and the plurality winner does not get a seat. In case all exceptional constituencies are represented by competing candidates of the plurality winner, the super-apportionment has reached its end.

In the remaining case, when there exist exceptional constituencies where no competitor of the plurality winner obtains a seat via a State list, Option S makes the following adjustments to guarantee local representation. It increases the house size, one by one, and re-calculates the proportionality seats until all constituencies are represented.

In summary, Option S gently combines proportional representation with the election of persons in such a way that, through the super-apportionment adjustments, political parties are allocated a number of seats in proportion to their nationwide count of list votes. At the same time it guarantees each constituency representation in the *Bundestag*. In most cases the constituency representative is the plurality winner, in exceptional cases it may be a competitor of the plurality winner. In any case, all 299 constituencies send at least one of the constituency candidates into the *Bundestag*. Thus Option S complies with the local representation principle of Section 1 II of the Federal Election Act.

6.4.1.3 Application to the 2009 Election

The abstract rules of Option S are illustrated using the electoral data of the 2009 *Bundestag* election. The CDU counted 173 plurality winners in constituencies nationwide. Starting out with a house size of 598 seats, the initial proportional representation calculation apportions 173 proportionality seats to the CDU. Hence all CDU plurality winners are carried by the CDU proportionality seats, a regular case. The other parties also constitute a regular case, except the CSU.

The CSU is an exceptional case. Her candidates emerged as the plurality winners in all Bavarian constituencies (45). However, the initial proportional representation calculation awards the CSU fewer proportionality seats (42), see Table 6.2.

Hence a discrepancy number comes into being, $45 - 42 = 3$. To carry out the gentle combination, the CSU plurality winners are sorted by decreasing constituency vote counts. The 42 strongest plurality winners are awarded a seat. The last three plurality winners, with rank score 43, 44, and 45, are further analyzed, see Table 6.3.

Rank 43 is occupied by the plurality winner of constituency 227 (Deggendorf), with 49,398 constituency votes (53% of the constituency votes in that constituency).

Table 6.2 Gentle combination 2009 for 598 seats. Initial super-apportionment calculation

Party	List votes	Quotient	Proportionality seats
CDU	11,828,277	173.4	173
SPD	9,990,488	146.497	146
FDP	6,316,080	92.6	93
Linke	5,155,933	75.6	76
Grüne	4,643,272	68.1	68
CSU	2,830,238	41.501	42
Sum [divisor]	40,764,288	[68,196]	598

Each 68,196 list votes account for about one proportionality seat.

Table 6.3 Gentle combination 2009 for 598 seats. Three CSU exceptional constituencies with rank score 43–45

Rank score	Constituency vote	%	Parties with successful competitors	Number and name of constituency
43	49,398	53	–	227 Deggendorf
44	48,943	37	SPD, Linke	244 Nürnberg-Nord
45	47,519	39	SPD	245 Nürnberg-Süd

Rank 44 is attained by the plurality winner of constituency 244 (Nürnberg-Nord), with 48,943 constituency votes (37%). Rank 45 features the plurality winner of constituency 245 (Nürnberg-Süd), with 47,519 constituency votes (39%). It turns out that constituencies 244 and 245 are represented by competitors of other parties who succeeded to obtain a seat via their party list. For this reason the plurality winners of constituencies 244 and 245 do not get a seat.

In constituency 227 (Deggendorf) none of the other candidates managed to obtain a seat via their party list. In order not to leave this exceptional constituency unrepresented, the initial house size of 598 seats is increased, one by one, until the plurality winner or another Deggendorf candidate obtains a seat. When the house size reaches 609 seats, the proportionality seats of the FDP are raised. The raise is handed on to the FDP list in Bavaria. The additional seat carries the FDP candidate of constituency 227 (Deggendorf) into the *Bundestag*. All three constituencies which initially were exceptional are now represented by a successful competitor, whence the plurality winners in these constituencies fall short of obtaining a seat in parliament. The final super-apportionment is shown in Table 6.4.

6.4.2 *Gentle Sub-Apportionments*

Sub-apportionment calculations follow the same strategy of honoring regular cases. Special regulations apply only in exceptional cases.

6.4.2.1 Regular Cases

A regular case is any situation where all plurality winners of a party in all constituencies of a State are carried by the proportionality seats of that party in that State. For regular cases, the gentle combination terminates and reproduces what the current law would do.

Table 6.4 Gentle combination 2009 for 609 seats. Final calculation for the super-apportionment

Party	List votes	Quotient	Proportionality seats	Plurality winners
CDU	11,828,277	177.0	177	173
SPD	9,990,488	149.49	149	64
FDP	6,316,080	94.51	95	0
Linke	5,155,933	77.1	77	16
Grüne	4,643,272	69.48	69	1
CSU	2,830,238	42.3	42	42
Sum [divisor]	40,764,288	[66,830]	609	296

Each 66,830 list votes account for about one proportionality seat.

6.4.2.2 Exceptional Cases

An exceptional case occurs in the presence of a party which in some State features more plurality winners than are carried by the proportionality seats. In such a case the plurality winners of that party in that State are ranked by decreasing constituency votes. The plurality winners with the stronger constituency vote counts are allocated a seat until the number of proportionality seats is exhausted.

For the few remaining exceptional constituencies Option S examines whether another candidate moves into the *Bundestag* via his or her party list. If there is a successful candidate from another party, the plurality winner is passed over and does not obtain a seat.

The final constellation to be considered is that there exists an exceptional constituency in which none of the candidates who compete with the plurality winner has been successful to obtain a seat via his or her State list. It is only now that Option S creates an "overhang seat of a new kind". The house size increases by the number of these overhang seats of a new kind.

These additional seats are created not exclusively under the conditions of a plurality system, but are justified by the principle of regional representation. The interrelation with the electoral system *in toto* gives overhang seats of a new kind a quality quite different from the character of overhang seats of the current kind. Because of this different character, and because of the identifiability of the overhang seats of a new kind, the decision of the Court that overhang seats cease to exist when the seat holder leaves the *Bundestag* becomes obsolete.

6.4.2.3 Application to the 2009 Election

In the 2009 election four of the five sub-apportionments constitute a regular case that all plurality winners of a party are carried by their parties' proportionality seats. The CDU constitutes an exceptional case.

As shown in Table 6.4, the house size of 609 seats allocates 177 seats to the CDU. With the sub-apportionment calculation, it transpires that there are 20 exceptional constituencies, in seven States. In these seven States, the CDU plurality winners are ranked by decreasing constituency vote counts. The lowest-ranked plurality winners identify the exceptional constituencies, see Table 6.5.

The third column in Table 6.5 shows that 13 of the 20 exceptional constituencies are represented by other candidates, while in seven constituencies no candidate obtained a *Bundestag* seat. For these seven exceptional constituencies Option S creates overhang seats of a new kind. Table 6.6 exhibits the final tally for the CDU lists. The nationwide super-apportionment and the per-party sub-apportionments yield a *Bundestag* that grows from 598 seats (initial house size), passes through 609 seats (super-apportionment increase), and terminates with 616 seats (by creating seven overhang seats of a new kind).

Table 6.5 Gentle combination 2009 for 609 seats. Twenty CDU overhang seats with associated exceptional constituencies

State/rank score	Constituency votes	%	Parties with successful competitors	Number and name of constituency
SH-9	48,136	39	–	006 Plön/Neumünster
MV-5	38,102	33	–	018 Neubrandenburg/.../Uecker-Randow
MV-6	34,633	29	SPD, FDP	013 Schwerin/Ludwigslust
SN-13	45,876	34	Linke	163 Chemnitz
SN-14	44,147	41	–	152 Nordsachsen
SN-15	42,704	33	SPD	153 Leipzig I
SN-16	41,101	29	SPD, Grüne	154 Leipzig II
TH-7	40,063	29	Grüne	192 Gotha/Ilm-Kreis
RP-12	53,705	46	FDP, Grüne	203 Bitburg
RP-13	50,035	39	–	211 Pirmasens
BW-29	54,172	47	–	286 Schwarzwald-Baar
BW-30	54,169	45	–	293 Bodensee
BW-31	53,872	38	FDP, Linke, Grüne	271 Karlsruhe-Stadt
BW-32	53,829	43	SPD	261 Esslingen
BW-33	50,967	42	–	288 Waldshut
BW-34	48,662	43	SPD	292 Biberach
BW-35	48,518	34	SPD	258 Stuttgart I
BW-36	48,137	36	FDP, Linke, Grüne	275 Mannheim
BW-37	44,002	35	SPD, Linke, Grüne	259 Stuttgart II
SL-4	45,748	32	SPD	296 Saarbrücken

6.5 Merits of the Gentle Combination

The possibility that a constituency may not be represented by the plurality winner is novel from the point of view of both voters and candidates. Yet this change, gentle as it is, comes along with multiple benefits. Firstly, by-elections may be avoided by substituting a deceased or resigned constituency candidate with a list candidate (5.1). An operationalization of negative voting weights, such as during the 2005 by-election in constituency 160 (Dresden I), would no longer be possible. Secondly, overhang seats of a new kind decrease to about a third of the overhang seats of the old kind (5.2). Thirdly, the overhang seats of a new kind are justified by guaranteeing the electorates their regional representation (5.3).

Fourthly, the gentle combination enhances the competitive character of the election. It provides a motivation for all – voters, candidates, and parties – to design constituencies that are as much as possible of the same size (5.4). Fifthly, smaller parties get a role to play even though they cannot profit from overhang seats. They can devise strategies to avoid them by identifying prospective exceptional

Table 6.6 Gentle combination 2009 for 609 seats. Sub-apportionment calculation for the CDU state lists

State	List votes	Quotient	Proportionality seats	Overhang seats
Schleswig-Holstein	518,457	7.8	8	1
Mecklenburg-Vorpommern	287,481	4.3	4	1
Hamburg	246,667	3.7	4	
Niedersachsen	1,471,530	22.0	22	
Bremen	80,964	1.2	1	
Brandenburg	327,454	4.9	5	
Sachsen-Anhalt	362,311	5.4	5	
Berlin	393,180	5.9	6	
Nordrhein-Westfalen	3,111,478	46.6	47	
Sachsen	800,898	12.0	12	1
Hessen	1,022,822	15.3	15	
Thüringen	383,778	5.7	6	
Rheinland-Pfalz	767,487	11.49	11	1
Baden-Württemberg	1,874,481	28.1	28	3
Saarland	179,289	2.7	3	
Sum [divisor]	11,828,277	[66,800]	177	7

Each 66,800 list votes account for about one proportionality seat.

constituencies, nominating candidates who may obtain a list seat, and thereby outmaneuvering the plurality winner (5.5). Sixthly, the gentle combination may enhance the transparency of voter behavior (5.6). Finally, Option S can be advanced to an Option W that definitely annihilates negative voting weights, at the expense of losing some of the benefits mentioned (5.7).

6.5.1 Avoidance of By-Elections

The gentle combination may be employed to avoid by-elections that otherwise would become necessary. Every by-election that takes place on a day different from the main election day is subject to different circumstances. These effects are independent from the occurrence of negative voting weights. Four of the seven by-elections since 1961 had to be conducted a couple of weeks after the main election day, such as the one during the 2005 election in constituency 160 (Dresden I).

For the avoidance of by-elections, Option S offers the possibility to profit from the novel combination of the proportional representation component with the election of persons. In the light of the current law it is alien to substitute a list candidate in place of a constituency candidate. However, since the gentle combination combines the two components, it seems possible to arrange for such a substitution. This would provide a rather pragmatic solution to the avoidance of by-elections.

For the 2005 election, where the NPD candidate in constituency 160 (Dresden I) deceased during the campaigning period, the first candidate from the NPD list in the Free State of Saxony who did not stand in some constituency would have been the one to substitute. This would have been a candidate whose home town was listed as

Coburg, Bavaria. However, on place eight there was a candidate from Dresden I who also did not campaign in any other constituency and who could have functioned as a substitute.[1] With a substitution of the deceased candidate, the by-election could have taken place on the main election day. The problem of negative voting weights would not have surfaced, and the voters in Dresden I would not have been affected by the peculiarities of the electoral system in any other way than the other voters in the rest of Saxony.

6.5.2 Reduction of Overhang Seats

At present 97 of the 299 constituencies in Germany are solely represented by the plurality winner. In the other 202 constituencies, additional candidates also obtained a seat in the *Bundestag*. Based on these numbers we predict that roughly a third of the constituencies are solely represented by their plurality winners, in which cases overhang seats of a new kind may become necessary. Hence switching from the current law to the proposed gentle combination, about a third of the overhang seats may persist.

A cumulative retrospective evaluation of past elections confirms the estimate. The period from 1980 through 2005 saw a total of 60 overhang seats. When reevaluating the elections with the gentle combination, 18 of these 60 overhang seats would have persisted.

6.5.3 Representativeness of Constituencies

The gentle combination secures at least one representative from every constituency. In allocating a seat to the plurality winner, the current law exclusively refers to constituency votes, that is, to the system component that covers the election of persons. In contrast, the gentle combination refers to the electoral system as a whole and merges the results from the election of persons and from the proportional representation component. In doing so, the gentle combination follows the goal that is set out in the Federal Election Act. The gentle combination interprets the Act in a wider sense that not only the plurality winner is legitimized to represent the constituency, but also any other candidate who obtains a *Bundestag* seat through his or her party's State list.

From a legal point of view the constituency votes lose some of their directness, since their effect is evaluated jointly with the results of the list votes. This follows the predominant view that candidates in a constituency stand as an individual, while candidates on a (closed) party list run on behalf of a political party. Factually,

[1] We gratefully acknowledge assistance from Renate Recknagel of the Statistical Office of the Free State of Saxony for retrieving the information on the 2005 party lists.

constituency candidates are also nominated by political parties, and their party affiliation is not less visible than the party affiliation of list candidates. As a matter a fact, Members of the *Bundestag* who obtain a seat via their party list also act as a representative of their constituency, run a constituency bureau, and maintain a high level of visibility in their constituency.

6.5.4 Equality of Constituency Sizes

The gentle combination refers exceptional cases to the absolute counts of constituency votes, not to relative vote shares. Hence the sizes of the constituencies play a central role and should be as equilibrated as possible. Of course, constituencies should always be of the same size, but the gentle combination emphasizes that unequal sizes entail unequal consequences.

For instance, the gentle combination would have led to twenty exceptional cases during the 2009 elections. As shown in Tables 6.3 and 6.5 the exceptional constituencies extend from the nationwide smallest constituency 227 (Deggendorf) with 189,600 population, to the 149th smallest constituency 192 (Gotha/Ilm-Kreis) with a population of 248,381. All exceptional constituencies belong to the smaller half of the constituencies nationwide. Six exceptional constituencies are among the very small constituencies with less than 212,676 population, deviating from the average 250,222.4 by more than 15%.[2]

A motivation to create constituencies of equal size is reinforced only when using the (absolute) constituency vote counts, and does not come to bear when referring to the (percentage) constituency vote shares. The 2009 plurality winner in the constituency 227 (Deggendorf) attracted 53% and hence more than half of the constituency votes. This result loses some of its appeal when noting that in the nationwide smallest constituency it is easier to meet a rather homogeneous electorate than elsewhere.[3]

---

[2]We have $74{,}816{,}496/299 = 250{,}222.4$. (We are grateful to Manfred Thoma from the Federal Statistical Office for supplying the census data as of 31 December 2008.) – Other countries measure the size of a constituency by the electorate (Max-Planck-Institut für ausländisches öffentliches Recht und Völkerrecht 1997). – For the 2009 election the Constituency Commission produced two reports (Bundestagsdrucksachen 16/4,300 and 16/6,286, on the basis of the census data of 31 December 2006). The constituency boundaries that were implemented by the decision of the *Bundestag* on 24 January 2008 deviate from both reports. Neither the parliamentary debate nor the accompanying printed matters advance reasons for the deviations, nor were the sizes of the actual constituencies documented by numbers. – An example of blatant unequal constituency geometry is analyzed in Arnim von (2003).

[3]As a member of the European Commission of Democracy and Law (Venice-Commission) Germany recommends since 2002 on an international level that constituencies should be drawn such as not to deviate more than 10% and 15% from the average [www.venice.coe.int/docs/2002/CDL-AD(2002)023rev-e.pdf]. However, on a national level the current ranges of 15% and 25% persist (Section 3 III 3 Federal Election Act).

6.5.5 Encouragement of Political Competition

The gentle combination opens up new possibilities for smaller parties to help avoiding overhang seats. This may be illustrated using the 2009 election data. There are 16 exceptional constituencies where overhang seats are avoided because the constituencies are represented by candidates other than the plurality winners. Smaller parties which cannot hope for overhang seats are encouraged to nominate their candidates in prospective exceptional constituencies high up on their list, in order to increase their chances of winning a list seat. As shown in Sect. 6.5.4 it suffices to focus on the 150 constituencies that are smaller in size than the average. Should the prospective speculation become reality such that the smaller parties' candidates win seats, the plurality winner from the larger party does not advance into the *Bundestag* and overhang seats of a new kind are avoided.

From the candidate's point of view the gentle combination creates a legitimization that reaches out beyond her or his constituency. A constituency candidate competes not only against the candidates of the other parties in that constituency, but with their party friends who run in other constituencies of the same State. In order not to miss a seat, he or she must strive for not being ranked too low in terms of constituency vote counts. This ranking is determined by the voters. In contrast, the nomination on a State list is a matter of party committees.

6.5.6 Transparency of the Balloting Process

The gentle combination reinforces the effects of constituency votes and of list votes in a way agreeing with the intentions of the electoral system. For instance, when in Baden-Württemberg many voters cast their constituency vote for the CDU constituency candidates and their list vote for the FDP list, they help the CDU candidate to acquire a top position in the constituency vote ranking and to obtain a seat. Hence the effect of the split-votes is exactly what it is supposed to be, in that the constituency votes support the constituency candidate as a person.

At the same time the proportional re-weighting through split-votes is less predictable under the gentle combination than under the current law. Those plurality winners who do not obtain a sufficient number of constituency votes must fear the possibility that in exceptional constituencies other candidates may represent the constituencies. Vote splitting may exercise a detrimental effect in all those other constituencies where the plurality winner does not obtain sufficiently many constituency votes (Fürnberg and Knothe 2009; Gschwend 2007; Jesse 1988; Schoen 1999; Winkler 1999).

6.5.7 Option W: Representativity and Proportionality

Since the gentle combination does not avoid overhang seats in all cases, there still remains a certain possibility that overhang seats emerge and imply negative voting weights. If this residual probability is considered unacceptable, then Option S may be extended to Option W. Option W definitely puts an end to negative voting weights, while ensuring that each constituency is represented by at least one member of the *Bundestag*.

Option W follows the strict rule, both in the super-apportionment and in the ensuing sub-apportionments, to identify as many plurality winners with fewer constituency vote counts and with successfully competing candidates as are needed to avoid overhang seats.

Thus Option W realizes the principle of local representation, since every constituency is represented by its plurality winner or by another successful constituency candidate. The necessary calculations turn out to be less laborious, since super- and sub-apportionments need to be determined only once. The house size of 598 seats is met precisely.

We illustrate Option W with the data of the 2009 election. The super-apportionment yields the nationwide seat numbers as shown in Table 6.2 and coincides with the initial calculation for the gentle combination Option S. There are 45 CSU plurality winners, while the proportionality seats for the CSU amount to only 42. Upon inspection the plurality winners that are left without a seat turn out to be those from constituency 229 (Passau), constituency 244 (Nürnberg-Nord), and constituency 245 (Nürnberg-Süd). In contrast, the plurality winner in constituency 227 (Deggendorf) does get a seat since none of his competitors acquired a seat over their party list. The plurality winner in the constituency 229 (Passau) is ranked 42, with 54,275 constituency votes (47%), but the constituency is represented also through a deputy from the FDP. The other constituencies mentioned are listed in Table 6.3.

6.6 Concluding Remarks

In this paper we have presented a novel Option S, to combine the principles of proportional representation and of the election of persons in the Federal Election Act. The new option stays close to the current electoral system while unwanted effects such as negative voting weights are reduced.

The further one moves away from the current system, the greater becomes the leeway for the required amendment. It remains to be seen whether some of the options of the present paper, or some other options that entail changes of a more substantial sort, are going to be adopted by the *Bundestag*.

At the time of writing it is not yet known which, if any, of these or other proposals currently contemplated by the Bundestag will ultimately be adopted and sanctioned by the German Federal Constitutional Court.

References

Arnim von, H. H. (2003). Der Zuschnitt der hessischen Wahlkreise ist verfassungswidrig. *Deutsches Verwaltungsblatt, 118*, 553–624.

Arnim von, H. H. (2008). Verfassungswidrigkeit des Bundeswahlgesetzes aufgrund des "negativen Stimmgewichts" – Anmerkungen zum Urteil des Bundesverfassungsgerichts vom 3. Juli 2008. *Recht und Politik, 44*, 136–138.

Behnke, J. (2010a). Negatives Stimmgewicht, Erfolgswert und Überhangmandate – einige Anmerkungen. *Kritische Vierteljahresschrift für Gesetzgebung und Rechtswissenschaft, 93*, 3–28.

Behnke, J. (2010b). Überhangmandate und negatives Stimmgewicht: Zweimannwahlkreise und andere Lösungsvorschläge. *Zeitschrift für Parlamentsfragen, 2*, 247–260.

Fürnberg, O. & Knothe, D. (2009). Wahlsieg ohne Stimmenmehrheit: Auswirkungen von verstärktem "Lagersplitting" auf Mandatsverteilung und Koalitionsoptionen. *Zeitschrift für Parlamentsfragen, 40*, 56–74.

Gschwend, T. (2007). Ticket-splitting and strategic voting in mixed electoral systems. *European Journal of Political Research, 46*, 1–23.

Holste, H. (2009). Die Bundestagswahl und das verfassungswidrige Wahlgesetz. *Recht und Politik, 45*, 152–156.

Isensee, J. (2010). Funktionsströrung im Wahlsystem: das negative Stimmgewicht – Denkbare Lösungen eines Dilemmas. *Deutsches Verwaltungsblatt, 125*, 269–277.

Jesse, E. (1988). Split-voting in the Federal Republic of Germany: An analysis of the Federal elections from 1953 to 1987. *Electoral Studies, 7*, 109–124.

König, M. (2009). Paradoxie des negativen Stimmgewichts. *Zeitschrift für Gesetzgebung, 24*, 177–195.

Max-Planck-Institute für ausländisches öffentliches Recht und Völkerrecht. (1997). Wahlkreiseinteilung in westlichen europäischen Demokratien, den USA und Kanada, Zeitschrift für ausländisches öffentliches Recht und Völkerrecht. *Heidelberg Journal of International Law, 57*, 633–674.

Meyer, H. (1994). Der Überhang und anderes Unterhaltsames aus Anlaß der Bundestagswahl 1994. *Kritische Vierteljahresschrift für Gesetzgebung und Rechtswissenschaft, 77*, 312–362.

Meyer, H. (2009). Lösungsmöglichkeiten nach dem Wahlrechtsurteil des BVerfG vom 3. Juli 2008. *Deutsches Verwaltungsblatt, 124*, 137–146.

Nohlen, D. (2009). Erfolgswertgleichheit als fixe Idee oder: Zurück zu Weimar? Zum Urteil des Bundesverfassungsgerichts über das Bundeswahlgesetz vom 3. Juli 2008. *Zeitschrift für Parlamentsfragen, 40*, 179–195.

Prittwitz von, V. Anders wählen: Vom widersprüchlichen Als-ob zur personalisierten Verhältniswahl, forthcoming. http://www.volkervonprittwitz.de/anders_waehlen.htm

Pukelsheim, F. (2008). Bundeswahlgesetz – Nächste Etappe. *Deutsches Verwaltungsblatt, 123*, 889–897.

Pukelsheim, F. & Rossi, M. (2010). Wahlsystemnahe Optionen zur Vermeidung negativer Stimmgewichte. *JuristenZeitung, 19*, 922–929.

Roth, G. (2008). Negatives Stimmgewicht und Legitimationsdefizite des Parlaments. *Neue Zeitschrift für Verwaltungsrecht, 27*, 1199–1201.

Schoen, H. (1999). Split-ticket voting in German Federal elections, 1953–90: An example of sophisticated balloting? *Electoral Studies, 18*, 473–496.

Schreiber, W. (2009). BWahlG – Kommentar zum Bundeswahlgesetz, 8. Auflage.

Winkler, M. (1999). Gleichheit der Wahl, "Stimmensplitting" und Überhangmandate. *Recht und Politik – Vierteljahreshefte für Rechts- und Verwaltungspolitik, 35*, 244–247.

Chapter 7
The Value of Research Based on Simple Assumptions about Voters' Preferences

William V. Gehrlein and Dominique Lepelley

7.1 Introduction

Many people have found it to be very interesting to think about strange and counterintuitive outcomes that might possibly be observed when a group of voters takes on the task of selecting a winning candidate from a set of available candidates. Books have been written to describe many of these paradoxical outcomes and to categorize them according to the types of unusual behaviors that they display. The categories of voting paradoxes that are defined by Nurmi (1999) are used in this current study. The most famous of these paradoxical voting outcomes is Condorcet's paradox, or the Condorcet effect, which is named after the renowned 18th century French mathematician-philosopher who formally described this phenomenon. We address this particular voting paradox at this point, so that it can be used as a basis for further discussion. Other voting paradoxes will be developed in detail later in the study.

A description of the phenomenon that is known as Condorcet's paradox begins with the definition of a given possible combination of voters' preferences for three candidates $\{A, B, C\}$ in an election. Voters are assumed to have complete and rational preference rankings on the candidates, and the six possible preference rankings that voters might have on the three candidates are listed in Fig. 7.1.

The n_i terms in Fig. 7.1 denote the number of voters who have the associated preference rankings on the candidates. That is, n_3 voters have Candidate B as most preferred, Candidate C as least preferred, and Candidate A as middle-ranked. With a total of n voters, $n = \sum_{i=1}^{6} n_i$. A *voting situation* is denoted by $\boldsymbol{n}$, and it defines

W.V. Gehrlein (✉)
University of Delaware, Newark, Delaware USA
e-mail: wvg@udel.edu

D. Lepelley
University of La Réunion, Saint-Denis, La Réunion, France
e-mail: dominique.lepelley@univ-reunion.fr

D.S. Felsenthal and M. Machover, *Electoral Systems*, Studies in Choice and Welfare,
DOI 10.1007/978-3-642-20441-8_7,

A	A	B	C	B	C
B	C	A	A	C	B
C	B	C	B	A	A
n_1	n_2	n_3	n_4	n_5	n_6

Fig. 7.1 A voting situation from a three-candidate election

a specific combination of voters' preference rankings, $(n_1, n_2, n_3, n_4, n_5, n_6)$, that could be observed.

Borda (1784) developed a procedure that extends the basic principle of majority rule in two-candidate elections to scenarios that involve three candidates, by looking at the basic majority rule relation as applied to pairs of candidates. Let AMB denote the event that Candidate A defeats B by *pairwise majority rule* (*PMR*) when only Candidates A and B are considered. By ignoring the relative position of C in the possible preference ranking for any of the individual voter's rankings in Fig. 7.1, we see that AMB if $n_1 + n_2 + n_4 > n_3 + n_5 + n_6$. By using the same basic logic, we then find that AMC if $n_1 + n_2 + n_3 > n_4 + n_5 + n_6$ and BMC if $n_1 + n_3 + n_5 > n_2 + n_4 + n_6$. Candidate A would be the *pairwise majority rule winner* (*PMRW*) if both AMB and AMC, which would make it an exceptionally good candidate for selection as the most preferred candidate according to the voters' preference rankings in the associated voting situation. The *pairwise majority rule loser* (*PMRL*) is then defined in the obvious way.

Condorcet (1785a) makes very strong arguments that the PMRW should always be chosen as the winner of an election, which has resulted in this principle being commonly referred to as the *Condorcet criterion.* But, Condorcet then continued on with an analysis of PMR relationships to make a fascinating discovery with his famous example of a voting situation with 60 voters on three candidates, as shown in Fig. 7.2:

Condorcet notes that we have a voting situation in this example that results in what he called a "*contradictory system*", and such an outcome has come to be widely known as representing *Condorcet's paradox.* In particular, we find that by using PMR comparisons with the voting situation in Fig. 7.2: *AMB* (33–27), *BMC* (42–18), and *CMA* (35–25). We therefore have a cycle in the PMR relations on the three candidates, so that no candidate emerges as being superior to each of the remaining candidates. Given Condorcet's strong arguments that the PMRW should always be selected as the winner, we are left with a difficult question in this case. In particular, "Which candidate should be selected as the winner?" No matter which candidate we select in this example, a majority of voters would prefer one of the other candidates for selection.

Condorcet wrote at length about the possibility that these cyclical majorities on pairs of candidates might occur, and he made some attempts to assess the likelihood that such outcomes might happen [Condorcet (1785b, c)]. A great deal of effort has been expended since Condorcet's early work to identify other voting paradoxes and to obtain probability representations for the likelihood that various

$$\begin{array}{ccccc} A & B & B & C & C \\ B & A & C & A & B \\ C & C & A & B & A \\ n_1 = 23 & n_3 = 2 & n_4 = 17 & n_5 = 10 & n_6 = 8\,. \end{array}$$

Fig. 7.2 A voting situation showing a PMR cycle from Condorcet (1785a)

voting paradoxes, including Condorcet's paradox, might be observed in election settings. The basic motivation behind this work has been to determine if any of these possible paradoxical events might actually pose a real threat to the legitimacy of elections. There have been significant advances in recent years in the modeling techniques that have been employed to develop these probability representations, and our objective here is to survey some of the results that have been obtained from the most elementary models that have been used. The primary goal is to show that significant results can indeed be obtained with analysis that is based on these elementary models, and we focus on outcomes for three-candidate elections.

The remainder of the paper is organized as follows. The next section discusses the classic assumptions that have been used to develop representations for the probability that various voting paradoxes are observed, while pointing out the differences and links that exist between these models. The third section develops arguments to support the relevance of the conclusions that are reached from the probability representations that are obtained with the assumptions that are based on these models. Attention is focused on the probability that Condorcet's paradox is observed. Sections 7.4, 7.5 and 7.6 extend this type of analysis to a number of other voting paradoxes that have been developed in the literature. The final section then summarizes the conclusions that can be reached from our analysis.

7.2 Calculating Probabilities for Observing Voting Paradoxes

The general procedure for calculating the probability that a voting paradox might be observed is quite direct. If we consider the example of Condorcet's paradox from the immediately preceding section, it is sufficient to enumerate all possible voting situations for a specified n, and identify the subset of all possible voting situations for which a PMR cycle exists. Then, the probability of observing Condorcet's paradox would be obtained by summing the probabilities that the individual voting situations in that subset will be observed. The outcome will obviously be completely driven by the specific mechanism that determines the probability with which each specific voting situation is observed.

Three probability models have formed the bulk of the traditional basis for assigning probabilities to voting situations: the dual culture condition (DC), the impartial culture condition (IC) and the impartial anonymous culture condition (IAC). We begin by describing each of these models. While doing this, some subtle differences between these models will be pointed out, along with the resulting impact that these differences will have on the characteristics of the voting situations that are obtained from them. Once that is complete, these models will then be analyzed to determine what they can tell us about the probability that a number of different voting paradoxes might be observed.

7.2.1 Dual Culture Condition

Specific voting situations are not obtained directly with the DC model. Instead, DC describes the probability that specific *voter preference profiles*, or *voter profiles*, will be observed. A voter profile identifies the specific preference ranking that each individual voter has for the candidates, so that each individual voter's preferences are not anonymous in a profile. However, once a voter profile has been established, it is very easy to accumulate the preferences of voters according to the possible preference rankings to obtain the associated voting situation for that profile. Since the preferences of specific voters can not be identified in a voting situation, the individual voter's preferences are anonymous in a voting situation.

The probability that any specific voter preference profile will be observed can be considered to be the result of a process that randomly generates n individual voter's preference rankings on the candidates. In this situation, we let $\boldsymbol{p}$ denote a six-dimensional vector $(p_1, p_2, p_3, p_4, p_5, p_6)$ for the three-candidate case, where p_i denotes the probability that a randomly selected voter from the population of potential voters will have the corresponding preference ranking on candidates that is shown in Fig. 7.3.

That is, a randomly selected voter will have a probability p_1 of having the linear preference ranking with Candidate A being most preferred, Candidate C being least preferred and Candidate B being middle-ranked. A very critical assumption is made at this point, in that each voter's preference ranking on candidates is assumed to be arrived at independently of the preferences of all other voters.

Following the traditional methods that are used in any analysis of this type of probability modeling, we start with an urn that contains some total number of balls, with each ball being one of six different colors. Each color corresponds to one of the six possible preference rankings on the three candidates. The proportions of the total number of balls of each color that are in the urn are equal to their associated probabilities for the specified $\boldsymbol{p}$. Then, balls are sequentially drawn at random from the urn over n different trials, with the selected ball being returned to

A	A	B	C	B	C
B	C	A	A	C	B
C	B	C	B	A	A
p_1	p_2	p_3	p_4	p_5	p_6

Fig. 7.3 Voter preference ranking probabilities for a three-candidate election

the urn after its color is noted on each draw. The random selection of balls is being done with replacement during the experiment so that the probability of observing any particular possible preference ranking for an individual voter does not change from draw to draw. The color of the ball that is drawn during the ith step of this sequential drawing is used to assign the associated preference ranking to the ith voter before the ball is placed back in the urn.

As noted previously, the voting situation, $\boldsymbol{n}$, that results from any given voter preference profile with its identifiable voters can be obtained simply by determining the number of voters in the voter preference profile that have each of the six possible preference rankings. The probability that any given $\boldsymbol{n}$ will be observed from the identifiable voters in such a randomly generated voter preference profile with this urn model is then given directly by the multinomial probability $n! \prod_{i=1}^{6} \frac{p_i^{n_i}}{n_i!}$.

The DC assumption represents a special case of $\boldsymbol{p}$ vectors such that the probability that a randomly selected voter will have any preference ranking on the candidates is the same as the probability that the same voter will have the dual, or inverted, preference ranking on the candidates, with $p_1 = p_6$, $p_2 = p_5$ and $p_3 = p_4$ in Fig. 7.3. In order to describe the context of DC, let $A \succ B$ denote the outcome that Candidate A is preferred to B in a specific voter's preference ranking on candidates.

Let $\Delta(A, B)$ denote the difference between the sum of the p_i values for preference rankings with $A \succ B$ and $B \succ A$. The same definition is extended in the obvious fashion to all pairs of candidates, so that:

$$\begin{aligned}
\Delta(A, B) &= p_1 + p_2 + p_4 - p_3 - p_5 - p_6 \\
\Delta(A, C) &= p_1 + p_2 + p_3 - p_4 - p_5 - p_6 \\
\Delta(B, C) &= p_1 + p_3 + p_5 - p_2 - p_4 - p_6. \qquad (7.1)
\end{aligned}$$

When each voter's preference ranking is independent of all of the other voters' preference rankings, a randomly selected voter will be more likely to have a preference ranking with $A \succ B$ than with $B \succ A$ whenever $\Delta(A, B) > 0$.

The law of large numbers requires that a randomly generated voting situation with $n \to \infty$ must have AMB if $\Delta(A, B) > 0$ for any pair of candidates like A and B. As a result, Candidate A will be the PMRW with probability approaching one whenever both $\Delta(A, B) > 0$ and $\Delta(A, C) > 0$, B will be the PMRW with probability approaching one if both $\Delta(A, B) < 0$ and $\Delta(B, C) > 0$, and C will be the PMRW with probability approaching one if both $\Delta(A, C) < 0$ and $\Delta(B, C) < 0$. There will be a PMR cycle $AMBMCMA$ with probability approaching one with $n \to \infty$ if each of $\Delta(A, B) > 0$, $\Delta(B, C) > 0$ and $\Delta(A, C) < 0$, and the reverse PMR cycle with $AMCMBMA$ will exist with probability approaching one if each of $\Delta(A, C) > 0$, $\Delta(B, C) < 0$ and $\Delta(A, B) < 0$.

By selectively constructing $\boldsymbol{p}$ to define the likelihood that randomly generated voting situations are observed from the urn experiment described above, it is easy to contrive situations as $n \to \infty$ for which either a PMRW must exist with near certainty, or for which a PMR cycle must exist with near certainty. However, there is a *complete balance* over all pairs of candidates on an expected value basis for a randomly selected voter when $\Delta(A, B) = \Delta(A, C) = \Delta(B, C) = 0$. When some $\boldsymbol{p}$ has this complete balance, it is neither intentionally forcing a PMRW to exist nor intentionally forcing a PMR cycle to exist. Moreover, it is easy to show that such a complete balance of individual voter's preferences on all pairs of candidates only exists with $\boldsymbol{p}$ vectors that meet the restriction of DC.

All of this leads to the conclusion that any results that are obtained with the assumption of DC represent a somewhat extreme case in which no candidate has any expected advantage whatsoever when the preferences on *pairs of candidates* are examined for a voter that is randomly selected from the population of voters. It is very important to emphasize that this balance of preferences applies to *individual voter's preferences on pairs of candidates* with DC. It does not preclude the possibility that some candidates might be ranked as most preferred, or least preferred, with greater likelihood than some other candidate in the preference ranking of a randomly selected voter. For example, DC applies to the case with $p_1 = p_6 = 1/2 - 2\varepsilon$ and $p_2 = p_3 = p_4 = p_5 = \varepsilon$, for small $\varepsilon > 0$, so that both Candidates A and C will frequently be ranked as either most preferred, or least preferred, in a randomly selected voter's preference ranking. Candidate B would therefore frequently be the middle-ranked candidate. However, the extreme case with $\varepsilon = 0$ would produce a scenario in which a PMRW must exist.

Another observation from the analysis of the assumption of complete balance of individual voter's preferences follows from a consideration of the resulting proportions of voters with preferences on pairs of candidates in a voting situation. That is, the proportion of voters with $A \succ B$ in a random voting situation will approach one-half with certainty as $n \to \infty$ if $\Delta(A, B) = 0$. The relative margins of all PMR wins and losses on pairs of candidates in voting situations will therefore be relatively small with a complete balance of preferences for individual voters. As a result, this will lead to an environment that is conducive to the occurrence of voting paradoxes that involve PMR cycles in voting situations. When the assumption of DC is being utilized, it can therefore be expected that exaggerated estimates will be obtained for the likelihood that voting paradoxes that involve PMR cycles will be observed in the resulting voting situations. But, it is important to stress that the DC assumption is neither forcing a PMRW to exist nor forcing a PMR cycle to exist as $n \to \infty$.

7.2.2 Impartial Culture Condition

The *impartial culture condition* (*IC*) is a refinement of DC which assumes that $p_i = 1/m!$ in an m-candidate election, so that each possible preference ranking on the candidates is equally likely to represent the preferences of a randomly selected voter. Since IC is a special case of DC, the preferences of any given voter are assumed to be independent of all other voters' preferences, and there is a complete expected balance of preferences on pairs of candidates for a randomly selected voter. The additional restriction of IC beyond DC requires that there is also a complete balance on the expected ranking position for all candidates, so that all candidates are equally likely to be most preferred, least preferred or middle ranked for a randomly selected voter. All of these assumptions make IC the "purest" assumption, since no candidate will have any advantage whatsoever when it is compared to any other candidates in the preference rankings of a randomly selected voter.

7.2.3 Impartial Anonymous Culture Condition

The *impartial anonymous culture condition* (*IAC*) is not based on the use of any particular $\boldsymbol{p}$ to generate a random voter preference profile that will then be used to obtain a random voting situation. Instead, the concept of IAC is based directly on the assumption that each possible voting situation with n voters is equally likely to be observed, with the probability of observing any voting situation being equal to $120/\prod_{i=1}^{5}(n+i)$. IAC also produces an expected balance of preferences on pairs of candidates. However, this balance does not apply to the preferences of specific individual voter's with IAC, it applies over all possible voting situations with anonymous voters. This balance follows from partitioning the set of all possible voting situations into pairs. To form a pair of voting situations in the partition, each voting situation is matched with the unique voting situation that interchanges voter preference rankings according to: $n_1 \leftrightarrow n_6$, $n_2 \leftrightarrow n_5$, and $n_3 \leftrightarrow n_4$.

This transformation matches every voting situation with its *dual voting situation*, which effectively reverses the preference ranking on candidates for every voter. Thus, for any two candidates, A and B, the number of voters with $A \succ B$ in one of the voting situations will have the same number of voters with $B \succ A$ in the matching voting situation in the partition. Since both voting situations are equally likely to be observed under IAC, there is an expected balance between the number of voters with $A \succ B$ and with $B \succ A$ within the pair of voting situations. This observation extends to all of the pairs of voting situations in the partition, since all voting situations are equally likely to be observed with IAC. In the event that $n_1 = n_6$, $n_2 = n_5$, and $n_3 = n_4$, the interchange of rankings matches the voting situation with itself. In this case, the difference in the number of rankings with $A \succ B$ and with $B \succ A$ is not cancelled out over a pair of equally likely voting situations, but within this particular voting situation.

If a voting situation is selected at random from the set of all possible voting situations with IAC, it is therefore equally likely that either AMB or BMA will be observed for all possible pairs of candidates. Estimates for the likelihood that voting paradoxes that involve PMR cycles will be observed can therefore be expected to be exaggerated with IAC. However, it is important to stress that the IAC assumption is neither forcing a PMRW to exist nor forcing a PMR cycle to exist. The DC assumption also requires that it is equally likely that AMB or BMA for all possible pairs of candidates in a voting situation, since it is based on the more restrictive requirement that it is equally likely to have $A \succ B$ or $B \succ A$ for each individual voter. IAC does not directly specify anything about the preferences of any individual voter.

The assumption that each voting situation has an equal likelihood of being observed as its dual voting situation is not sufficient on its own to conclude that the resulting likelihood of observing a paradox involving PMR cycles will be maximized. Consider the subset of all voting situations with $n_3 = n_4 = 0$, and add the IAC-like assumption that each such voting situation is equally likely to be observed. The use of the same transformation for voting situations from above will show that it is equally likely to have AMB or BMA over this particular subset of voting situations. When $n_3 = n_4 = 0$, Candidate A is never the middle-ranked candidate in voters' preferences, and it follows directly that A must either be the PMRW or the PMRL with an odd number of voters. Thus, the probability that a PMR cycle exists must be zero, despite the fact that it is equally likely to have AMB or BMA over this particular subset of voting situations.

The IAC assumption represents a very simple concept, and it can take on some other equivalent interpretations. For example, Berg (1985) shows an interesting connection between IC and IAC through a discussion of *Pólya–Eggenberger* $(P\text{–}E)$ probability models. These models are best described in the context of generating random voter preference profiles by drawing colored balls from an urn, following earlier discussion. The experiment starts with balls of six different colors being placed in the urn. For each possible individual preference ranking, there are A_i balls of the particular color that corresponds to the ith possible individual preference ranking, and the A_i values vary according the components of the predetermined $\boldsymbol{p}$. A ball is drawn at random and the corresponding individual preference ranking is assigned to the first voter. The ball is then replaced, just as in the original experiment, but now α additional balls of the same color as the drawn ball are also placed into the urn. A second ball is then drawn, the corresponding ranking for its color is assigned to the second voter, and the ball is replaced along with α additional balls of the same color as the drawn ball. The process is repeated over n trials to obtain an individual preference ranking for each of the n voters. When $\alpha > 0$, the color of the ball that is drawn for the first voter will have an increased likelihood of representing the color of the ball that is drawn for the second voter, and so on.

These P–E-based *contagion models* create an increasing degree of dependence among the voters' preferences as α increases. However, there is no dependence among voters' preferences for the particular case with $\alpha = 0$. The special case of a P–E model in which $A_i = 1$ for $1 \le i \le 6$ is obviously identical to the assumption of IC when $\alpha = 0$. The particularly interesting observation from Berg (1985) is that the same special case of a P–E model is equivalent to IAC when $\alpha = 1$, so that IAC inherently requires the presence of some degree of dependence among voters' preferences in voting situations.

Another interesting connection between IAC and the *uniform culture condition* (*UC*) is developed in Gehrlein (1981). We have described how the probability that a voting paradox will be observed can be calculated for a specified $\boldsymbol{p}$ that describes the probability that n individual independent voters will have the preference rankings in Fig. 7.3. UC assumes that each such $\boldsymbol{p}$ with $\sum_{i=1}^{6} p_i = 1$ is equally likely to be observed. For any voting paradox, different probabilities will be obtained for observing that paradox with different $\boldsymbol{p}$. But, if we consider the expected value of the probability that this voting paradox will be observed over all possible $\boldsymbol{p}$ with UC, the result will be identical to the probability that is obtained for n voters with IAC.

7.3 Relevance of DC, IC and IAC Based Probability Models

It was mentioned previously that an extensive amount of research has been conducted to develop probability representations for the likelihood that various voting paradoxes will occur with the assumptions of DC, IC and IAC; and it is obviously of interest to discuss the relevance of the probability estimates that result from such studies. This is particularly true since a number of recent studies have clearly raised this issue after performing empirical analysis to reach the conclusion that the distribution of voters' preferences in most election results do not correspond to anything like DC, IC or IAC. The most notable empirical studies of this type include Regenwetter et al. (2006) and Tideman and Plassmann (2008). We shall see that there are in fact many very good reasons to explain why it is indeed very relevant to consider the results that are obtained with such probability models.

7.3.1 General Arguments

A number of general arguments that support investigations that are based the use of assumptions like DC, IC and IAC to develop probability representation are summarized in Gehrlein and Lepelley (2004), given the fact that we have already determined that they are likely to represent scenarios that exaggerate the probability that paradoxical voting events that involve PMR relationships will occur:

1. They are very useful when large amounts of relevant empirical data are not available, which is typically the case when analyzing elections.
2. They can show that some paradoxical events are very unlikely to be observed. That is, if we use conditions that tend to exaggerate the likelihood of observing paradoxes to find that the probability for some paradox is small with such calculations, then this paradox is assuredly very unlikely to be observed in reality.
3. They can suggest the relative impact that paradoxical events can have on different types of voting situations. For example, different voting rules can be compared on the basis of their relative likelihood of electing the PMRW.
4. By using such probability models to obtain closed form representations, it is easy to observe the impact of varying specific parameters of voting situations or voter preference profiles, which is more difficult to do with other approaches.

5. The probability representations that are obtained are directly reproducible and verifiable with mathematical analysis, which is not as simple to do with other approaches.
6. Analysis of this type can be useful to find out if the relative probabilities of paradoxical outcomes on various voting mechanisms behave in a consistent fashion over a number of different assumptions about the likelihood that voting situations or voter preference profiles are observed.

With very few exceptions, actual elections are only conducted with one voting rule being used, and it typically is not at all easy to compare the resulting election outcome to what else might have happened if some different voting rule had been used. In fact, it is not always straightforward to determine exactly what actually did happen in an election, based only on typical election results. Fishburn (1980) considers the restrictions under which it is possible simply to determine whether or not the PMRW has been selected as the winner of an election, based only on the reported vote counts from the election. It is assumed that voters have weak ordered preferences on candidates, and assumptions are established to define admissible voting behavior. The severity of these restrictions leads Brams and Fishburn (1983a, p. 95) to conclude that

> Because of the varieties of strategies that are allowed and the paucity of detail about how people voted, the likelihood of concluding that the winner is a (PMRW)... is often small if not zero.

As a result, other factors about voting behavior must typically be assumed with some model that reconstructs the preferences of voters from the reported ballot outcomes in an election, simply to determine which candidate was the PMRW, and to determine what might have happened if a different voting rule had been used. The significant difficulties that can arise from making such assumptions in these models that reconstruct voters' preferences are pointed out in the conclusion of an empirical study by Regenwetter et al. (2002, p. 461)

> Similarly, we conclude from the analysis of four ... data sets ... that even the most basic and subtle changes in modeling approaches can affect the outcome on any analysis of voting or ballot data against the Condorcet criterion.

This conclusion was reached when actual data sets were examined to determine the resulting PMR ranking on candidates with the use of a very basic and plausible model, and it was found that very different rankings could be obtained with very minor changes in a preference threshold parameter in their model. It can therefore be concluded that completely abandoning theoretical models to pursue empirical studies that are based only on the voting results from typical elections can often be expected to lead to a new set of problems regarding the validity of the results.

We now proceed to develop some of the types of basic results that can be obtained by analyzing probability representations that are obtained with the simple assumptions of DC, IC and IAC.

7.3.2 Results from the DC Assumption

As a specific example of the some of the types of analyses that are suggested in the list of general arguments that is presented above, we consider some results that follow from probability representations that Condorcet's Paradox is observed with the assumption of DC. Let $P^S_{PMRC}(3, n, DC)$ denote the probability that a Strict PMR cycle, or an occurrence of Condorcet's paradox, is observed in a three-candidate election with an odd number n voters for a specified $\boldsymbol{p}$ vector from the subspace of DC. A Strict PMR relationship indicates that no PMR ties exist on any of the pairs of candidates. A representation for $P^S_{PMRC}(3, n, DC)$ follows directly from related work in Gehrlein and Fishburn (1976a) as given in (7.2)

$$P^S_{PMRC}(3, n, DC) =$$

$$1 - \sum_{m_1=0}^{\frac{n-1}{2}} \sum_{m_2=0}^{\frac{n-1}{2}-m_1} \sum_{m_3=0}^{\frac{n-1}{2}-m_1} \frac{n!}{m_1!m_2!m_3!m_4!} \left\{ \begin{array}{l} \left(\frac{1}{2} - p_3\right)^{n-m_2-m_3} p_3^{m_2+m_3} + \\ \left(\frac{1}{2} - p_1\right)^{n-m_2-m_3} p_1^{m_2+m_3} + \\ \left(\frac{1}{2} - p_2\right)^{n-m_2-m_3} p_2^{m_2+m_3} \end{array} \right\}, \quad (7.2)$$

for odd n with $m_4 = n - m_1 - m_2 - m_3$.

The limiting case for large electorates as $n \to \infty$ is addressed in Fishburn and Gehrlein (1980) to lead to a representation for $P^S_{PMRC}(3, \infty, DC)$ in (7.3)

$$P^S_{PMRC}(3, \infty, DC) = \frac{1}{4} - \frac{1}{2\pi} \sum_{j=1}^{3} Sin^{-1}(1 - 4p_j). \quad (7.3)$$

Computed values of $P^S_{PMRC}(3, \infty, DC)$ that are obtained from (7.3) are listed in Table 7.1 for each $p_1, p_2 = .000(.025).500$. The range of values in Table 7.1 is truncated since it is obvious from (7.3) that $P^S_{PMRC}(3, \infty, DC)$ is invariant to permutations of p_1, p_2 and p_3.

The limiting probability values in Table 7.1 show that $P^S_{PMRC}(3, \infty, DC)$ goes to zero if any of p_1, p_2 or p_3 is equal to zero, it is also proved that $P^S_{PMRC}(3, \infty, DC)$ is maximized for the special case of IC, with $p_i = 1/6$ for $1 \leq i \leq 6$, for $\boldsymbol{p}$ that are consistent with the assumption of DC. It has already been concluded that DC can be expected to produce exaggerated estimates of the probability that paradoxical outcomes that involve PMR relationships will be observed. By adding the fact that that $P^S_{PMRC}(3, \infty, DC)$ is maximized with IC suggests that $P^S_{PMRC}(3, n, IC)$ estimates are very likely to produce significant overestimates of the likelihood with which Condorcet's paradox can be expected to be observed.

Table 7.1 Computed values of $P_{PMRC}^{S}(3, \infty, DC)$

p_1	p_2										
	0.000	0.025	0.050	0.075	0.100	0.125	0.150	0.175	0.200	0.225	0.250
0.000	0.000	0.000	0.000	0.000	0.000	0.000	0.000	0.000	0.000	0.000	0.000
0.025	0.000	0.041	0.048	0.051	0.053	0.054	0.055	0.055	0.056	0.056	0.056
0.050	0.000	0.048	0.057	0.062	0.065	0.068	0.069	0.070	0.070	0.071	0.070
0.075	0.000	0.051	0.062	0.069	0.073	0.075	0.077	0.078	0.079	0.079	0.078
0.100	0.000	0.053	0.065	0.073	0.077	0.080	0.082	0.083	0.083	0.083	0.082
0.125	0.000	0.054	0.068	0.075	0.080	0.083	0.085	0.086	0.086	0.085	0.083
0.150	0.000	0.055	0.069	0.077	0.082	0.085	0.087	0.088	0.087	0.085	0.082
0.175	0.000	0.055	0.070	0.078	0.083	0.086	0.088	0.088	0.086	0.083	0.078
0.200	0.000	0.056	0.070	0.079	0.083	0.086	0.087	0.086	0.083	0.079	0.070
0.225	0.000	0.056	0.071	0.079	0.083	0.085	0.085	0.083	0.079	0.071	0.056
0.250	0.000	0.056	0.070	0.078	0.082	0.083	0.082	0.078	0.070	0.056	0.000
0.275	0.000	0.056	0.070	0.077	0.080	0.080	0.077	0.070	0.056	0.000	
0.300	0.000	0.055	0.069	0.075	0.077	0.075	0.069	0.055	0.000		
0.325	0.000	0.055	0.068	0.073	0.073	0.068	0.055	0.000			
0.350	0.000	0.054	0.065	0.069	0.065	0.054	0.000				
0.375	0.000	0.053	0.062	0.062	0.053	0.000					
0.400	0.000	0.051	0.057	0.051	0.000						
0.425	0.000	0.048	0.048	0.000							
0.450	0.000	0.041	0.000								
0.475	0.000	0.000									
0.500	0.000										

It follows directly from (7.3) that

$$P^{S}_{PMRC}(3, \infty, IC) = \frac{1}{4} - \frac{3}{2\pi} Sin^{-1}\left(\frac{1}{3}\right) \approx .088. \tag{7.4}$$

This result indicates that a significant overestimate of the probability that Condorcet's paradox will be observed in a three-candidate election is approximately nine percent. We can therefore conclude that such observations should actually be infrequent phenomena, which follows the logic of the second general argument in the list that is given above. Moreover, this outcome is completely consistent with many empirical results that indicate that while Condorcet's paradox is not a commonly observed election outcome, it does occasionally occur. A thorough survey of these empirical studies is Gehrlein and Lepelley (2011).

In general, $P^{S}_{PMRC}(m, n, IC)$ values should be viewed as an upper bound on $P^{S}_{PMRC}(m, n, \boldsymbol{p})$ when $\boldsymbol{p}$ vectors are not biased either to produce a PMR cycle or to produce a PMRW. They have never been intended to produce estimates of the probability that Condorcet's paradox would ever be observed in any actual voting scenario, but they can tell us a great deal about the likelihood of extreme cases.

The calculated $P^{S}_{PMRC}(3, \infty, DC)$ values in Table 7.1 indicate that there is a great deal of variability over the range of $\boldsymbol{p}$ vectors in DC, and it is natural to wonder if there is some natural underlying explanation for this variation. Many studies have been conducted to evaluate the impact that various measures of the consistency of voters' preferences in a population will have on the probability that a PMRW exists. It is intuitively appealing to speculate that paradoxical voting outcomes should become less likely to be observed as a population of voters tends to have preferences that are more mutually consistent. This degree of the consistency of voters' preferences can be defined in the context of *social homogeneity*. The preferences of a population of voters would be totally homogeneous if every member of that society had exactly the same preference ranking on the candidates. The opposite extreme is a situation that reflects a situation like IC, where the individual voters have preferences that are completely dispersed over all possible preference rankings on the candidates.

Simple measures of the amount of dispersion among the p_i terms in $\boldsymbol{p}$ vectors have been used as a gauge of the amount of social homogeneity that exists among voters' preferences in a population. Abrams (1976) considers one such measure of homogeneity for three-candidate elections, with

$$H(\mathbf{p}) = \sum_{i=1}^{6} p_i^2. \tag{7.5}$$

$H(\boldsymbol{p})$ is maximized when $p_i = 1$ for some ranking, so that all voters will have identical preference rankings on candidates, and it is minimized with the assumption of IC, with $p_i = 1/6$ for all $1 \leq i \leq 6$. Increased values of $H(\boldsymbol{p})$ will generally tend to reflect increased levels of homogeneity for a population of voters. With a large value of $H(\boldsymbol{p})$, we would expect an increased likelihood of observing random voting situations from such a population that have voters' preferences that are clustered around one, or a few, of the possible linear rankings on candidates. As $H(\boldsymbol{p})$ increases, intuition therefore suggests that $P^S_{PMRC}(3, n, \boldsymbol{p})$ should also be expected to decrease.

Fishburn and Gehrlein (1980) show that $P^S_{PMRC}(3, \infty, DC)$ decreases as $H(\boldsymbol{p})$ increases for $\boldsymbol{p}$ vectors in DC when $H(\boldsymbol{p})$ is changed by keeping one of p_1, p_2 or p_3 fixed while changing the other two. Of course, p_4, p_5 and p_6 must also change accordingly to keep $\boldsymbol{p}$ in accord with the definition of DC. An expected negative relationship is also found between $H(\boldsymbol{p})$ and $P^S_{PMRC}(3, n, \boldsymbol{p})$ for general $\boldsymbol{p}$ with independent voters, but this relationship does tend to deteriorate as the number of voters gets very large.

An important conclusion can be reached from these two observed relationships that exist between $H(\boldsymbol{p})$ and $P^S_{PMRC}(3, n, \boldsymbol{p})$ for $\boldsymbol{p}$ vectors with independent voters. In particular, the impact of any possible dependence among voters' preferences is completely eliminated as a potential component of an explanation of the source of this relationship. This provides an example application of the fourth item in the list of general arguments for developing such representations above, since it allows for an analysis of the impact of varying just one specific parameter of voting situations. That is, a relationship exists between $H(\boldsymbol{p})$ and $P^S_{PMRC}(3, n, \boldsymbol{p})$ when the direct impact of dependence among voters' preferences is excluded.

7.3.3 Results from the IAC Assumption

A representation for the probability $P^S_{PMRC}(3, n, IAC)$ that Condorcet's paradox is observed for odd n with the assumption of IAC is directly obtainable from a result in Gehrlein and Fishburn (1976b), with

$$P^S_{PMRC}(3, n, IAC) = \frac{(n-1)(n+7)}{16(n+2)(n+4)}, \quad \text{for odd } n. \tag{7.6}$$

Two interesting observations can be made by considering the limiting result as $n \to \infty$ in (7.6), with $P^S_{PMRC}(3, \infty, IAC) = 1/16$. The first of these observations goes back to a consideration of the relationship between the assumptions of IAC and UC that was discussed above. That is, if all possible $\boldsymbol{p}$ vectors are equally likely to be observed as $n \to \infty$, then the expected value of $P^S_{PMRC}(3, \infty, \boldsymbol{p})$ is 1/16, or about six percent. This again verifies that the IC assumption, which leads to $P^S_{PMRC}(3, \infty, IC)$ equal to about nine percent, gives an exaggerated estimate of the probability that Condorcet's paradox will be observed. Since the proportion of all possible $\boldsymbol{p}$ vectors that meet the restrictions of DC is of measure zero, this observation and previous discussion jointly lead to an alternative form of this conclusion. That is, only 1/16 of all possible $\boldsymbol{p}$ vectors will result in an observation of Condorcet's paradox as $n \to \infty$, while 15/16 of all possible $\boldsymbol{p}$ vectors will result in the existence of a PMRW.

The second general observation that stems from (7.6) follows from a comparison of the limiting results that are obtained for the probability that Condorcet's paradox is observed with IC and with IAC. The limiting probability is approximately nine percent with IC, while it is reduced to approximately six percent with IAC. Both of these assumptions were found to result in an expected balance in PMR comparisons on all pairs of candidates, so it is natural to wonder what else remains to explain the difference. The earlier discussion of P–E probability models indicated that the difference between IC and IAC stems from the fact that IAC introduces a degree of dependence among voters' preferences while IC does not do so. As a result, the presence of a degree of dependence among voters' preferences can be isolated as a cause for reducing the probability that Condorcet's paradox will be observed. This observation must be balanced with an understanding that some link must be expected to exist between social homogeneity and the degree of dependence among voters' preferences.

We have said a great deal about how these different probability models can be used to analyze the likelihood that Condorcet's paradox might be observed, and to isolate different parameters of voting situations that will have an impact on this probability. Attention will now be focused on what has been discovered much more recently by applying these same techniques in the consideration of other voting paradoxes.

7.4 Incompatibility Paradoxes

Incompatibility paradoxes occur in voting situations when there are multiple definitions as to which candidate should be viewed as being the "best" possible candidate among the set of available candidates, and where these definitions cannot be satisfied simultaneously by a voting rule. Condorcet's paradox reflects one such incompatibility paradox. Two other incompatibility paradoxes are Borda's paradox and Condorcet's other paradox, and we consider the relative likelihood that each will be observed.

7.4.1 *Borda's Paradox*

Borda (1784) presented another early example of a voting paradox. The background of Borda's paradox is associated with the use of a weighted scoring rule (WSR) to determine the winner of an election. A WSR is defined in terms of weights $(1, \lambda, 0)$, with $1 \geq \lambda \geq 0$ for a three-candidate election. Each voter assigns a score of one to their most preferred candidate, a score of zero to their least preferred candidate and a score of λ to their middle-ranked candidate. The WSR winner is the candidate that receives the most total points from all voters. The most commonly use WSR is *plurality rule* (*PR*) with $\lambda = 0$, such that each voter gives a score of one to their most preferred candidate. Let APB denote the outcome that Candidate A beats B by PR. A variation of this voting rule is *negative plurality rule* (*NPR*) with $\lambda = 1$, such that each voter casts a vote for their two more preferred candidates. This is equivalent to having each voter cast a vote against some candidate, where the candidate with the fewest negative votes is declared the winner. Let ANB denote the outcome that Candidate A beats B by NPR.

Borda presented an example voting situation in Fig. 7.4 for 21 voters.

If PR is used with the voting situation that is shown in Fig. 7.4, we have the outcomes APB (8–7), APC (8–6) and BPC (7–6) to give a linear ranking by PR, with $APBPC$. A very different and very paradoxical result is observed with the use of PMR. Here, BMA (13–8), CMA (13–8) and CMB (13–8) to give a linear PMR ranking, with $CMBMA$. With this particular voting situation, PR and PMR completely reverse the election rankings on the three candidates. This specific phenomenon is referred to as representing an occurrence of a *strict Borda paradox*.

Borda was particularly concerned about the fact that the PMRL could be chosen as the winner by PR, leading to his suggestion that PR should never be used. Given this primary source of concern, a less stringent *strong Borda paradox* is defined as a situation in which PR elects the PMRL, without necessarily having a complete reversal in PR and PMR rankings. Both forms of Borda's paradox can obviously be observed with other voting rules than PR. Borda proposed a procedure that he referred to as "*election by order of merit*", that has come to be widely known as *Borda rule* (*BR*), to deal with this type of situation, and BR is equivalent to a WSR with $\lambda = 1/2$. For a general voting situation, as described in Fig. 7.1, with n voters and three candidates, the *Borda score* for Candidates A, B and C under BR would respectively be $BS(A)$, $BS(B)$ and $BS(C)$ with:

$$\begin{aligned} BS(A) &= (n_1 + n_2) + (n_3 + n_4)/2 \\ BS(B) &= (n_3 + n_5) + (n_1 + n_6)/2 \\ BS(C) &= (n_4 + n_6) + (n_2 + n_5)/2. \end{aligned} \tag{7.7}$$

For the particular example in Fig. 7.4, we obtain $BS(C) = 13$, $BS(B) = 10.5$, and $BS(A) = 8$. If we let ABB denote the event that Candidate A beats B by BR, we get a linear ranking on the candidates, with $CBBBA$. This ranking of candidates by BR is now in the reverse order of the ranking by PR, and it is in perfect agreement with the ranking that was obtained by PMR. It has since been proved that BR can never elect the PMRL as the unique winner, so it is completely resistant to the possibility of exhibiting both a strict Borda paradox and a strong Borda paradox.

A	A	B	C
B	C	C	B
C	B	A	A
$n_1 = 1$	$n_2 = 7$	$n_5 = 7$	$n_6 = 6$.

Fig. 7.4 An example voting situation from Borda (1784)

However, every WSR other than BR can exhibit both of these this phenomena, and representations have been obtained for the associated limiting probabilities for each in Diss and Gehrlein (2009).

Let $P_{StBP}^{WSR(\lambda)}(3,\infty,IC^{*})$ denote the conditional limiting probability as $n \to \infty$ that a strict Borda paradox is observed with a WSR that uses weights $(1,\lambda,0)$ under the IC* assumption, which adds the additional restriction that a strict PMRW must exist to the IC assumption. When there are only three candidates, a requirement that a strict PMRW exists is equivalent to a requirement that a strict PMR ranking exists for odd n. Then, $P_{StBP}^{WSR(\lambda)}(3,\infty,IAC^{*})$ is defined in the same fashion. It is proved that $P_{StBP}^{WSR(\lambda)}(3,\infty,IC^{*}) = P_{StBP}^{WSR(1-\lambda)}(3,\infty,IC^{*})$, and the same relationship is also valid with IAC. Computed values of both $P_{StBP}^{WSR(\lambda)}(3,\infty,IC^{*})$ and $P_{StBP}^{WSR(\lambda)}(3,\infty,IAC^{*})$ are listed in Table 7.2 for each $\lambda = .00(.05).50$.

The results from Table 7.2 indicate that $P_{StBP}^{WSR(\lambda)}(3,\infty,IC^{*})$ and $P_{StBP}^{WSR(\lambda)}(3,\infty,IAC^{*})$ both decrease as λ increases for the interval $0 \le \lambda \le .5$, so that the likelihood of the outcome is maximized by both PR and NPR. However, these probabilities are typically less than one percent in all cases. Given that the IC and IAC scenarios can be expected to exaggerate the probability that paradoxical events that involve PMR relationships will be observed, it can easily be concluded that actual observations of a strict Borda paradox should be very rare events, which is completely consistent with empirical studies. A survey of these empirical studies is given in Gehrlein and Lepelley (2011). Since these probabilities are so small, no really significant differences can be observed between the cases of IC and IAC from Table 7.2.

The definition of a strong Borda paradox specifies conditions that are not as stringent as the requirements for a strict Borda paradox, so it is obvious that it should have a greater probability of being observed. Computed values of the limiting conditional probabilities $P_{SgBP}^{WSR(\lambda)}(3,\infty,IC^{*})$ and $P_{SgBP}^{WSR(\lambda)}(3,\infty,IAC^{*})$ that a strong Borda paradox is observed with IC and IAC respectively are listed in Table 7.3 for each $\lambda = .00(.05).50$. As before, these IC* and IAC* representations are conditional on the existence of a PMRW.

Table 7.2 Computed values of $P_{StBP}^{WSR(\lambda)}(3,\infty,IC^{*})$ and $P_{StBP}^{WSR(\lambda)}(3,\infty,IAC^{*})$ ^

λ	$P_{StBP}^{WSR(\lambda)}(3,\infty,IC^{*})$	$P_{StBP}^{WSR(\lambda)}(3,\infty,IAC^{*})$
0.00	0.0126	0.0111
0.05	0.0100	0.0091
0.10	0.0077	0.0073
0.15	0.0057	0.0056
0.20	0.0039	0.0040
0.25	0.0024	0.0027
0.30	0.0013	0.0016
0.35	0.0006	0.0008
0.40	0.0002	0.0003
0.45	0.0000	0.0000
0.50	0.0000	0.0000

^From Diss and Gehrlein (2009)

Table 7.3 Computed values of $P_{SgBP}^{WSR(\lambda)}(3,\infty,IC^*)$ and $P_{SgBP}^{WSR(\lambda)}(3,\infty,IAC^*)$ ∧

λ	$P_{SgBP}^{WSR}(\lambda)$ $(3,\infty,IC^*)$	$P_{SgBP}^{WSR}(\lambda)$ $(3,\infty,IAC^*)$	λ	$P_{SgBP}^{WSR(\lambda)}$ $(3,\infty,IC^*)$	$P_{SgBP}^{WSR(\lambda)}$ $(3,\infty,IAC^*)$
0.00	0.0371	0.0296	0.50	0.0000	0.0000
0.05	0.0303	0.0242	0.55	0.0001	0.0002
0.10	0.0238	0.0192	0.60	0.0007	0.0013
0.15	0.0179	0.0146	0.65	0.0021	0.0033
0.20	0.0126	0.0105	0.70	0.0046	0.0061
0.25	0.0081	0.0070	0.75	0.0081	0.0096
0.30	0.0046	0.0042	0.80	0.0126	0.0136
0.35	0.0021	0.0021	0.85	0.0179	0.0178
0.40	0.0007	0.0007	0.90	0.0238	0.0223
0.45	0.0001	0.0001	0.95	0.0303	0.0269
0.50	0.0000	0.0000	1.00	0.0371	0.0315

∧ From Diss and Gehrlein (2009)

Similar to observations from Table 7.2 for a strict Borda paradox, it is seen that $P_{SgBP}^{WSR(\lambda)}(3,\infty,IC^*) = P_{SgBP}^{WSR(1-\lambda)}(3,\infty,IC^*)$, but this symmetry relationship is no longer valid for a strong Borda paradox with IAC. The probabilities in Table 7.3 are obviously greater than the associated probabilities in Table 7.2, and they are maximized with the use of NPR for both IC and IAC, with PR having a marginally smaller probability than NPR for IAC. However, all of these probabilities remain less than four percent in all cases. This indicates that observations of a strong Borda paradox should be unlikely events, which is consistent with results from empirical studies that show that they do occasionally occur. The increase in dependence among voters' preferences that is inherent to the IAC assumption reduces the already small probabilities of observing a strong Borda paradox with the assumption of IC for all $0 \leq \lambda \leq .5$. But, there are some instances in which the IAC probabilities are actually greater than the associated IC probabilities when $\lambda > .5$.

The results in Table 7.3 are taken directly from Diss and Gehrlein (2009), but they could also have been obtained from an earlier representation in Tataru and Merlin (1997) for IC*.

Since we have already concluded that IC and IAC based probabilities can be expected to exaggerate the likelihood of observing paradoxes that are based on PMR relationships, neither of these two forms of Borda's paradox can be viewed as posing a significant threat to typical voting scenarios with a small number of candidates. This conclusion follows, despite the fact that neither IC nor IAC is ever expected to mirror the reality of any given election.

7.4.2 Condorcet's Other Paradox

Condorcet (1785d) gives the example voting situation in Fig. 7.5 to show a phenomenon that has come to be known as *Condorcet's other paradox*.

A	A	B	C	B	C
B	C	A	A	C	B
C	B	C	B	A	A
$n_1 = 30$	$n_2 = 1$	$n_3 = 29$	$n_4 = 10$	$n_5 = 10$	$n_6 = 1$.

Fig. 7.5 A voting situation from Condorcet (1785d)

Condorcet notes that AMB (41–40) and AMC (61–20) in this voting situation, so that Candidate A is the PMRW, and then goes on to compute $Score(A, \lambda)$ and $Score(B, \lambda)$ for Candidates A and B when the WSR with weights $(1, \lambda, 0)$ is used, and:

$$\begin{aligned} Score(A, \lambda) &= 1^*31 + \lambda^*39 + 0^*11 \\ Score(B, \lambda) &= 1^*39 + \lambda^*31 + 0^*11. \end{aligned} \tag{7.8}$$

In order for Candidate A to be elected by this WSR, we must therefore have:

$$\begin{aligned} &Score(A, \lambda) > Score(B, \lambda) \\ &31 + 39\lambda > 39 + 31\lambda \\ &8\lambda > 8 \\ &\lambda > 1. \end{aligned} \tag{7.9}$$

This contradicts the basic definition of a WSR, so that no WSR, including BR, can elect the PMRW in this example voting situation, which is Condorcet's other paradox. This observation led Condorcet to the conclusion that no WSR should ever be used to determine the winner of an election.

It is of definite interest to obtain some estimate of the relative probability with which this paradox might be observed, since it has a highly significant impact on the relevance of using a WSR. Merlin et al. (2002) obtain a limiting conditional representation as $n \to \infty$ for the probability that a similar event is observed in a three-candidate election, given that a PMRW exists. They consider the probability that a given candidate that is not the PMRW will be the winner over the range of all possible WSR's with $1 \geq \lambda \geq 0$. With the assumption of IC, this limiting probability is estimated to be 0.01808. Given that IC will tend to create voting situations that have a PMRW with relatively small PMR margins over other candidates, this gives an estimate for scenarios in which Condorcet's other paradox should be much more likely to be observed. And, we find that this probability is still small for this scenario that is expected to exaggerate it.

Gehrlein and Lepelley (2009) obtain a different representation for this limiting conditional probability and find a very similar numerical result with IC. Moreover, a limiting representation is also found with IAC, and the resulting conditional probability is reduced to $19/1{,}620 = 0.01173$. So, the already small IC related probability is further reduced with the introduction of some degree of dependence among voters' preferences with IAC. More relaxed conditions are also introduced to consider probabilities that are more closely associated with the pure definition of Condorcet's other paradox, but very little change resulted in the associated probabilities that have just been given. It therefore follows that there is very little reason to expect that Condorcet's other paradox would ever be observed in any realistic three-candidate election.

7.5 Monotonicity Paradoxes

Monotonicity paradoxes represent situations in which some reasonable definition has been established to determine which candidate should be viewed as being the "best" available candidate, and where a voting rule has been selected and that voting rule is not monotonic. *Monotonicity* of a voting rule requires consistency of election outcomes as voters' preferences change. That is, increased support (decreased support) for a candidate in voters' preferences should not be detrimental (beneficial) to that candidate in the election outcome. The No Show Paradox is one specific type of a monotonicity paradox.

The *no show paradox* is developed in Brams and Fishburn 1983b, with an example in which some subset of voters chooses not to participate in an election, and then prefers the resulting winner to the winner that would have been selected if they had actually participated in the election. The winner of an election is determined by *negative plurality elimination rule* (*NPER*) in a three-candidate election in this example. In the first stage, voters cast votes according to NPR. The candidate that receives the fewest number of votes is then eliminated, and the ultimate winner is selected in the second stage by using PMR on the remaining two candidates.

Consider a voting situation with 21 voters and three candidates, as shown in Fig. 7.6.

In the first stage of NPR voting, Candidates A, B, and C receive 15, 14 and 13 votes respectively. Candidate C is therefore eliminated in the first stage of voting and then BMA by a vote of 11–10 in the second stage, to select B as the overall election winner.

Voters with the linear preference ranking $A \succ B \succ C$ would not get their most preferred candidate in this situation, since B is the ultimate election winner. But, suppose that two of these particular voters had not participated in this election for some reason. The resulting voting situation for the 19 remaining voters is shown in Fig. 7.7.

A	A	B	C	B	C
B	C	A	A	C	B
C	B	C	B	A	A
$n_1 = 3$	$n_2 = 5$	$n_3 = 5$	$n_4 = 2$	$n_5 = 3$	$n_6 = 3$.

Fig. 7.6 An example voting situation from Brams and Fishburn (1983b)

A	A	B	C	B	C
B	C	A	A	C	B
C	B	C	B	A	A
$n_1 = 1$	$n_2 = 5$	$n_3 = 5$	$n_4 = 2$	$n_5 = 3$	$n_6 = 3$.

Fig. 7.7 The modified example voting situation from Brams and Fishburn (1983b)

Table 7.4 Probability values for $P_{NSP}^{VR}(3, \infty, IC)$ and $P_{NSP}^{VR}(3, \infty, IAC)$

VR	$P_{NSP}^{VR}(3, \infty, IC)$	$P_{NSP}^{VR}(3, \infty, IAC)$
PER	0.0558	0.0408
NPER	0.1623	0.0425
BER	0.0502	0.0243

In the first stage of NPR voting with this modified voting situation, Candidates A, B, and C respectively receive 13, 12 and 13 votes. Candidate B is eliminated in the first stage and then AMC by a vote of eleven to eight in the second stage. Since the winner in this modified voting situation is A, the two voters with linear preferences $A \succ B \succ C$ who did not participate will now have their most preferred candidate chosen as the winner. These two voters have therefore obtained a more preferred outcome from the election with NPER as a result of not participating in the election, which violates the definition of monotonicity.

Probability representations for the limiting probability $P_{NSP}^{VR}(3, \infty, IC)$ that the no show paradox is observed with the assumption of IC are obtained in Lepelley and Merlin (2001) for thee voting rules (VR). The analysis includes NPER, as described above, along with plurality elimination rule (PER) and Borda elimination rule (BER). PER and BER operate in the same fashion as NPER, by using PR and BR respectively in the initial stage to determine which candidate is eliminated in the first round of voting. Limiting representations for $P_{NSP}^{VR}(3, \infty, IAC)$ are also obtained for both PER and NPER. A representation for $P_{NSP}^{BER}(3, \infty, IAC)$ is obtained in Wilson and Pritchard (2007). All numerical results are summarized in Table 7.4.

Occurrences of a monotonicity paradox are very often associated to the presence of a PMR cycle in voting situations. Consequently, it should be expected that the introduction of some degree of homogeneity or dependence in voters' preferences will considerably reduce the vulnerability of WSR runoff systems to these paradoxes. This expectation is clearly shown to exist in Table 7.4, where the $P_{NSP}^{VR}(3, \infty, IC)$ probabilities are significantly greater than their associated $P_{NSP}^{VR}(3, \infty, IAC)$ probabilities, particularly for NPER. With the exception of the entry for $P_{NSP}^{NPER}(3, \infty, IC)$, all probabilities remain less likely than the probability that Condorcet's paradox will be observed with IC and IAC. The no show paradox should therefore have a relatively low probability of being observed, particularly with PER and BER.

The impact of using assumptions like IC for these probability calculations can also be considered from the fact that PMR is used on the second stage of all of these elimination rules. The use of IC will tend to support the generation of voting situations for large n such that there will be a relatively close PMR comparison in this second stage, so there will be a good chance of either of the two candidates being selected as the winner in the second stage, resulting in an exaggerated chance that the outcome in the second stage might be changed with the removal of some subset of voters' preferences from the election.

7.6 Choice Set Variance Paradoxes

Choice set variance paradoxes represent situations in which a series of propositions are put before voters, where each individual issue will be approved or disapproved by majority rule voting. A paradoxical result then arises when the overall final election outcome on the propositions represents a result that is somehow inconsistent with the underlying preferences of the voters. We consider two such paradoxes in the form of Ostrogorski's paradox and the majority paradox.

7.6.1 *Ostrogorski's Paradox*

Suppose that there are m independent issues that are to be presented to n voters and that each individual issue will be approved or disapproved by majority rule voting. There are two parties, R and L, that have opposing positions on each of the issues. Each voter therefore has a position that is in agreement with either Party R or Party L on each individual issue, but each voter does not necessarily agree with the position of the same party on every issue. A voter is considered to be a member of Party R (Party L) if their individual position on issues is in agreement with Party R (Party L) over a majority of the issues that are being considered. The outcome of voting on each issue will be determined to be in agreement Party R, or Party L, based on the majority rule outcome of voting on that issue.

Consider the example in Fig. 7.8 where each of five voters has preferences on three different issues.

The results in Fig. 7.8 indicate for example that Voter 1 has preferences on Issues 1 and 2 that are in agreement with Party L, while this voter has preferences that

	Voter					Position
Issue	1	2	3	4	5	Winner
1	L	L	R	R	R	R
2	L	R	L	R	R	R
3	R	L	L	R	R	R
Party Membership	L	L	L	R	R	

Fig. 7.8 An example voting situation from Bezembinder and Van Acker (1980)

are in agreement with Party R on Issue 3. Since Voter 1 is in agreement with Party L on a majority of issues by a 2–1 margin, this voter is listed as having a membership affiliation with Party L. Using this same logic, three of the five voters have a membership affiliation with Party$L$, to make it the *majority party* (*MP*) by a 3–2 margin. However, given the preferences of the voters on the issues, the position of Party R will win by a 3–2 majority margin on every issue. So, the position of Party R wins on every issue, while Party L is the MP.

Deb and Kelsey (1987) define this very contrary outcome as a *strict Ostrogorski paradox*, and it was first discussed in Ostrogorski (1902) . A less restrictive outcome of a *weak Ostrogorski paradox* occurs when Party R (Party L) is the MP, while a majority of election outcomes on issues are in agreement with the position of Party L (Party R).

Probability representations for the likelihood that various forms of Ostrogorski's paradox are observed are developed in Gehrlein and Merlin (2009a) with an application of the IC assumption. That is, each possible assignment of voters' preferences on the m issues, according to party positions, is assumed to be equally likely to be observed. This will tend to result in voting situations in which there is a small relative margin of victory for the determination of the MP as $n \to \infty$. Such a balanced outcome will make it easier for paradoxical outcomes to be observed on majority rule votes on the issues, compared to scenarios in which most voters are expected to have the same party membership.

Representations are obtained for the limiting probability $P^{\infty}_{MP}(m,k,IC)$ as $n \to \infty$ that the majority rule outcomes on exactly k issues are in agreement with the MP positions in an m-candidate election. It follows that $P^{\infty}_{MP}(m,0,IC)$ is the probability that a strict Ostrogorski paradox will be observed, and that these results become less paradoxical as k increases for a given m. Computed values of all possible $P^{\infty}_{MP}(m,k,IC)$ are listed in Table 7.5 for each $m = 2,3,4$.

Given the completely balanced nature of the IC assumption, the maximum agreement values in Table 7.5 occur for k values near $(m+2)/2$. Since we know that these probabilities are expected to produce exaggerated estimated of paradoxical outcomes, it is clear that the likelihood of observing an extreme strict Ostrogorski paradox is very small. Moreover, there is strong evidence to suggest that strong versions of a weak Ostrogorski paradox can also be expected to be relatively rare. While less stringent occurrences of a weak Ostrogorski paradox will have greater

Table 7.5 Probability values of $P^{\infty}_{MP}(m,k,IC)$

k	m		
	2	3	4
0	0.0000	0.0104	0.0005
1	0.5000	0.2187	0.0594
2	0.5000	0.5312	0.3750
3		0.2396	0.4406
4			0.1245

probabilities of being observed, it can also be pointed out that such outcomes are not very paradoxical. It is also found that creating a bias toward situations in which individual voter's preferences on issues that are more uniformly consistent with the position of either Party R or Party L, has a significant impact on reducing the probability that Ostrogorski's paradox will be observed.

Following the discussion above regarding the possibility of contriving situations in which Condorcet's paradox must occur; it is also possible to do the same type of thing with Ostrogorski's paradox.

7.6.2 Majority Paradox

The Majority Paradox is similar in nature to Ostrogorski's paradox. With Ostrogorski's paradox, we were concerned about the number of majority rule outcomes on issues that were in agreement with the MP. With the majority paradox, we are concerned instead about the number of majority rule outcomes on issues that are in agreement with the *overall majority party* (*OMP*). Party R (Party L) is the OMP if there are more R (L) entries than L (R) entries in the mn different party position associations for preferences of the voters over all of the issues. The example in Fig. 7.8 shows the fifteen different preference agreements that the five voters have with the parties on the three issues, with nine Party R agreements and six Party L agreements.

The party membership of each voter in Fig. 7.8 has no impact on the definition of the majority paradox; we simply note that Party R is the OMP in this example since it beats Party L by a 9–6 margin in the set of all voters' preferences on issues. The *majority paradox* occurs if the OMP is selected as the winner in a minority of elections on issues. This definition of the majority paradox is equivalent to the *referendum paradox* in Nurmi (1999), as studied by Feix et al. (2004). There cannot be a strict majority paradox, since if any party is the winner by majority rule for every issue, then that same party must also be the OMP.

Representations are obtained in Gehrlein and Merlin (2009b) for the limiting probability $P^{\infty}_{OMP}(m,k,IC)$ as $n \to \infty$ that the majority rule outcomes on exactly k issues are in agreement with the OMP positions in an m-candidate election with the same IC assumption that was used in the discussion of representations for

Table 7.6 Probability values of $P^{\infty}_{OMP}(m,k,IC)$

k	m		
	2	3	4
0	0.0000	0.0000	0.0000
1	0.5000	0.1623	0.0417
2	0.5000	0.5877	0.3750
3		0.2500	0.4583
4			0.1250

Ostrogorski's paradox. Computed values of all possible $P_{OMP}^{\infty}(m, k, IC)$ are listed in Table 7.6 for each $m = 2, 3, 4$.

The computed majority paradox probabilities in Table 7.6 are similar to the Ostrogorski paradox probabilities that were observed in Table 7.5, so similar conclusions can be drawn. That is, there is strong evidence to suggest that extreme versions of a majority paradox can be expected to be quite rare. While less stringent occurrences of a majority paradox will have greater probabilities of being observed, such outcomes are not really very paradoxical.

7.7 Conclusions

We have seen that the classic assumptions for producing probability representations for the likelihood that voting paradoxes will be observed do have valid uses. By utilizing the fact that these classic assumptions will tend to exaggerate the probability of observing paradoxes that involve PMR relationships, we have been able to show that the probability of observing any extreme paradoxical results in an election is very small for a number of different paradoxes. It is also consistently observed that the introduction of a degree of dependence among voters' preferences will further reduce these already small probabilities. However, there were some minor aberrations in this last observation for paradoxes that involve WSR's. The classic assumptions are also shown to be useful for isolating the effects that different parameters can have on these probabilities.

We also note that extensions of these classic assumptions can lead to even more dramatic results, as shown in Gehrlein and Lepelley (2011). For example, by introducing a more sophisticated measurable parameter from voting situations, it is possible to develop a metric of how close any voting situation is to the greatly studied condition of perfectly single-peaked preferences. These more sophisticated measurable parameters are found to have a very significant impact on the probabilities that various voting paradoxes are observed, to further strengthen argument four in the list of reasons to support the use of these types of models.

The authors acknowledge very helpful input from an anonymous reviewer.

References

Abrams, R. (1976). The voter's paradox and the homogeneity of individual preference orders. *Public Choice, 26*, 19–27.

Berg, S. (1985). Paradox of voting under an urn model: the effect of homogeneity. *Public Choice, 47*, 377–387.

Bezembinder, T., & Van Acker, P. (1980). The Ostrogorski paradox and its relation to nontransitive choice. Unpublished paper.

de Borda, J (1784). A paper on elections by ballot. In F. Sommerlad, I. McLean (1989, Eds.), pp. 122–129.

Brams, S.J., & Fishburn, P.C. (1983a). *Approval voting*. Boston: Birkhäuser.

Brams, S.J., & Fishburn, P.C. (1983b). Paradoxes of preferential voting. *Mathematics Magazine, 56*, 207–214.

Condorcet, M. (1785a). An essay on the application of probability theory to plurality decision making: an election between three candidates. In F. Sommerlad, I. McLean (Eds., 1989), *The political theory of Condorcet* (pp. 69–80). Oxford: University of Oxford Working Paper.

Condorcet, M. (1785b). An essay on the application of probability theory to plurality decision making: elections. In F. Sommerlad, I. McLean (Eds., 1989), *The political theory of Condorcet* (pp. 81–89). Oxford: University of Oxford Working Paper.

Condorcet, M. (1785c). An essay on the application of probability theory to plurality decision making: Part five. In F. Sommerlad, I. McLean (Eds. 1989), *The political theory of Condorcet* (pp. 109–118). Oxford: University of Oxford Working Paper.

Condorcet, M. (1785d). An essay on the application of probability theory to plurality decision making: hypothesis eleven. In F. Sommerlad, I. McLean (Eds., 1989), *The political theory of Condorcet* (pp. 90–108). Oxford: University of Oxford Working Paper.

Deb, R., & Kelsey, D. (1987). On constructing a generalized Ostrogorski paradox: necessary and sufficient conditions. *Mathematical Social Sciences, 14*, 161–174.

Diss, M., & Gehrlein, W.V. (2009). Borda's Paradox and weighted scoring rules. *Social Choice and Welfare, 19*, 193–206 (2009). doi:10.1007/s00355-010-0522-1.

Feix, M.R., Lepelley, D., Merlin, V.R., Rouet, J.L. (2004). The probability of conflicts in a U. S. Presidential type election. *Economic Theory, 23*, 227–257.

Fishburn, P.C. (1980). Deducing majority candidates from election data. *Social Science Research, 9*, 216–224.

Fishburn, P.C., & Gehrlein, W.V. (1980). Social homogeneity and Condorcet's paradox. *Public Choice, 35*, 403–420.

Gehrlein, W.V. (1981). The expected probability of Condorcet's paradox. *Economics Letters, 7*, 33–37.

Gehrlein, W.V., & Fishburn, P.C. (1976a). The probability of the paradox of voting: a computable solution. *Journal of Economic Theory, 13*, 14–25.

Gehrlein, W.V., & Fishburn, P.C. (1976b). Condorcet's paradox and anonymous preference profiles. *Public Choice, 26*, 1–18.

Gehrlein, W.V., & Lepelley, D. (2004). Probability calculations in voting theory: an overview of recent results. In M. Wiberg (ed.), *Reasoned choices: essays in honor of Hannu Nurmi* (pp. 140–160). Turku: The Finnish Political Science Association.

Gehrlein, W.V., & Lepelley, D. (2009). A note on Condorcet's other paradox. *Economics Bulletin, 29*, 2000–2007.

Gehrlein, W.V., & Lepelley, D. (2011). *Voting paradoxes and group coherence: the Condorcet efficiency of voting rules*. Berlin: Springer.

Gehrlein, W.V., & Merlin, V. (2009a). On the probability of the Ostrogorski paradox. Unpublished manuscript.

Gehrlein, W.V., & Merlin, V. (2009b). The probability of the majority paradox. Unpublished manuscript.

Lepelley, D., & Merlin, V. (2001). Scoring run-off paradoxes for variable electorates. *Economic Theory, 17*, 53–80.

Merlin, V., Tataru, M., Valognes, F. (2002). On the likelihood of Condorcet's profiles. *Social Choice and Welfare, 19*, 193–206.

Nurmi, H. (1999). *Voting paradoxes and how to deal with them*. Berlin: Springer.

Ostrogorski, M. (1902). *La démocratie et l'organisation des partis politiques*. Paris: Calmann Levy.

Regenwetter, M., Grofman, B., Marley, A.A.J. (2002). On the model dependence of majority preference relations reconstructed from ballot or survey data. *Mathematical Social Sciences, 43*, 451–466.

Regenwetter, M., Grofman, B., Marley, A., Tsetlin, I. (2006). *Behavioral social choice*. Cambridge: Cambridge University Press.

Sommerlad, F., & McLean, I. (1989). *The political theory of Condorcet*. University of Oxford Working Paper.

Tataru, M., & Merlin, V. (1997). On the relationship of the Condorcet winner and positional voting rules. *Mathematical Social Sciences, 34*, 81–90.

Tideman, T.N., & Plassmann, F. (2008). *Evaluating voting methods by their probability of success: an empirical analysis*. Virginia Polytechnic Institute and State University. Unpublished manuscript.

Wilson, M.C., & Pritchard, G. (2007). Probability calculations under the IAC hypothesis. *Mathematical Social Sciences, 54*, 244–256.

Chapter 8
The Impact of Group Coherence on the Condorcet Ranking Efficiency of Voting Rules

William V. Gehrlein and Dominique Lepelley

8.1 Introduction

Recent developments in voting theory, based on both probabilistic and empirical considerations, have led to the conclusion that Condorcet's paradox should be a rare event in actual election settings with a small number of candidates, as soon as voters display any significant level of group mutual coherence, i.e. as soon as voters tend, in one way or another, to have similar preferences (see e.g. Gehrlein 2011, and for an empirical point of view, Regenwetter et al. 2006). In the light of this conclusion, the Condorcet criterion, which requires that the pairwise majority rule winner (PMRW) – or Condorcet winner – should be elected when such a candidate exists, appears as being very relevant. It is therefore of particular interest to investigate the propensity of common voting rules to be in agreement with pairwise majority rule (PMR) *when group coherence is taken into consideration*.

Five voting rules are considered in this study, where attention is restricted to three-candidate elections: the plurality rule (PR), the negative plurality rule (NPR), the Borda rule (BR), the plurality elimination rule (PER) and the negative plurality elimination rule (NPER). PR, BR and NPR are the most often used examples of (single stage) scoring rules: A scoring rule is defined in terms of weights $(1, \lambda, 0)$, with $1 \geq \lambda \geq 0$ for a three-candidate election; each voter assigns a score of one to their most preferred candidate, a score of zero to their least preferred candidate and a score of λ to their middle-ranked candidate, with $\lambda = 0$ for PR, $\lambda = 1/2$ for BR, $\lambda = 1$ for NPR, and the scoring rule winner is the candidate that receives the greatest

W.V. Gehrlein
University of Delaware
e-mail: wvg@udel.edu

D. Lepelley (✉)
University of La Réunion
e-mail: dominique.lepelley@univ-reunion.fr

D.S. Felsenthal and M. Machover, *Electoral Systems*, Studies in Choice and Welfare,
DOI 10.1007/978-3-642-20441-8_8,

Fig. 8.1 The six possible linear preference rankings on three candidates

A	A	B	C	B	C
B	C	A	A	C	B
C	B	C	B	A	A
$n1$	n_2	n_3	n_4	n_5	n_6

score from all voters[1]. PER and NPER are two-stage scoring rules (associated respectively with PR and NPR) in which the candidate with the lowest score is eliminated at the first stage and the winner is selected at the second stage by using PMR. Notice that, as we only consider three-candidate elections, PER coincides here with the very often used "Two-round Majority" and also with "Alternative Vote" (if we assume sincere voting); on the other hand, NPER corresponds –under the same assumption of sincere voting– to "Coombs Rule". All these voting rules share the distinguishing feature of violating the Condorcet criterion and can be analyzed in a similar framework.[2]

A number of studies have been conducted to compute the *Condorcet efficiency* of these voting rules, with the notion of Condorcet efficiency of a voting rule VR being defined as the conditional probability that the PMRW be elected under VR, given that a PMRW exists. In these studies, it is implicitly assumed that the objective of the collective choice process is to determine only the winning candidate. We consider in this paper a somewhat different context in which the objective is to obtain a complete ranking of the candidates and we define the *Condorcet ranking efficiency* of VR as the conditional probability that candidate rankings are identical for both PMR and VR, given that a (strict) PMR ranking exists. The notion of Condorcet ranking efficiency is clearly more demanding than the notion of Condorcet efficiency: in three-candidate elections, it is required that the voting rule under consideration not only ranks the PMRW in first position but also ranks the pairwise majority rule loser (PMRL or Condorcet loser) in third (and last) position.

Let n be the number of voters. In an election with three candidates A, B, and C, there are six possible complete preference orders that voters may have, as shown in Fig. 8.1.

Here, n_i denotes the number of voters that have the associated complete preference ranking on the candidates. Any given combination of $n_i's$ such that $\sum_{i=1}^{6} n_i = n$ is referred to as a voting situation and reflects the voters' opinion. A well known assumption in voting theory is the so-called impartial anonymous culture (IAC) assumption, which supposes that every possible voting situation is equally likely to be observed. Let $CRE(VR, n)$ be the Condorcet ranking efficiency of VR in three-candidate elections with n voters when IAC is assumed. Cervone et al. (2005) give a general representation for the limiting Condorcet ranking

[1]Equivalently, NPR can be implemented by asking each voter to cast a negative vote against his or her least preferred candidate and the candidate who receives the fewest number of votes is the election winner.

[2]Some other rules such as approval voting or range voting need additional assumptions to be analyzed.

efficiency of scoring rules as $n \to \infty$ with IAC, from which it follows that:

$$CRE(PR, \infty) = CRE(NPR, \infty) = 303/540 = .5611, \tag{8.1}$$

$$CRE(BR, \infty) = 111/135 = .8222. \tag{8.2}$$

Using recent developments for obtaining IAC probability representations (see Gehrlein, 2002; Lepelley et al. 2008), it is easy to show that:

$$CRE(PER, \infty) = 17/27 = .6296, \tag{8.3}$$

$$CRE(NPER, \infty) = 119/135 = .8815. \tag{8.4}$$

The aim of the present paper is to analyze the impact that various degrees of group coherence in voters' preferences might have on the Condorcet ranking efficiency of these five common voting rules[3]. We begin by introducing various measures of group coherence.

8.2 Measuring the Level of Group Coherence in Voters' Preferences

Given a voting situation, we can evaluate the level of group coherence in voters' preferences associated with this voting situation in various ways. Two sets of measures are used in our study.

8.2.1 Weak Measures of Group Coherence

The first measure of group coherence we consider is directly inspired from Black's single-peakedness. We introduce a parameter b, that measures the minimum number of times that some candidate is bottom ranked, or is least preferred, in the preferences of the n voters in a voting situation, to serve as a simple measure of the proximity of a voting situation to representing perfectly single-peaked preferences in a three-candidate election, where

$$b = Min\,\{n_1 + n_3, n_2 + n_4, n_5 + n_6\}\,. \tag{8.5}$$

[3]This paper completes two recent studies by the same authors and H. Smaoui (Gehrlein et al. 2008, 2010) that consider the impact of group coherence on the Condorcet efficiency of these voting rules.

If b is equal to zero for a voting situation with three candidates, it means that the n voters agree that some candidate is never ranked as least preferred, so the voting situation represents the condition in which voters have perfectly single-peaked preferences. This would happen, for example if $n_1 + n_3 = 0$, where the definitions from Fig. 8.1 indicate that this requires that candidate C is never the least preferred candidate for any voter in the associated voting situation. When b is maximized at $n/3$, a voting situation reflects very disperse preferences of voters over candidates to reflect a situation that is very far removed from perfect single-peakedness.

As parameter b increases in voting situations, the preferences of voters in a voting situation become more removed from the condition of perfect single-peakedness. Another perspective on this issue is that a voting situation with a small parameter b reflects a situation in which there is some candidate that very few voters think is the worst of the three candidates. The electorate would be somewhat united by their *weak* support of, or lack of complete opposition to, the election of such a candidate. In that sense, this candidate can be viewed as a *weak positively unifying candidate* that voters would not generally think of as reflecting the worst possible outcome if that candidate were to be elected.

Following the development of parameter b above, parameter t measures the proximity of a voting situation to meeting the condition of perfectly single-troughed preferences (see Vickery 1960), with

$$t = Min\{n_1 + n_2, n_3 + n_5, n_4 + n_6\}. \tag{8.6}$$

The definition of $n_i's$ in Fig. 8.1 are used to define parameter t as the minimum number of times that some candidate is top-ranked as the most preferred candidate in the voters' preference rankings, so that a voting situation is perfectly single-troughed if $t = 0$, and the value of t then reflects the relative proximity of a voting situation to the condition of perfect single-troughedness. Any candidate that very few voters rank as the most preferred candidate in a voting situation can be viewed as a *weak negatively unifying candidate* since none of the voters would generally think of the election of this candidate as reflecting the best possible outcome. The electorate would be weakly unified by their opposition to, or lack in complete support of, the election of such a candidate.

According to Ward (1965), a candidate is said to be *perfectly polarizing* if this candidate is never middle ranked, or ranked at the center, of any voter's preference ranking. That is, every voter will either consider this candidate to be either the most preferred or the least preferred. The definition of $n_i's$ in Fig. 8.1 are used to define parameter c to reflect the proximity of a voting situation to the condition of perfect polarization, with

$$c = Min\{n_3 + n_4, n_1 + n_6, n_2 + n_5\}. \tag{8.7}$$

If $c = 0$, some candidate is perfectly polarizing, since all voters will rank that candidate as either least preferred or most preferred, and the value of c measures the proximity of a voting situation to the condition of perfect polarization. Any candidate that very few voters rank in the middle of their preference ranking can generally be viewed as a *weak polarizing candidate*.

8.2.2 Strong Measures of Group Coherence

Stronger measures of group coherence are developed in Gehrlein (2011), and each of these measures is a more restrictive variation of parameters b, t, and c. A weak positively unifying candidate was defined as some candidate that is ranked as least preferred by a small proportion of voters in a voting situation, and the proximity of a voting situation to having a perfect weak positively unifying candidate is measured by parameter b. A candidate would more strongly reflect the notion of being a positively unifying candidate by being ranked as most preferred by a large proportion of the voters in a voting situation. *Parameter* t^* is defined accordingly from the definition of the $n_i's$ in Fig. 8.1, with

$$t^* = Max\{n_1 + n_2, n_3 + n_5, n_4 + n_6\}. \tag{8.8}$$

If $t^* = n$, the same candidate is ranked as most preferred by all voters, making it a perfect *strong positively unifying candidate*, and parameter t^* is used as a measure of the proximity of a voting situation to this condition.

The same basic logic can be used to strengthen the definition the proximity of a voting situation to having perfect weak negatively unifying candidate, as measured by parameter t. *Parameter* b^* is defined accordingly by

$$b^* = Max\{n_5 + n_6, n_2 + n_4, n_1 + n_3\}. \tag{8.9}$$

If $b^* = n$, the same candidate is ranked as least preferred by all voters, making it a perfect *strong negatively unifying candidate*, and parameter b^* is used as a measure of the proximity of a voting situation to this condition.

Parameter c measured the proximity of a voting situation to the condition of perfect weak polarization. The strong measure that is associated with this parameter is parameter c^*, with

$$c^* = Max\{n_3 + n_4, n_1 + n_6, n_2 + n_5\}. \tag{8.10}$$

If $c^* = n$, the same candidate is middle-ranked in the preferences of all voters, so that this candidate is neither extremely liked nor extremely disliked by any voter, making it a perfect *strong centrist candidate*, and parameter c^* is used as a measure of the proximity of a voting situation to this condition.

We have finally six different measures of group mutual coherence at our disposal, with an increasing level of mutual coherence when b, t and c move from $n/3$ to 0 and when b^*, t^* and c^* move from $n/3$ to n.

8.3 Obtaining Condorcet Ranking Efficiency Representations

To determine the impact that our measures of group coherence have on the probability that candidate rankings are identical for both PMR and a given voting rule, we develop representations for the conditional probability of that event, given that voting situations have specified values of these measures.

These probability representations are based on a direct extension of the IAC assumption: for any particular X in $\{b, t, c, b^*, t^*, c^*\}$, we assume that only voting situations for which parameter X has a specified value can be observed, and that each of these possible voting situations is equally likely to be observed.

Using some techniques recently developed in the literature (see Gehrlein 2005 and Lepelley et al. 2008), it is possible to derive some representations for the desired probabilities as functions of n and X. Unhappily, these representations are often of a very complicated nature, which makes them of limited value. For this reason, we only focus here on the limiting probabilities as n tends to infinity. Let $\alpha_X = X/n$; thus, for example, α_b denotes the minimal proportion of voters that rank a candidate in last position. The limiting representations we present in this paper express the Condorcet ranking efficiency of the voting rules as a function of the proportion α_X as $n \to \infty$, with α_X in [0,1/3] for $X = b, t, c$ and α_X in [1/3,1] for $X = b^*, t^*, c^*$.[4]

Let $CRE(VR, n/X)$ denote the Condorcet ranking efficiency of *VR* with n voters, given a specified value of parameter X. As we consider five voting rules and six distinct group coherence parameters, 30 representations are a priori needed. However, some representations do not need to be determined as a result of the following observations.

Consider first the weak measures of group coherence. We have:

Lemma 1. *(i)* $CRE(PR, n/t) = CRE(NPR, n/b), CRE(NPR, n/t) = CRE(PR, n/b)$ *and* $CRE(BR, n/t) = CRE(BR, n/b)$.
(ii) $CRE(PR, n/c) = CRE(NPR, n/c)$.

In order to illustrate the argument on which these results are based, consider a voting situation such that (8.1) A is the PMRW and C the PMRL; (8.2) A is the PR Winner and C the PR Loser; (8.3) $b = k$, where k is an integer between 0 and $n/3$. To this voting situation, it is possible to associate an equally likely dual voting situation obtained by inverting the preference of each of the voters, with the following 1–1 mapping: $n_1 \leftrightarrow n_6$, $n_2 \leftrightarrow n_5$ and $n_3 \leftrightarrow n_4$. It is easy to check that, in this dual voting situation (8.1) C is the PMRW and A the PMRL; (8.2) C is the NPR winner and A the NPR loser; (8.3) $t = k$. Consequently, the Condorcet ranking efficiency of PR given a specified value of b will be equal to the Condorcet ranking efficiency of NPR given the same specified value of t.

[4]For more details on this approach, see Gehrlein et al. (2010).

Furthermore, it can be noticed that the Condorcet ranking efficiency of the two-stage scoring rules PER and NPER can be obtained from the Condorcet efficieny of NPR and PR (respectively). Let $CE(VR, n/X)$ denote the Condorcet efficiency of VR, given a specified value of X.

Lemma 2. *(i)* $CRE(PER, n/b) = CE(NPR, n/t)$ *and* $CRE(NPER, n/b) = CE(PR, n/t)$.
(ii) $CRE(PER, n/t) = CE(NPR, n/b)$ *and* $CRE(NPER, n/t) = CE(PR, n/b)$.
(iii) $CRE(PER, n/c) = CE(NPR, n/c)$ *and* $CRE(NPER, n/c) = CE(PR, n/c)$.

The proof techniques that are used to obtain these results are very similar to the ones used to prove Lemma 1. Lemma 2 allows us to use the Condorcet efficiency results given in Gehrlein et al. (2008) to obtain the Condorcet ranking efficiency of PER and NPER with weak measures of group coherence.

Similar observations can be made for the strong measures of coherence.

Lemma 3. *(i)* $CRE(PR, n/t^*) = CRE(NPR, n/b^*)$, $CRE(NPR, n/t^*) = CRE(PR, n/b^*)$ *and* $CRE(BR, n/t^*) = CRE(BR, n/b^*)$.
(ii) $CRE(PR, n/c^*) = CRE(NPR, n/c^*)$.

Lemma 4. *(i)* $CRE(PER, n/b^*) = CE(NPR, n/t^*)$ *and* $CRE(NPER, n/b^*) = CE(PR, n/t^*)$.
(ii) $CRE(PER, n/t^*) = CE(NPR, n/b^*)$ *and* $CRE(NPER, n/t^*) = CE(PR, n/b^*)$.
(iii) $CRE(PER, n/c^*) = CE(NPR, n/c^*)$ *and* $CRE(NPER, n/c^*) = CE(PR, n/c^*)$.

Thanks to Lemma 4, the Condorcet ranking efficiency of PER and NPER for strong measures of coherence can be deduced from the recent work by Gehrlein et al. (2010), where the Condorcet efficiency of PR and NPR is derived for parameters b^*, t^* and c^*.

Finally, our preliminary observations allow us to reduce the number of needed representations from 30 to 10. These 10 (limiting) representations are given in Appendix.

8.4 Results

The representations given in Appendix are used to obtain computed values of $CRE(VR, \infty/\alpha_X)$ for $VR \in \{PR, NPR, BR, PER, NPER\}$ and for various values of α_X. The results are listed in Table 8.1–8.6. Table 8.1–8.3 allow to analyze the relationship between Condorcet ranking efficiency of voting rules and weak measures of group mutual coherence (parameters b, t, c), whereas Table 8.1–8.6 deal with strong measures of group coherence (parameters b^*, t^*, c^*). Although these numerical values have been obtained for infinitely many voters, they can be considered as giving a good approximation for finite cases as soon as the number of voters is higher than about 100.

Table 8.1 Computed values of $CRE(VR, \infty/\alpha_b)$ for $VR \in \{PR, NPR, BR, PER, NPER\}$

α_b	PR	NPR	BR	PER	NPER
0.00	0.5764	0.7500	0.8333	0.6528	1.0000
0.02	0.5759	0.7279	0.8334	0.6463	0.9925
0.04	0.5730	0.7065	0.8335	0.6385	0.9850
0.06	0.5678	0.6855	0.8337	0.6296	0.9772
0.08	0.5606	0.6647	0.8338	0.6199	0.9689
0.10	0.5518	0.6438	0.8336	0.6098	0.9598
0.12	0.5418	0.6225	0.8331	0.5997	0.9494
0.14	0.5315	0.6003	0.8318	0.5907	0.9370
0.16	0.5222	0.5767	0.8295	0.5841	0.9218
0.18	0.5163	0.5509	0.8252	0.5826	0.9020
0.20	0.5178	0.5213	0.8177	0.5897	0.8750
0.22	0.5327	0.4858	0.8055	0.6112	0.8357
0.24	0.5717	0.4394	0.7905	0.6578	0.7736
0.26	0.6491	0.3756	0.7862	0.7430	0.6737
0.28	0.7269	0.3111	0.7973	0.8213	0.5698
0.30	0.7885	0.2515	0.8166	0.8783	0.4731
0.32	0.8257	0.1983	0.8303	0.9104	0.3857
1/3	0.8333	0.1667	0.8333	0.9163	0.3460

Table 8.2 Computed values of $CRE(VR, \infty/\alpha_t)$ for $VR \in \{PR, NPR, BR, PER, NPER\}$

α_t	PR	NPR	BR	PER	NPER
0.00	0.7500	0.5764	0.8333	0.7500	0.8611
0.02	0.7279	0.5759	0.8334	0.7353	0.8674
0.04	0.7065	0.5730	0.8335	0.7210	0.8728
0.06	0.6855	0.5678	0.8337	0.7071	0.8774
0.08	0.6647	0.5606	0.8338	0.6933	0.8811
0.10	0.6438	0.5518	0.8336	0.6796	0.8839
0.12	0.6225	0.5418	0.8331	0.6657	0.8857
0.14	0.6003	0.5315	0.8318	0.6515	0.8861
0.16	0.5767	0.5222	0.8295	0.6367	0.8850
0.18	0.5509	0.5163	0.8252	0.6209	0.8818
0.20	0.5213	0.5178	0.8177	0.6037	0.8768
0.22	0.4858	0.5327	0.8055	0.5840	0.8717
0.24	0.4394	0.5717	0.7905	0.5601	0.8710
0.26	0.3756	0.6491	0.7862	0.5279	0.8826
0.28	0.3111	0.7269	0.7973	0.4839	0.8965
0.30	0.2515	0.7885	0.8166	0.4311	0.9080
0.32	0.1983	0.8257	0.8303	0.3732	0.9151
1/3	0.1667	0.8333	0.8333	0.3433	0.9166

It is worth noticing that, in many cases, the results do not correspond to what intuition suggests: it could be expected that the Condorcet ranking efficiency of voting rules would monotonically increase when the group coherence in voters' preferences increases. It turns out that this pattern of behavior is actually rather rare

Table 8.3 Computed values of $CRE(VR, \infty/\alpha_c)$ for $VR \in \{PR, NPR, BR, PER, NPER\}$

α_c	PR	NPR	BR	PER	NPER
0.00	0.4097	0.4097	0.8333	0.4236	0.8611
0.02	0.4147	0.4147	0.8386	0.4351	0.8662
0.04	0.4231	0.4231	0.8419	0.4501	0.8702
0.06	0.4349	0.4349	0.8433	0.4685	0.8733
0.08	0.4502	0.4502	0.8428	0.4906	0.8755
0.10	0.4689	0.4689	0.8405	0.5165	0.8769
0.12	0.4909	0.4909	0.8364	0.5463	0.8776
0.14	0.5162	0.5162	0.8307	0.5800	0.8780
0.16	0.5447	0.5447	0.8236	0.6178	0.8783
0.18	0.5768	0.5768	0.8156	0.6590	0.8793
0.20	0.6128	0.6128	0.8076	0.7022	0.8818
0.22	0.6518	0.6518	0.8009	0.7459	0.8864
0.24	0.6921	0.6921	0.7970	0.7883	0.8924
0.26	0.7302	0.7302	0.7867	0.8275	0.8974
0.28	0.7630	0.7630	0.7996	0.8608	0.9002
0.30	0.7891	0.7891	0.8044	0.8863	0.9023
0.32	0.8059	0.8059	0.8085	0.9016	0.9043
1/3	0.8095	0.8095	0.8095	0.9046	0.9047

Table 8.4 Computed values of $CRE(VR, \infty/\alpha_{b*})$ for $VR \in \{PR, NPR, BR, PER, NPER\}$

α_{b*}	PR	NPR	BR	PER	NPER
0.33	0.8333	0.1667	0.8333	0.9162	0.3233
0.35	0.8220	0.2053	0.8285	0.9074	0.3833
0.40	0.7029	0.3323	0.7642	0.7990	0.5366
0.45	0.5495	0.4797	0.7023	0.6388	0.7161
0.50	0.4074	0.6875	0.7292	0.4676	1.0000
0.55	0.4647	0.6534	0.8541	0.5259	1.0000
0.60	0.5300	0.6250	0.9106	0.5918	1.0000
0.65	0.6034	0.6010	0.9327	0.6636	1.0000
0.70	0.6792	0.5804	0.9464	0.7321	1.0000
0.75	0.7500	0.5625	0.9583	0.7917	1.0000
0.80	0.8125	0.5469	0.9688	0.8438	1.0000
0.85	0.8676	0.5331	0.9779	0.8897	1.0000
0.90	0.9167	0.5208	0.9861	0.9306	1.0000
0.95	0.9605	0.5099	0.9934	0.9671	1.0000
1.00	1.0000	0.5000	1.0000	1.0000	1.0000

since it is only observed with NPR and NPER for parameter b, with PR and PER for parameter t and with NPER for parameter b^*. The somewhat paradoxical opposite behavior, where the Condorcet ranking efficiency monotonically decreases when the level of group coherence increases, appears to be more frequent: this behavior occurs with NPER for parameter t, with PR, NPR, PER and NPER for parameter c, and with PR, NPR, PER and NPER for parameter c^*. In all other combinations of voting rules and scenarios that we have considered, the behavior of the Condorcet

Table 8.5 Computed values of $CRE(VR, \infty/\alpha_{t*})$ for $VR \in \{PR, NPR, BR, PER, NPER\}$

α_{t*}	PR	NPR	BR	PER	NPER
0.33	0.1667	0.8333	0.8333	0.3208	0.9166
0.35	0.2053	0.8220	0.8285	0.3955	0.9143
0.40	0.3323	0.7029	0.7642	0.5640	0.8864
0.45	0.4797	0.5495	0.7023	0.6770	0.8432
0.50	0.6875	0.4074	0.7292	0.6875	0.8148
0.55	0.6534	0.4647	0.8541	0.6534	0.8750
0.60	0.6250	0.5300	0.9106	0.6250	0.9118
0.65	0.6010	0.6034	0.9327	0.6009	0.9326
0.70	0.5804	0.6792	0.9464	0.5803	0.9464
0.75	0.5625	0.7500	0.9583	0.5625	0.9583
0.80	0.5469	0.8125	0.9688	0.5469	0.9688
0.85	0.5331	0.8676	0.9779	0.5331	0.9779
0.90	0.5208	0.9167	0.9861	0.5208	0.9861
0.95	0.5099	0.9605	0.9934	0.5099	0.9934
1.00	0.5000	1.0000	1.0000	0.5000	1.0000

Table 8.6 Computed values of $CRE(VR, \infty/\alpha_{c*})$ for $VR \in \{PR, NPR, BR, PER, NPER\}$

α_{c*}	PR	NPR	BR	PER	NPER
0.33	0.8095	0.8095	0.8095	0.9047	0.9047
0.35	0.8040	0.8040	0.8080	0.8999	0.9040
0.40	0.7494	0.7494	0.7969	0.8475	0.8972
0.45	0.6850	0.6850	0.7908	0.7738	0.8889
0.50	0.6179	0.6179	0.7900	0.6988	0.8691
0.55	0.5392	0.5392	0.8075	0.6181	0.8536
0.60	0.4738	0.4738	0.8359	0.5339	0.8591
0.65	0.4128	0.4128	0.8671	0.4452	0.8802
0.70	0.3458	0.3458	0.8950	0.3571	0.9052
0.75	0.2713	0.2713	0.9188	0.2766	0.9267
0.80	0.2024	0.2024	0.9394	0.2063	0.9453
0.85	0.1422	0.1422	0.9574	0.1450	0.9616
0.90	0.0892	0.0892	0.9733	0.0909	0.9759
0.95	0.0421	0.0421	0.9874	0.0429	0.9886
1.00	0.0000	0.0000	1.0000	0.0000	1.000

ranking efficiency is not monotonic when the degree of group mutual coherence increases.

It is also of interest to remark that the Condorcet ranking efficiencies of PR and NPR can be very different when some measures of group coherence are taken in consideration, despite the fact that, on average, PR and NPR display the same level of Condorcet ranking efficiency ($CRE(PR, \infty) = CRE(NPR, \infty) = 303/540 = 0.5611$, as seen in Sect. 8.1). To illustrate, Table 8.1 shows that NPR continuously decreases from $3/4$ to 1/6 as α_b increases, to reflect voting situations that are farther removed from having a perfect positively unifying candidate, whereas the results for PR slowly decrease over the range $0 \leq \alpha_b \leq .21$, and then continue to increase significantly up to 5/6 as α_b continues to increase to 1/3.

We now turn to the overall comparison of the five voting rules. In what follows, we will say that voting rule VR1 dominates voting rule VR2 for parameter X if $CRE(VR1, \infty/\alpha_X) \geq CRE(VR2, \infty/\alpha_X)$ for every possible value of α_X and there exists some value of α_X for which $CRE(VR1, \infty/\alpha_X) > CRE(VR2, \infty/\alpha_X)$.

The results from Table 8.1–8.6 show that:

- BR dominates PR and NPR for every X in $\{b, t, c, b^*, t^*, c^*\}$, i.e. for all weak and strong measures of group mutual coherence we have considered.
- Similarly, the two-stage voting rules (PER and NPER) dominate the corresponding single stage voting rules (PR and NPR) for every X in $\{b, t, c, b^*, t^*, c^*\}$.
- NPER dominates each of the four other voting rules for $X \in \{t, c, c^*\}$. Moreover, NPER dominates PR, NPR and PER for $X = t^*$.[5]
- BR dominates PER for $X \in \{t, t^*\}$.

Thus, BR and NPER clearly exhibit the best performances on the basis of Condorcet ranking efficiency. But the most striking observation from Table 8.1–8.6 is that BR remains relatively stable over the entire range of the various parameters we have introduced and it turns out that the Condorcet ranking efficieny of BR is never lower than 0.70. When BR and NPER are compared, it can be seen that $CRE(NPER, \infty/\alpha_X)$ is higher than $CRE(BR, \infty/\alpha_X)$ in many cases but the margin of dominance for NPER over BR remains always rather low. Moreover, the possibility exists in which NPER could behave very poorly with large values of parameter b or with low values of parameter b^*, reflecting scenarios in which voting situations are far removed either from having a perfect weak positively unifying candidate or from having a perfect strong negatively unifying candidate. In these circumstances, the Condorcet ranking efficiency of NPER can be lower than 0.35

8.5 Conclusion

The results that are observed for Condorcet ranking efficiency in the presence of group coherence are very similar to the results obtained previously, in the same context, for Condorcet efficiency when the objective is to select a single winner (see Gehrlein et al. 2008; Gehrlein et al. 2010). The Condorcet ranking efficiency of BR remains somewhat stable across the complete range of all measures of group mutual coherence. BR dominates both PR and NPR for all weak and strong measures of group mutual coherence, particularly for parameter c and parameter c^*. While PER does display superior performance to BR over a small range of some parameters, it very frequently exhibits extremely poor performance on the basis of Condorcet ranking efficiency and it is not a viable option for consideration. The efficiency of

[5] NPER does not dominate BR for $X = t^*$ because $CRE(BR, \infty/\alpha_{t*}) > CRE(NPER, \infty/\alpha_{t^*})$ for $0.62 \leq \alpha_{t*} \leq 0.65$.

NPER is very often superior to that of BR, but there are ranges in which NPER performs very poorly for both parameter b and parameter b^*, while BR does not do so. Since we cannot exclude the possibility that voters are obtaining preference rankings with some model that will fall into the ranges in which NPER performs very poorly, the Borda compromise suggested in Gehrlein et al. (2008) still has a good foundation for Condorcet ranking efficiencies.

Appendix A: Condorcet Ranking Efficiency Representations

A.1 Limiting Condorcet Ranking Efficiency of PR for Parameter b

$$CRE(PR, \infty/\alpha_b) = \frac{83 - 243\alpha_b - 828\alpha_b^2 + 2898\alpha_b^3}{144\left(1 - 3\alpha_b - 4\alpha_b^2 + 11\alpha_b^3\right)}, \quad \text{for} \quad 0 \le \alpha_b \le 1/6$$

$$\frac{-121 + 4800\alpha_b - 30240\alpha_b^2 + 74304\alpha_b^3 - 66528\alpha_b^4}{3456\alpha_b\left(1 - 3\alpha_b - 4\alpha_b^2 + 11\alpha_b^3\right)}, \quad \text{for} \quad 1/6 \le \alpha_b \le 1/4$$

$$\frac{46 - 549\alpha_b + 2457\alpha_b^2 - 3267\alpha_b^3}{54\left(1 - 6\alpha_b + 18\alpha_b^2 - 18\alpha_b^3\right)}, \quad \text{for} \quad 1/4 \le \alpha_b \le 1/3$$

A.2 Limiting Condorcet Ranking Efficiency of NPR for Parameter b

$$CRE(NPR, \infty/\alpha_b) = \frac{3\left(1 - 2\alpha_b\right)\left(2 - 5\alpha_b - 6\alpha_b^2\right)}{8\left(1 - 3\alpha_b - 4\alpha_b^2 + 11\alpha_b^3\right)}, \quad \text{for} \quad 0 \le \alpha_b \le 1/4$$

$$\frac{3\left(1 - 2\alpha_b\right)^2\left(1 - \alpha_b\right)}{4\left(1 - 6\alpha_b + 18\alpha_b^2 - 18\alpha_b^3\right)}, \quad \text{for} \quad 1/4 \le \alpha_b \le 1/3$$

A.3 Limiting Condorcet Ranking Efficiency of BR for Parameter b

$$CRE(BR, \infty/\alpha_b) = \frac{40 - 120\alpha_b - 152\alpha_b^2 + 369\alpha_b^3}{48\left(1 - 3\alpha_b - 4\alpha_b^2 + 11\alpha_b^3\right)}, \quad \text{for} \quad 0 \le \alpha_b \le 1/6$$

$$\frac{1 + 136\alpha_b - 264\alpha_b^2 - 1472\alpha_b^3 + 2772\alpha_b^4}{192\alpha_b\left(1 - 3\alpha_b - 4\alpha_b^2 + 11\alpha_b^3\right)}, \quad \text{for} \quad 1/6 \le \alpha_b \le 1/5$$

$$\frac{41-664\alpha_b+5736\alpha_b^2-21472\alpha_b^3+27772\alpha_b^4}{192\alpha_b\left(1-3\alpha_b-4\alpha_b^2+11\alpha_b^3\right)}, \quad \text{for} \quad 1/5 \leq \alpha_b \leq 1/4$$

$$\frac{-77+520\alpha_b-840\alpha_b^2-1568\alpha_b^3+4484\alpha_b^4}{96(3\alpha_b-1)\left(1-6\alpha_b+18\alpha_b^2-18\alpha_b^3\right)}, \quad \text{for} \quad 1/4 \leq \alpha_b \leq 2/7$$

$$\frac{525-6792\alpha_b+33768\alpha_b^2-75264\alpha_b^3+62744\alpha_b^4}{96(1-3\alpha_b)\left(1-6\alpha_b+18\alpha_b^2-18\alpha_b^3\right)}, \quad 2/7 \leq \alpha_b \leq 3/10$$

$$\frac{5-43\alpha_b+153\alpha_b^2-177\alpha_b^3}{4\left(1-6\alpha_b+18\alpha_b^2-18\alpha_b^3\right)}, \quad 3/10 \leq \alpha_b \leq 1/3$$

A.4 Limiting Condorcet Ranking Efficiency of BR for Parameter *c*

$$CRE(BR, \infty/\alpha_c) = \frac{5(16-48\alpha_c-96\alpha_c^2+309\alpha_c^3)}{6\left(16-54\alpha_c-28\alpha_c^2+139\alpha_c^3\right)}, \quad \text{for} \quad 0 \leq \alpha_c \leq 1/6$$

$$\frac{-2+448\alpha_c-1632\alpha_c^2-672\alpha_c^3+5133\alpha_c^4}{30\alpha_c\left(16-54\alpha_c-28\alpha_c^2+139\alpha_c^3\right)}, \quad \text{for} \quad 1/6 \leq \alpha_c \leq 1/5$$

$$\frac{-27+948\alpha_c-5382\alpha_c^2+11828\alpha_c^3-10492\alpha_c^4}{30\alpha_c\left(16-54\alpha_c-28\alpha_c^2+139\alpha_c^3\right)}, \quad \text{for} \quad 1/5 \leq \alpha_c \leq 1/4$$

$$\frac{3+468\alpha_c-2502\alpha_c^2+4148\alpha_c^3-2812\alpha_c^4}{30(3\alpha_c-1)\left(1-29\alpha_c+63\alpha_c^2-39\alpha_c^3\right)}, \quad \text{for} \quad 1/4 \leq \alpha_c \leq 2/7$$

$$\frac{23-151\alpha_c+693\alpha_c^2-933\alpha_c^3}{6\left(-1+29\alpha_c-63\alpha_c^2+39\alpha_c^3\right)}, \quad 2/7 \leq \alpha_c \leq 1/3$$

A.5 Limiting Condorcet Ranking Efficiency of PR for Parameter *c*

$$CRE(PR, \infty/\alpha_c) = \frac{236-702\alpha_c+1716\alpha_c^2-6421\alpha_c^3)}{36\left(16-54\alpha_c-28\alpha_c^2+139\alpha_c^3\right)}, \quad \text{for} \quad 0 \leq \alpha_c \leq 1/8$$

$$\frac{3+140\alpha_c+450\alpha_c^2-4428\alpha_c^3+5867\alpha_c^4}{36\alpha_c\left(16-54\alpha_c-28\alpha_c^2+139\alpha_c^3\right)}, \quad \text{for} \quad 1/8 \leq \alpha_c \leq 1/6$$

$$\frac{-1+1488\alpha_c-4860\alpha_c^2+10584\alpha_c^3-30894\alpha_c^4}{216\alpha_c\left(16-54\alpha_c-28\alpha_c^2+139\alpha_c^3\right)}, \quad \text{for} \quad 1/6 \leq \alpha_c \leq 1/5$$

$$\frac{89-312\alpha_c+8640\alpha_c^2-34416\alpha_c^3+25356\alpha_c^4}{216\alpha_c\left(16-54\alpha_c-28\alpha_c^2+139\alpha_c^3\right)}, \quad \text{for} \quad 1/5 \leq \alpha_c \leq 1/4$$

$$\frac{-163+2136\alpha_c+3456\alpha_c^2-45936\alpha_c^3+62220\alpha_c^4}{216(3\alpha_c-1)\left(1-29\alpha_c+63\alpha_c^2-39\alpha_c^3\right)}, \quad \text{for} \quad 1/4 \le \alpha_c \le 3/10$$

$$\frac{263-2727\alpha_c+12933\alpha_c^2-17685\alpha_c^3}{54\left(-1+29\alpha_c-63\alpha_c^2+39\alpha_c^3\right)}, \quad 3/10 \le \alpha_c \le 1/3$$

A.6 Limiting Condorcet Ranking Efficiency of PR for Parameter b^*

$$CRE(PR,\infty/\alpha_{b*}) = \frac{-3591\alpha_{b*}^3+4401\alpha_{b*}^2-1737\alpha_{b*}+208}{54(18\alpha_{b*}^3-18\alpha_{b*}^2+6\alpha_{b*}-1)}, \quad \text{for} \quad 1/3 \le \alpha_{b*} \le 3/8$$

$$\frac{61776\alpha_{b*}^4-102816\alpha_{b*}^3+60480\alpha_{b*}^2-14448\alpha_{b*}+1141}{864(1-3\alpha_{b*})(18\alpha_{b*}^3-18\alpha_{b*}^2+6\alpha_{b*}-1)}, \quad \text{for} \quad 3/8 \le \alpha_{b*} \le 5/12$$

$$\frac{20304\alpha_{b*}^4-33696\alpha_{b*}^3+17280\alpha_{b*}^2-2448\alpha_{b*}-109}{24(3\alpha_{b*}-1)(18\alpha_{b*}^3-18\alpha_{b*}^2+6\alpha_{b*}-1)}, \quad \text{for} \quad 5/12 \le \alpha_{b*} \le 1/2$$

$$\frac{432\alpha_{b*}^4-1728\alpha_{b*}^3+2592\alpha_{b*}^2-1392\alpha_{b*}+149}{3456\alpha_{b*}(\alpha_{b*}-1)^3}, \quad \text{for} \quad 1/2 \le \alpha_{b*} \le 2/3$$

$$\frac{32\alpha_{b*}^4-96\alpha_{b*}^2+80\alpha_{b*}-15}{128\alpha_{b*}(1-\alpha_{b*})^3}, \quad \text{for} \quad 2/3 \le \alpha_{b*} \le 3/4$$

$$\frac{7\alpha_{b*}^{-}3}{4\alpha_{b*}}, \quad \text{for} \quad 3/4 \le \alpha_{b*} \le 1.$$

A.7 Limiting Condorcet Ranking Efficiency of BR for Parameter b^*

$$CRE(BR,\infty/\alpha_{b*}) = \frac{15\alpha_{b*}^3+9\alpha_{b*}^2-11\alpha_{b*}+1}{4(18\alpha_{b*}^3-18\alpha_{b*}^2+6\alpha_{b*}-1)}, \quad \text{for} \quad 1/3 \le \alpha_{b*} \le 3/8$$

$$\frac{3961\alpha_{b*}^4-6180\alpha_{b*}^3+3582\alpha_{b*}^2-906\alpha_{b*}+84}{12(1-3\alpha_{b*})(18\alpha_{b*}^3-18\alpha_{b*}^2+6\alpha_{b*k}-1)}, \quad \text{for} \quad 3/8 \le \alpha_{b*} \le 2/5$$

$$\frac{1453\alpha_{b*}^4-2640\alpha_{b*}^3+1836\alpha_{b*}^2-588\alpha_{b*}+72}{24(3\alpha_{b*}-1)(18\alpha_{b*}^3-18\alpha_{b*}^2+6\alpha_{b*}-1)}, \quad \text{for} \quad 2/5 \le \alpha_{b*} \le 1/2$$

$$\frac{621\alpha_{b*}^4-1680\alpha_{b*}^3+1692\alpha_{b*}^2-744\alpha_{b*}+118}{48\alpha_{b*}(\alpha_{b*}-1)^3}, \quad \text{for} \quad 1/2 \le \alpha_{b*} \le 2/3$$

$$\frac{9\alpha_{b*}^{-}1}{8\alpha_{b*}}, \quad \text{for} \quad 2/3 \le \alpha_{b*} \le 1.$$

A.8 Limiting Condorcet Ranking Efficiency of NPR for Parameter b^*

$$CRE(NPR, \infty/\alpha_{b*}) = \frac{87\alpha_{b*}^3 - 99\alpha_{b*}^2 + 31\alpha_{b*} - 3}{8(18\alpha_{b*}^3 - 18\alpha_{b*}^2 + 6\alpha_{b*} - 1)}, \quad \text{for} \quad 1/3 \leq \alpha_{b*} \leq 1/2$$

$$\frac{5\alpha_{b*}^{+}3}{16\alpha_{b*}}, \quad \text{for} \quad 1/2 \leq \alpha_{b*} \leq 1.$$

A.9 Limiting Condorcet Ranking Efficiency of BR for Parameter c^*

$$CRE(BR, \infty/\alpha_{c^*}) = \frac{-525\alpha_{c^*}^3 + 765\alpha_{c^*}^2 - 335\alpha_{c^*} + 31}{6(123\alpha_{c^*}^3 - 99\alpha_{c^*}^2 + 25\alpha_{c^*} - 5)}, \quad \text{for} \quad 1/3 \leq \alpha_{c^*} \leq 3/8$$

$$\frac{41911\alpha_{c^*}^4 - 56004\alpha_{c^*}^3 + 28746\alpha_{c^*}^2 - 7404\alpha_{c^*} + 831}{90(3\alpha_{c^*} - 1)(123\alpha_{c^*}^3 - 99\alpha_{c^*}^2 + 25\alpha_{c^*} - 5)}, \quad \text{for} \quad 3/8 \leq \alpha_{c^*} \leq 3/7$$

$$\frac{71\alpha_{c^*}^4 - 10684\alpha_{c^*}^3 + 11706\alpha_{c^*}^2 - 3884\alpha_{c^*} + 391}{60(1 - 3\alpha_{c^*})(123\alpha_{c^*}^3 - 99\alpha_{c^*}^2 + 25\alpha_{c^*} - 5)}, \quad \text{for} \quad 3/7 \leq \alpha_{c^*} \leq 1/2$$

$$\frac{1991\alpha_{c^*}^4 - 4604\alpha_{c^*}^3 + 4026\alpha_{c^*}^2 - 1724\alpha_{c^*} + 351}{60(17\alpha_{c^*} - 1)(1 - \alpha_{c^*})^3}, \quad \text{for} \quad 1/2 \leq \alpha_{c^*} \leq 2/3$$

$$\frac{1249\alpha_{c^*}^{-}289}{60(17\alpha_{c^*} - 1)}, \quad \text{for} \quad 2/3 \leq \alpha_{c^*} \leq 1.$$

A.10 Limiting Condorcet Ranking Efficiency of PR for Parameter c^*

$$CRE(PR, \infty/\alpha_{c^*}) = \frac{-17901\alpha_{c^*}^3 + 22653\alpha_{c^*}^2 - 9135\alpha_{c^*} + 1055}{54(123\alpha_{c^*}^3 - 99\alpha_{c^*}^2 + 25\alpha_{c^*} - 5)}, \text{ for } 1/3 \leq \alpha_{c^*} \leq 3/8$$

$$\frac{21390\alpha_{c^*}^4 - 42696\alpha_{c^*}^3 + 27540\alpha_{c^*}^2 - 6456\alpha_{c^*} + 409}{108(1 - 3\alpha_{c^*})(123\alpha_{c^*}^3 - 99\alpha_{c^*}^2 + 25\alpha_{c^*} - 5)}, \quad \text{for} \quad 3/8 \leq \alpha_{c^*} \leq 5/12$$

$$\frac{20082\alpha_{c^*}^4 - 26424\alpha_{c^*}^3 + 15660\alpha_{c^*}^2 - 5544\alpha_{c^*} + 841}{108(3\alpha_{c^*} - 1)(123\alpha_{c^*}^3 - 99\alpha_{c^*}^2 + 25\alpha_{c^*} - 5)}, \quad \text{for} \quad 5/12 \leq \alpha_{c^*} \leq 3/7$$

$$\frac{70503\alpha_{c*}^4 - 112860\alpha_{c*}^3 + 71226\alpha_{c*}^2 - 21420\alpha_{c*} + 2542}{108(3\alpha_{c*} - 1)(123\alpha_{c*}^3 - 99\alpha_{c*}^2 + 25\alpha_{c*} - 5)}, \quad \text{for} \quad 3/7 \leq \alpha_{c*} \leq 1/2$$

$$\frac{1545\alpha_{c*}^4 - 4644\alpha_{c*}^3 + 5670\alpha_{c*}^2 - 3372\alpha_{c*} + 815}{108(17\alpha_{c*} - 1)(1 - \alpha_{c*})^3}, \quad \text{for} \quad 1/2 \leq \alpha_{c*} \leq 2/3$$

$$\frac{565\alpha_{c*}^4 - 1620\alpha_{c*}^3 + 1566\alpha_{c*}^2 - 540\alpha_{c*} + 27}{36(17\alpha_{c*} - 1)(\alpha_{c*} - 1)^3}, \quad \text{for} \quad 2/3 \leq \alpha_{c*} \leq 3/4$$

$$\frac{51(1 - \alpha_{c*})}{4(17\alpha_{c*} - 1)}, \quad \text{for} \quad 3/4 \leq \alpha_{c*} \leq 1.$$

References

Cervone, D., Gehrlein, W., Zwicker, W. (2005). Which scoring rule maximizes Condorcet efficiency under IAC? *Theory and Decision, 58*, 145–185.

Gehrlein, W. (2002). Obtaining representations for probabilities of voting outcomes with effectively unlimited precision integer arithmetic. *Social Choice and Welfare, 19*, 503–512.

Gehrlein, W. (2005). Probabilities of election outcomes with two parameters: the relative impact of unifying and polarizing candidates. *Review of Economic Design, 9*, 317–336.

Gehrlein, W. (2011). Strong measures of group coherence and the probability that a pairwise majority rule winner exists. *Quality and Quantity, 45*, 365–374.

Gehrlein W., Lepelley D., Smaoui H. (2008). The condorcet efficiency of voting rules with mutually coherent voter preferences: a Borda compromise. *Annales d'Economie et de Statistiques.*

Gehrlein, W., Lepelley, D., Smaoui, H. (2010). Condorcet efficiency of voting rules with strong measures of group coherence. Unpublished manuscript.

Lepelley, D., Louichi, A., Smaoui, H. (2008). On Ehrhart polynomials and probability calculations in voting theory. *Social Choice and Welfare, 30*, 363–383.

Regenwetter, M., Grofman, B., Marley, A., Tsetlin, I. (2006). *Behavioral social choice*. Cambridge: Cambridge University Press.

Vickery, W. (1960). Utility, strategy and social decision rules. *Quarterly Journal of Economics, 74*, 507–535.

Ward, B. (1965). Majority voting and alternative forms of public enterprise. In J. Margolis (Ed.), *Public economy of urban communities* (pp. 112–126). Baltimore: Johns Hopkins.

Chapter 9
Modeling the Outcomes of Vote-Casting in Actual Elections

T. Nicolaus Tideman and Florenz Plassmann

9.1 Introduction

How often do events of interest to voting theorists occur in actual elections? For example, what is the probability of observing a voting cycle – an outcome in which no candidate beats all other candidates in pairwise comparison by majority rule? When there is a candidate who beats all others in such pairwise comparisons – a Condorcet winner – what is the probability that a voting method chooses this candidate? What is the probability that voters have an incentive to vote strategically – that is, cast their votes in ways that do not reflect their true preferences? Voting theorists have analyzed these questions in great detail, using a variety of statistical models that describe different distributions of candidate rankings. But there has been no systematic effort to determine which statistical model comes closest to describing the distribution of rankings of candidates in actual elections. Thus we know how often various voting events occur under different statistical models, but not how often voting events occur in actual elections. This chapter provides a framework for answering this question.

We consider elections in which each voter is asked to submit a strict ranking of m candidates. We interpret the rankings submitted by all voters as the outcome of a statistical model of vote-casting that yields a vector with $m!$ components, representing the possible strict rankings of the m candidates. To assess the probabilities of voting events, we need to know the likelihoods of these vectors – are all vectors of rankings equally likely or are some more likely to occur than others? We identify a statistical model of vote-casting that comes very close to describing the distribution of vectors of rankings in actual three-candidate elections.

T.N. Tideman
Department of Economics, Virginia Tech, Blacksburg, VA 24061, USA

F. Plassmann (✉)
Department of Economics, Binghamton University, Binghamton, NY 13902, USA
e-mail: fplass@binghamton.edu

D.S. Felsenthal and M. Machover, *Electoral Systems*, Studies in Choice and Welfare,
DOI 10.1007/978-3-642-20441-8_9,

We find it intuitive to view the statistical model of vote-casting as a two-part process. One part describes the distribution of the expected shares of the voters who will report each of the $m!$ rankings. The other part describes the distribution of the observed shares of the voters who report each of the $m!$ rankings, given a vector that describes the expected shares of these rankings.

Voting theorists have proposed various models to describe possible distributions of the expected shares. However, these models were generally proposed for purposes other than describing rankings in actual elections, and we are not aware of a systematic investigation of whether any of these models is at all likely to describe rankings in actual vote-casting processes.

We consider, for the case of three candidates, nine models that have been proposed by others as well as three new models. We evaluate these 12 models with two sets of voting data. Our first data set was assembled by Nicolaus Tideman in 1987 and 1988, and it consists of individual ballot information for 87 elections that we transform into 883 three-candidate elections.[1] Our second data set consists of 913 three-candidate "elections" that we construct from the "thermometer scores" that are part of the surveys conducted by the American National Election Studies (ANES).[2] The results that we obtain from these two rather different data sets are very consistent; they indicate that the combination of a spatial model of voting to describe the distribution of the vector of expected shares and a multinomial model to describe the distribution of the vector of actual shares fits the observed results of three-candidate elections much better than any of the models that have so far been used in theoretical analyses of voting events, and well enough that it may be difficult to devise an alternative model that would fit actual election data significantly better.

While the spatial model fits the observed data very well, it is too complex to permit the type of calculation of the probabilities of voting events that theorists have undertaken so far (as, for example, in Gehrlein 2002). Thus we envisage using this model instead for Monte Carlo simulations, to generate data that have the same characteristics as data from actual elections. Such simulations would be unnecessary if there were enough data from actual elections in which voters rank the candidates to determine the frequencies of rare voting events with a reasonable degree of accuracy. But there are not nearly enough ranking data to undertake such a project. For example, Brams and Fishburn (2001) and Saari (2001) use data from a single election – the 1999 election for president of the Social Choice and Welfare Society – to illustrate and analyze the properties of different voting methods.

[1]These data have been analyzed previously by Feld and Grofman (1990 and 1992), Felsenthal et al. (1993), Felsenthal and Machover (1995), Tideman and Richardson (2000), Regenwetter et al. (2002), and Tideman (2006).

[2]ANES survey data have been used in several previous analyses of voting. For example, Chamberlin and Featherston (1986) use scores from ANES surveys administered in 1972, 1974, 1976, and 1978 to construct combinations of four candidates. Regenwetter et al. (2002 and 2003) analyze the thermometer scores of the three major candidates in the four ANES surveys administered in 1968, 1980, 1992, and 1996 to construct combinations of three candidates. Our method of constructing three-candidate "elections" is the same as that in these earlier analyses.

Chamberlin and Featherston (1986) analyze data from five presidential elections of the American Psychological Association, while Regenwetter et al. (2002) use data from 12 elections for positions in professional organizations. Tideman's data set of 87 elections is one of the largest data sets of elections with voters' rankings that voting analysts have used. Even our two data sets with 883 and 913 three-candidate elections cannot provide reliable information about the frequencies of rare voting events. But if we are able to infer the underlying model of vote-casting from these elections, then data simulated under this model can reveal the frequencies of voting events of interest.

The remainder of this chapter is organized as follows: in Sect. 9.2 we formalize the two-part model of vote casting and introduce the 12 statistical models of the distribution of the expected shares. In Sect. 9.3 we explain our strategy for assessing the accuracy of these models, and we describe our data and report the results of our statistical analysis in Sect. 9.4. In Sect. 9.5 we illustrate the practical relevance of our analysis: we use the different models to predict the frequencies of voting cycles ("Condorcet's paradox") in our two data sets and show that several popular models come nowhere close to predicting the observed frequencies. We conclude in Sect. 9.6.

9.2 A Statistical Model of Vote-Casting

Consider an election with m candidates in which each of n voters submits a strict ranking of the candidates. There are $m!$ possible strict rankings, and n_r voters submit ranking r, $r = 1, \ldots, m!$. Let the discrete random variable N_r describe the frequencies with which $n_r = 0, \ldots, n$ voters submit ranking r, with $\Sigma n_r = n$, and let $N = \{N_1, \ldots, N_{m!}\}$ be a random vector defined on the N_r. Define p_r as the expected share of ranking r among the n submitted ballots, with $\Sigma p_r = 1$ so that $p = \{p_1, \ldots, p_m!\}$ is a vector of length $m!$ of expected shares. Let P be a random vector of length $m!$ that is defined on the collection of feasible p. A statistical model of vote-casting consists of specifications of both N and P.

Previous analyses of the frequencies of voting events have focused predominantly on specifying P. The requirement that $\Sigma p_r = 1$ implies that the support of all permissible models of P is contained in the unit $(m! - 1)$-simplex. Some models assume that P is defined on the entire $m!$-simplex, while others assume that P is defined on a strict subset of the simplex. In the following subsection, we introduce 12 models of P and discuss and compare their properties. In Sect. 9.2.2, we introduce two intuitive contenders for the distribution of N for a given realization of P.

9.2.1 *Statistical Models of P*

Statistical models of P are mappings from the unit $(m! - 1)$-simplex to $[0, \infty]$. They differ in the probability density or probability mass that they assign to different

p as well as in the subsets of the unit $(m! - 1)$-simplex that form the support of the mapping. It is straightforward to describe the differences among the probability structures but less straightforward to illustrate the differences in support for general m. However, it is customary in the theoretical literature of voting events to restrict the number of candidates, and many theoretical results are available only for elections with three candidates. We continue this tradition and restrict our analysis in this chapter to elections with $m = 3$; this permits us to represent the corresponding five-simplex (or hexateron) as a three-dimensional octahedron, which we use to derive intuitive graphical illustrations of the differences in the support of different models. In this octahedron, each of the six vertices represents a vector p with one probability equal to 1 and the remaining five probabilities equal to zero, each of the 15 edges (including the three virtual edges connecting pairs of opposite vertices) represents a vector p with four probabilities equal to zero while the remaining two p_r sum to 1, and each of the 20 (real and virtual) faces represents a vector p with three probabilities equal to zero while the remaining three p_r sum to 1. The point at the center of the octahedron represents the vector of equal probabilities with each $p_r = 1/6$. Although such a three-dimensional representation of a five-dimensional space cannot distinguish among all vectors of probabilities permitted by the five-dimensional space, it is nevertheless sufficient to illustrate the differences in the support of all but one of the models that we analyze. To simplify the exposition, we label the three candidates A, B, and C, and order the six rankings {ABC, ACB, CAB, CBA, BCA, BAC} so that ABC is ranking 1, ACB is ranking 2, and so on.

We classify each statistical model that describes the distribution of P as being in one of four categories. Models in the first category either have support of the entire five-simplex and accept every feasible vector p as a potential source of elections or they have support of uniform lattices on the entire five-simplex. Models in categories 2 through 4 assign zero probability as the source of an election to all points in the simplex except for subsets of Lebesgue measure zero that contain the subspaces of permitted probability vectors. Models in category 2 are of zero dimensionality and consist of either a single point or a single set of symmetric points within the simplex. Models in this category are described by discrete probability distributions rather than continuous probability distributions. Models in categories 3 and 4 are of higher dimensionality; those in category 3 are specified by linear restrictions on the unit simplex, while those in category 4 impose non-linear restrictions on the unit simplex. We describe each model below and summarize the properties of all 12 models in Table 9.1.

9.2.1.1 Models Whose Support is the Entire Unit Simplex

Our first model, the impartial anonymous culture (IAC), was proposed in Kuga and Nagatani (1974) and Gehrlein and Fishburn (1976). This model assumes that all points within the five-simplex are equally likely. Figure 9.1a shows the support of IAC – the entire octahedron. Several voting theorists have used IAC to calculate the

Table 9.1 Comparison of the 12 models of P

#	Model	Number of dimensions in subspace(s)	Parameters per election to be calibrated to fit the model to this election	Parameters to be calibrated to simulate data from this model	First proposal (of which we are aware) as a description of P
A. Models whose support is the entire five-simplex:					
1.	IAC	5	5	0	Kuga and Nagatani (1974) Gehrlein and Fishburn (1976)
2.	$\mathrm{IAC}_b(k_b)$	5	5	1	Gehrlein (2004)
3.	$\mathrm{IAC}_t(k_t)$	5	5	1	Gehrlein (2006)
4.	$\mathrm{IAC}_c(k_c)$	5	5	1	Gehrlein (2006)
B. Models whose support is one or more zero-dimensional subspaces of the five-simplex:					
5.	IC	0	0	0	Campbell and Tullock (1965)
6.	UUP	0	5 (for all elections)	5	Chamberlin and Featherston (1986)
C. Models whose support is one or more more-than-zero-dimensional subspaces, defined by linear restrictions:					
7.	SPP ($\mathrm{IAC}_b(0)$)	3	4	0	Lepelley (1995)
8.	DC	3	2	2	Gehrlein (1978)
9.	EPSF	3	2	2	This paper
D. Models whose support is one or more more-than-zero-dimensional subspaces, defined by nonlinear restrictions:					
10.	Borda	1	1	1	Conitzer and Sandholm (2005)
11.	Condorcet	1	1	1	This paper
12.	Spatial model	4	4	4	This paper

(1) The four models in A can describe any set of observed vote-shares and the parameters equal the observed shares. We describe our strategy for simulating data under these models in the appendix
(2) The five parameters of UUP are calibrated from all elections simultaneously, and the model's fit to any individual election is assessed on the basis of these parameter values. The parameters of SPP, DC, EPSF, the Borda model, the Condorcet model, and the spatial model are calibrated for each election individually
(3) The five parameters of UUP are constants in simulations from UUP. To simulate from any of the other models with unknown parameters, we assign distributions to all parameters and draw pseudo random numbers from these distributions that we use as inputs into the density functions (9.6) and (9.7)

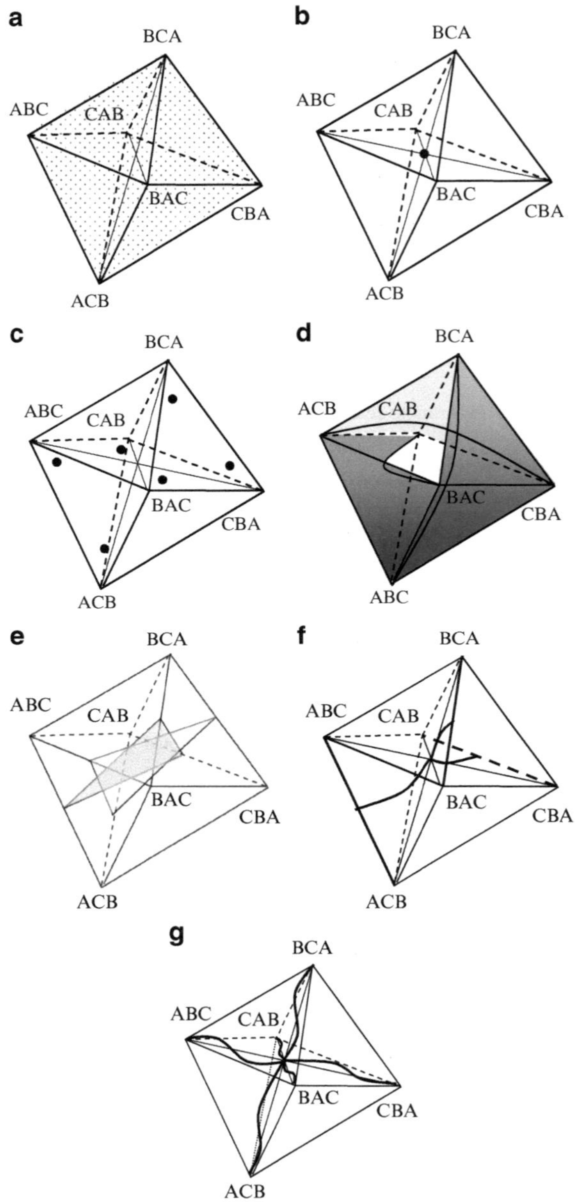

Fig. 9.1 (**a**) Three-dimensional representation of the support of IAC and variations. (**b**) Three-dimensional representation of the support of IC. (**c**) Three-dimensional representation of the support of UUP. (**d**) Three-dimensional representation of the support of SPP. (**e**) Three-dimensional representations of the support of DC and EPSF. (**f**) Three-dimensional representation of the support of the Borda model. (**g**) Three-dimensional representation of the support of the Condorcet model

probability that specific voting events will occur,[3] but they generally emphasize that they do not claim that equally likely probabilities of strict rankings describe vote-casting.[4] We are not aware of any published formal test of the statistical adequacy of the IAC assumption.

The assumption that all points within the five-simplex are equally likely permits expected vote shares that are unattainable for the number of voters in a specific election. For example, in an election with 10 voters it is possible to observe the shares $p = \{0.2, 0.2, 0.1, 0.1, 0.1, 0.1\}$ but not $p = \{0.25, 0.15, 0.1, 0.1, 0.1, 0.1\}$. A variation of IAC permits only those vote shares that yield integer values when multiplied by n. The support of this variation of IAC is not the entire five-simplex but rather a lattice on the five-simplex that becomes finer as n increases and fills the simplex for $n = \infty$. This variation of IAC yields slightly different results when the number of voters is small.

Several variations on IAC were developed specifically to examine the probability of observing Condorcet's paradox, and we consider three of them. Our second model, $\text{IAC}_b(k_b)$ was introduced in Gehrlein (2004). Assume that one candidate, say C, is ranked last no more often than either A or B, so that the relative frequency with which C is ranked last is a real number k_b between 0 and 1/3. $\text{IAC}_b(k_b)$ assumes that the probability of ABC ($=p_1$) is distributed uniformly on the interval $[0, k_b]$, and that the probability of BAC ($=p_6$) equals k_b minus the probability of ABC. The probabilities of the four other rankings are distributed uniformly on a subset of the tetrahedron formed by ACB, BCA, CBA, and CAB that includes all probability combinations for which $p_2+p_3 > k_b$ and $p_4+p_5 > k_b$ and whose components sum to $1 - k_b$. Thus while the components of P are determined simultaneously under IAC, they are determined sequentially in two steps under $\text{IAC}_b(k_b)$. Lepelley (1995) considers the case of this model when $k_b = 0$, so that the support of the model is limited to a subset of the unit simplex. We consider this limiting case $\text{IAC}_b(0)$ below as model 7.

Gehrlein (2006) describes two variations on $\text{IAC}_b(k_b)$ that he calls $\text{IAC}_t(k_t)$ and $\text{IAC}_c(k_c)$. $\text{IAC}_t(k_t)$ assumes that one candidate is ranked first no more often than the other two candidates, while $\text{IAC}_c(k_c)$ assumes that one candidate is ranked in the middle no more often than the other two candidates.[5] In both models, the probabilities of the six rankings are determined analogously to those in $\text{IAC}_b(k_b)$. We analyze $\text{IAC}_t(k_t)$ and $\text{IAC}_c(k_c)$ as models three and four. As with IAC, models 2–4 can be defined on either the entire five-simplex or the subset of the five-simplex

[3]For example, Saari (1990) uses this assumption to analyze the probability of strategic voting under different voting methods, Gehrlein (2002) uses it to analyze the probability of observing Condorcet's paradox, and Cervone et al. (2005) use it to analyze the probability that a Condorcet candidate, if it exists, will win the election.

[4]See, for example, Gehrlein (2002, p. 169) and Cervone et al. (2005, p.182).

[5]$\text{IAC}_b(k_b)$ measures the proximity of voter preferences to the case of single-peakedness (which occurs at $k_b = 0$), while the other two representations measure the proximities to "single-troughness" and "single-centeredness."

that contains only those vote-share vectors whose elements yield integer values when multiplied by n. The exact values of p in models 2–4 depend on the unknown parameters k_b, k_t, and k_c. In Appendix 7.1 we describe how we use our two data sets to estimate the parameters of these three models as well as of the models described below. Models 1–4 each have five degrees of freedom per election, which implies that they are able to describe perfectly every observable vector p.[6] Thus it is not possible to evaluate their accuracy through a likelihood of the realization of N given the most likely p, because every such likelihood is 1. In Sect. 9.3.1 we describe our strategy for assessing the accuracy of such models.

9.2.1.2 Models Whose Support is Composed of Zero-dimensional Subsets of the Unit Simplex

The most restrictive of the models that limit the support to a proper subset of the unit simplex is our fifth model, the impartial culture (IC). This model assumes that the support of P is a single point at the center of the simplex where all rankings are equally likely, or $p_r = 1/6$ for $r = 1, \ldots, 6$, so that IC has no degrees of freedom. Figure 9.1b shows IC's support as the octahedron's center. Its computational simplicity made IC popular in early Monte Carlo studies (see, for example, Campbell and Tullock 1965), but there is now considerable empirical evidence that IC does not describe actual elections (see Regenwetter et al. 2006, for a summary).

Our sixth model, which was proposed in Chamberlin and Featherston (1986) and which we call unique unequal probabilities (UUP), assumes that in every election, each candidate occupies a specifiable ranking niche (first, second, etc.), and that for each possible ranking of the candidates described by these niches, there is a constant probability that this ranking will be used by a voter. UUP restricts the support of P to a set of six points, one for each permutation of the rankings. Figure 9.1c shows the support of UUP – six points in symmetric locations in the octahedron. Unlike IC, UUP does not specify the values of the six probabilities, so it has five degrees of freedom that determine the six probabilities p_r. Probabilities corresponding to these five degrees of freedom, constant across elections, can be estimated. Note that UUP has five degrees of freedom to fit all elections, while models 1–4 have five degrees of freedom for each election. Because all six probabilities can be equal, UUP includes IC as a limiting case.

[6]This is true if the support is the entire 5-simplex as well as if the support is the subset of the 5-simplex that yields integer values when multiplied by n. It is worth emphasizing that our use of models 2–4 differs in spirit from William Gehrlein's original setup. Gehrlein defines these models for specific elections for which k_b, k_t, and k_c are observable, so that each model has only four degrees of freedom and its support is a proper subset of the 5-simplex. However, prior to applying these models to a specific election and thereby determining the parameter value, each model has five degrees of freedom and its support is the entire five-simplex.

9.2.1.3 Models Whose Support is more than Zero-dimensional and is Specified by Linear Restrictions on the Unit Simplex

Our seventh model is the limiting case of $\text{IAC}_b(k_b)$ with $k_b = 0$ that was proposed in Lepelley (1995). We refer to this model as SPP because it ensures single-peaked (group) preferences, making a voting cycle impossible. The support of SPP consists of all vectors p with either $p_4 = p_5 = 0$ (A is never ranked last), $p_2 = p_3 = 0$ (B is never ranked last), or $p_1 = p_6 = 0$ (C is never ranked last). Each of these restrictions eliminates two rankings, implying that the remaining possibilities can be viewed as a tetrahedron; if we could see in five dimensions, then the support of SPP would be three tetrahedrons, of which any two share an edge where four rankings are assigned zero probability. Figure 9.1d depicts this support in three dimensions.

Our eighth model, the Dual Culture (DC) proposed in Gehrlein (1978), assumes that the probabilities of opposite rankings are equal, that is, $p_1 = p_4$, $p_2 = p_5$, and $p_3 = p_6$. Our ninth model is a straightforward variation that assumes that the probability of each ranking is the same as that of the other ranking with the same first candidate, that is, $p_1 = p_2$, $p_3 = p_4$, and $p_5 = p_6$. We are not aware of any prior use of this model, which we call equal probabilities for same first (EPSF). Because neither model specifies the probabilities of any of the three pairs, both DC and EPSF have two degrees of freedom per election. Each set of three equalities specifies a plane in the five-simplex. Both the support of DC and the support of EPSF can be represented by a triangle determined by the midpoints of three edges of the octahedron. These triangles appear to be coplanar in the octahedron (see Fig. 9.1e), but in five dimensions their only common point is their centers where all probabilities equal 1/6. Thus both models include IC as a special case.

9.2.1.4 Models Whose Support is Curved Subsets of the Unit Simplex

As tenth and eleventh models, we investigate two models for which the Borda voting method and the Condorcet voting method, respectively, are maximum likelihood estimators. The "Borda model" supposes that there is a "best candidate" and evaluates, for every candidate, the evidence that this candidate is best. Such evidence is measured by the number of times a candidate is ranked second plus twice the number of times the candidate is ranked first. Conitzer and Sandholm (2005) show that this measure of the evidence is optimal only if the probabilities for both rankings with the best candidate first are equal, the probabilities for both rankings with the best candidate second are equal, and the probabilities for both rankings with the best candidate third are equal. That is, $p_1 = p_2$, $p_3 = p_6$, and $p_4 = p_5$ if A is best, $p_1 = p_4$, $p_2 = p_3$, and $p_5 = p_6$ if B is best, and $p_1 = p_6$, $p_2 = p_5$, and $p_3 = p_4$ if C is best. In addition, the probabilities for the pairs of rankings must follow a geometric sequence, so that

$$p_r = c_1 e^{\beta w_r} \tag{9.1}$$

where $w_r = 0, 1, 2$ denotes 1 minus the position that the "best candidate" occupies in ranking r (the ranking's contribution to the candidate's Borda score), β is a constant, and

$$c_1 = 1/2 \sum_{i=0}^{2} e^{i\beta} \tag{9.2}$$

to ensure $\Sigma p_r = 1$. Thus the Borda model has one degree of freedom (β) per election. Because the Borda model does not distinguish between rankings that rank a given candidate in the same position but differ in the positions they assign to the other candidates, its support includes the midpoints of the three edges of the octahedron joining rankings that list a given candidate first (where $\beta = \infty$), but the support includes none of the vertices. The three curved lines in Fig. 9.1f that start at the center of the octahedron (where $\beta = 0$; here the Borda model nests IC) and end midway between the pairs of vertices where a given candidate is first ($\beta = \infty$) depict the support of the Borda model for $0 < \beta < \infty$. The differences among the three equality restrictions imply that these three curved lines lie in three different planes.

Analogous to the derivation of the Borda model in Conitzer and Sandholm (2005), we define our eleventh model so that the Condorcet voting method is a maximum likelihood estimator of the ranking that is most favorable in terms of the statistical model that we assume has generated the election data (the "correct ranking"). While the Borda voting method assigns a score to each of the m candidates, the Condorcet voting method assigns a score to each of the $m!$ possible rankings.[7] Let r^* be the correct ranking and let $n^*_{rr} = 0, \ldots, 3$ be the number of pairs of candidates that are ranked the same in ranking r and ranking r^*. The Condorcet model specifies the components of p as

$$p_r = c_2 e^{\gamma n_{rr^*}} \tag{9.3}$$

where γ is a constant and

$$c_2 = 1 \Bigg/ \left(\sum_{i=0}^{3} f(i, 3) e^{i\gamma} \right) \tag{9.4}$$

to ensure $\Sigma p_r = 1$, where $f(i,\ 3)$ is the frequency distribution of Kendall's τ.[8] Like the Borda model, the Condorcet model has one degree of freedom (γ) per election.

The six corkscrew-shaped lines in Fig. 9.1g that start at the center of the octahedron (where the Condorcet model nests IC at $\gamma = 0$) and end at the six vertices ($\gamma = \infty$) depict the support of the Condorcet model for $0 < \gamma < \infty$. The

---

[7]Condorcet's explanation of his method (Condorcet, 1785) was opaque and contained errors; Kemeny (1959) proposed the same voting method in the twentieth century, and Young (1988) explained how Condorcet's intention could be understood despite his errors.

[8]See Kendall and Gibbons (1990, pp. 91 – 92).

actual support in the five-simplex is six elongated corkscrews that each span a three-space specified by two opposite vertices (for example, ABC and CBA) and the two points that are midway between pairs of vertices whose orderings both differ from one of the opposite vertices by a permutation of one pair of adjacent candidates (for example, BAC and ACB both differ from ABC by such permutations).

None of the 11 models discussed so far are based on a stated belief that the associated distributions of P might actually describe rankings in real elections.[9] IAC, IC, UUP, DC, and EPSF assume that various components of p are equally likely, for the sake of algebraic tractability. $\mathrm{IAC}_b(k_b)$, $\mathrm{IAC}_t(k_t)$, $\mathrm{IAC}_c(k_c)$ and SSP seek to describe rankings that have meaningful interpretations for the problem of defining probabilities of observing Condorcet's paradox. The Borda and Condorcet models are rationalizations of claims about how one ought to determine the winner in an election.

In contrast, our final model, the spatial model of voting, is based on plausible models of distributions of voter and candidate characteristics. As in other spatial models of voting, our spatial model assumes that voters care about the "attributes" of candidates; these attributes form a multi-dimensional "attribute space."[10] Every voter has an indifference map in attribute space, which contains an "ideal point" that describes the quantities of each attribute that the voter's ideal candidate would possess. Actual candidates also possess specifiable quantities of each attribute and therefore have locations in attribute space. We assume that attribute space has at least two dimensions and that the candidates are in "general position," where any slight change in the position of any one candidate does not change the dimensionality of the space that they span, so that the positions of the three candidates in attribute space span a two-dimensional "candidate plane" that is a subspace of attribute space.[11] Voters' indifference maps are defined in candidate space through their definitions in attribute space.

We follow Good and Tideman (1976) and assume that the positions of voters' ideal points in attribute space follow a spherical multivariate normal distribution, which implies that the distribution of "relative" ideal points in candidate space is bivariate normal. We further assume that every voter's utility loss from the choice of a particular candidate is the same increasing function of the distance between the candidate's location in candidate space and the voter's relative ideal point in candidate space, so that every voter's indifference surfaces are concentric spheres centered on the voter's ideal point.[12]

---

[9] See Gehrlein (2002, p. 197) for a discussion of the reasons for developing models 1–7.

[10] See Davis et al. (1970), and Enelow and Hinich (1984 and 1990).

[11] The case when all candidates' attributes lie in a single line requires special treatment because not all of the six possible rankings of the candidates occur, but it does not pose conceptual difficulties. See Good and Tideman (1976, pp. 380–381) for a description of the general case with $m > 3$.

[12] None of these assumptions is conceptually necessary and each could be replaced – at a cost of more complex calculations – if there is evidence that it does not represent election data sufficiently well. See Good and Tideman (1976) for a discussion.

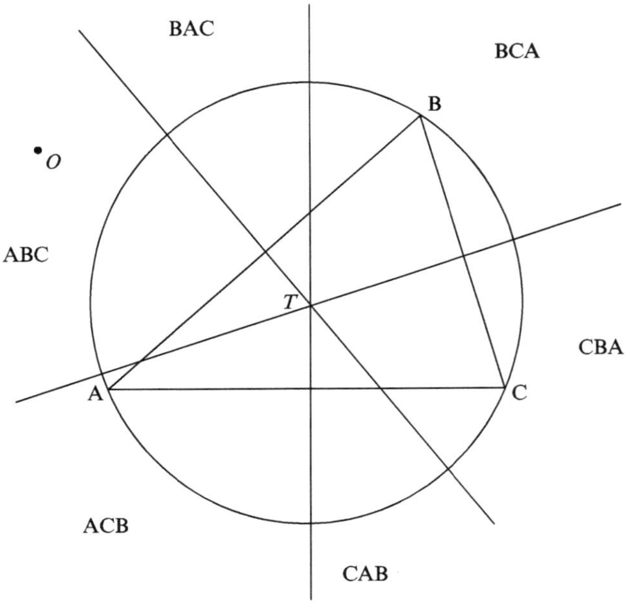

Fig. 9.2 Division of the candidate plane into six sectors by drawing the perpendicular bisectors of the three sides of the triangle formed by the candidates' locations, and the associated rank orders of the sectors. (The figure is taken from Good and Tideman 1976, p. 372.)

Suppose there is a set of candidates for which every voter submits a truthful ranking that reflects his ideal point, his indifference surfaces, and the positions of the candidates. To determine the vote share of each ranking, consider the triangle in the candidate plane that is formed by the locations of the three candidates, A, B, and C. We divide the candidate plane into six sectors by drawing the perpendicular bisectors of the three sides of this triangle. These bisectors intersect at the triangle's circumcenter, T. For the voters' ideal points in each sector, the distances to the locations of the three candidates have a unique rank order. These rank orders are indicated in Fig. 9.2, together with the mode of the circular bivariate normal distribution at O. The integral of the density function of this distribution over each sector is the expected value of the fraction of the voters who rank the candidates in the order corresponding to the sector's rank order.[13] These six integrals determine the probabilities p_r of the six rankings. Note that even though sectors that are opposite each other have the same angle, they do not have the same integral of the density function (and therefore do not imply the same p_r), unless O is not inside either of the sectors and the two lines that form the sectors come equally close to O.

[13]We use the algorithm described in DiDonato and Hageman (1980) to compute the integral of the bivariate normal distribution over each sector.

If O is exactly at the triangle's circumcenter T, then the spatial model coincides with our eighth model, DC, that assumes $p_1 = p_4$, $p_2 = p_5$, and $p_3 = p_6$. The spatial model has four degrees of freedom per election, and in Appendix 1 we explain how we parameterize the model.

The support of the spatial model in the five-simplex is sufficiently complex to make it difficult to represent it in two or three dimensions, but we can offer some insights. Every vertex and every edge of the five-simplex is included in the support of the spatial model. Of the 20 faces of the five-simplex, each specified by three vertices, 18 are included in the support of the spatial model. The two faces that are not included in the support of the spatial model are the ones that are defined by rankings that form a voting cycle: ABC, BCA, CAB and CBA, BAC, ACB. There are 15 (three-dimensional) cells in the five-simplex, each specified by four vertices and spanning a three-space on the "surface"of the five-simplex. Of these 15 cells, nine are included in the support of the model and six are not. The nine that are included in the support are the ones for which the two vertices that are not among the four in the defining set differ by a permutation of the positions of two candidates.

Within the five-simplex, the support of the spatial model consists of six curved hyper-cells (four-spaces), that is, one curved hyper-cell for each ordering of the candidates. Figure 9.3 offers some insights into the shape of these hyper-cells. Each triangle in the figure represents the subset of the five-simplex that is consistent with fixed shares for three rankings. The largest triangle imposes the restrictions $p_{\mathrm{ABC}} = p_{\mathrm{ACB}} = p_{\mathrm{CAB}} = 1/6$, and shows the intersection of the support of the spatial model with the two-simplex consistent with these restrictions (formed by $p_{\mathrm{CBA}} + p_{\mathrm{BCA}} + p_{\mathrm{BAC}} = 1/2$). For the middle triangle p_{CAB} is increased to 1/3 and the intersection of the support of the spatial model with the plane described by $p_{\mathrm{CBA}} + p_{\mathrm{BCA}} + p_{\mathrm{BAC}} = 1/3$ is plotted as the curve in this triangle. Finally, for the smallest triangle, p_{CAB}

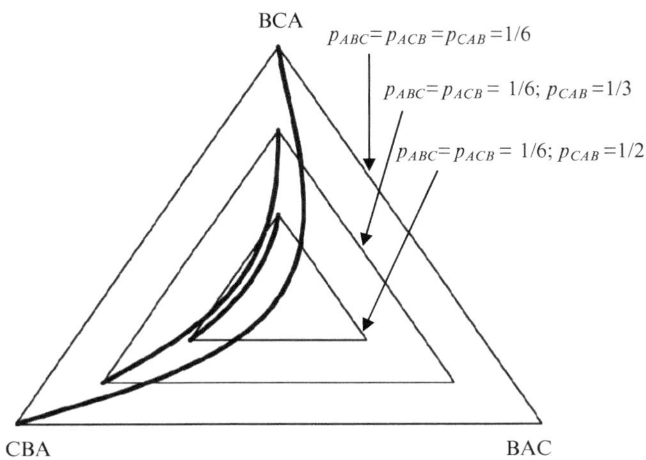

Fig. 9.3 Two-dimensional representations of the parts of the support of the spatial model that are consistent with specified values for the shares of ABC, ACB, and CAB

is increased to 1/2 and the intersection of the support of the spatial model with the plane described by $p_{\mathrm{CBA}} + p_{\mathrm{BCA}} + p_{\mathrm{BAC}} = 1/6$ is plotted as the curve in this triangle. The combination of the three curves suggests that, if we increased p_{CAB} continuously from 0 to 2/3, we would be looking at a saddle-shaped figure, from the front of the saddle. The saddle we show is symmetric because $p_{\mathrm{CBA}} = p_{\mathrm{BCA}}$, and it tilts to one side if $p_{\mathrm{CBA}} \neq p_{\mathrm{BCA}}$.

The spatial model nests DC (when the locations of T and O coincide) and thus IC, and we confirmed empirically that it also nests the Borda model but not the Condorcet model.

9.2.2 Statistical Models of *N*

Two intuitive contenders to describe the distribution of N for given n and p are the multinomial distribution and the multivariate Pólya distribution. The density function of the multinomial distribution is

$$f(n_1, \ldots, n_{m!}; n, p_1, \ldots, p_{m!}) = \frac{n!}{\prod_{r=1}^{m!} n_r!} \prod_{r=1}^{m!} p_r^{n_r}, \tag{9.5}$$

with first two moments $E[N_r] = np_r$, $Var[N_r] = np_r(1-p_r)$, and $Cov[N_r, N_s] = -np_r p_s$. An intuitive way of motivating the multinomial distribution is to assume that the vector p describes the probabilities with which a voter submits any of the six rankings, that these probabilities are the same for all voters, and that voters submit their votes independently, in which case the N_rs follow a multinomial distribution. Note that the multinomial distribution has no unknown parameters besides n and the p_rs, which implies that the model of P must provide a complete explanation of the expected vote shares as well as their variation.

The multivariate Pólya distribution can be motivated by relaxing the assumption that voters submit their votes independently. Consider the possibility that, if one voter submits a particular ranking r, the probability that the next voter will submit the same ranking increases while the probabilities that this voter will submit any of the other rankings decrease. If it is true that the probabilities of observing first ranking r and then ranking s are identical to observing first ranking s and then ranking r, then the probabilities behave like those associated with urns to which balls are being added. In such a case, the distribution of N is described by the multivariate Pólya distribution with density function

$$f(n_1, \ldots, n_{m!}; n, p_r, \ldots, p_{m!}, \delta) = \frac{n!}{\prod_{r=1}^{m!} n_r!} \frac{\Gamma(\delta)}{\Gamma(n+\delta)} \prod_{r=1}^{m!} \frac{\Gamma(n_r + \delta p_r)}{\Gamma(\delta p_r)} \tag{9.6}$$

(where Γ is the gamma function), whose first two moments are $E[N_r] = np_r$, $Var[N_r] = np_r(1 - p_r)\Psi$, and $Cov[N_r, N_s] = -np_r p_s \Psi$, where $\Psi =$

$(n + \delta)/(1 + \delta)$.[14] As δ approaches infinity, Ψ approaches 1, and the multivariate Pólya distribution converges to the multinomial distribution. This genesis of the multivariate Pólya distribution suggests that $1/\delta$ is a measure of the dependence between voters.

The multivariate Pólya distribution can be derived alternatively by relaxing the assumption that the vote-share vector of the multinomial distribution is deterministic. The assumption of a deterministic vote-share vector might not be acceptable if the random vector P is specified on a strict subset of the $m!$-simplex. Models 5–12 that we describe in Sect. 9.2.1 assume that P is defined on either spots or lines of zero width and thus assign zero probability to all points of the simplex except for the subsets of Lebesgue measure zero that contain the spots and lines of zero width. For example, our fifth model, IC, assumes that P describes a single spot at the center of the simplex, and IC therefore does not provide good descriptions of observed vote-casting processes with variances of the vote shares around the center of the simplex that differ from $5n/36$ in the three-candidate case. However, if the vote-share vector of the multinomial distribution is not deterministic but rather a draw from a random vector Q, then the resulting model of vote-casting can accommodate probability vectors that are outside the support of a particular model of P. A natural assumption is that Q follows an $m!$-variate Dirichlet distribution with parameter vector δp, where δ is inversely proportional to the variances of Q. Compounding the multinomial distribution with the Dirichlet distribution yields the multivariate Pólya distribution with density function (9.6). This genesis suggests that δ indicates how well a model of P that is defined on a strict subset of the $m!$-simplex describes observed vote-casting processes: the larger the value of δ, the smaller are the variances of Q and the less additional permitted but unexplained variation in the p_rs is necessary to fit the resulting model of vote-casting to actual elections.

9.3 Model Evaluation

9.3.1 Evaluation of Models with Fewer than Five Degrees of Freedom per Election

For models with fewer than five degrees of freedom per election (models 5–12, whose support is a proper subset of the five-simplex), we use likelihood calculations to assess how well the model can explain the vectors of vote shares that we observe in actual three-candidate elections, taking account of the fact that the models differ in their degrees of freedom. We derive the likelihoods by identifying, for each

[14]See Mosimann (1962, pp. 67–68).

election, the vector p that, according to the model of P, is most likely to have produced the observed numbers of votes for the six rankings, assuming that these rankings follow either a multinomial or a multivariate Pólya distribution. For a set of elections whose outcomes are independent of each other, the likelihood function is proportional to the product over all elections of density function (9.6) or (9.7). For each model, we estimate the unknown parameters (if any) by maximizing the likelihood function over the observed election data. For nested models (for example, the Borda model nested in the spatial model or DC nested in the spatial model), a likelihood ratio test indicates which model yields a better fit of the data, given their different degrees of freedom. Nested as well as non-nested model can be compared by the Akaike and Bayesian information criteria (AIC and BIC) that account for differences in the degrees of freedom.[15] Models with lower values of AIC or BIC use degrees of freedom more efficiently in describing the data than models with higher AIC or BIC.

9.3.2 *Evaluation of Models with Five Degrees of Freedom per Election*

The four models of P with five degrees of freedom per election and whose support is the entire five-simplex are able to describe perfectly any vector of observed vote shares. This makes it impossible to evaluate their accuracy on the basis of the likelihood functions described above.

To assess these models, we use their assumptions to simulate multiple "elections" and determine whether these simulated data have the same statistical properties as observed election data. For each model of P whose support is a proper subset of the simplex (models 5-12), we calculate the mean (multi-dimensional Euclidean) distance from an observed vector of vote shares to the closest vector that this model permits as the source of the election. We measure the mean distance σ as the mean square root of the sum of the squared differences,

$$\sigma = \frac{1}{E}\sum_{e=1}^{E}\sqrt{\sum_{r=1}^{6}(p_{re} - s_{re})^2}, \tag{9.7}$$

where E is the number of elections, s_{re} is the observed vote share of ranking r in election e, and p_{re} is the corresponding vote share predicted by the respective model. We place two subscripts on σ. The first indicates the source of the data, O for observed elections, and a number for a simulation with the corresponding model. The second subscript indicates which of models 5–12 was used to measure

[15]These criteria are determined as AIC $= -2\ln(L) + 2d$ and BIC $= -2\ln(L) + d\overline{\ln(N_e)}$, where d is the total number of degrees of freedom, L is the maximum value of the likelihood function, and $\overline{\ln(N_e)}$ is the mean value of the log of the number of voters in an election. Thus BIC imposes a heavier penalty for the use of degrees of freedom than AIC.

distance to the nearest outcome permitted by a model. We compare σ_{ij}, $i = 1, \ldots, 6, 8, \ldots, 12$, $j = 5, 6, 8, \ldots, 12$, computed from the simulated data, with the σ_{Oj}, $j = 5, 6, 8, \ldots, 12$, computed from observed elections. If σ_{ij}, for any j, differs significantly from σ_{Oj}, then this is evidence that the simulated data differ in significant ways from the observed data and that model i is unlikely to be a good statistical description of the process that generated the observed data.

9.4 Empirical Evaluation of the 12 Models

9.4.1 The Data

Our first data set consists of 84 elections that were administered by the Electoral Reform Society (ERS) and tabulated by Nicolaus Tideman in 1987 and 1988 and three elections that he included from another source. For each election, we have individual ballot information about the strict ranking of candidates provided by each voter (the ballots did not permit ties). The number of voters in these elections ranges from 9 to 3,422, with a mean of 410.5, and the number of candidates ranges from 3 to 29, with a mean of 8.7. Most voters ranked only some of the candidates in these elections. We use these ballots to construct all possible combinations of three candidates within an election, treating each combination as one election with three candidates. We use a ballot in such a three-candidate election only if the voter ranks all three candidates, which yields a total of 20,087 three-candidate elections with between 1 and 1,957 voters. We found that elections with too few voters contain mostly random noise and do not provide much information about the model of vote-casting. We therefore limit our analysis to elections with more than 350 voters, which leads us to use information from only 10 of the original 87 elections in our analysis.[16] Our ERS data set contains 883 three-candidate elections with between 350 and 1,957 voters, with a mean of 716.4 voters. (Had all voters in every election ranked all available candidates, this procedure would have yielded 5,348 three-candidate elections with a mean of 763.7 voters.)

The data from these 883 three-candidate elections are not independent, but this lack of independence is unlikely to affect our conclusions.[17] It is possible that such three-candidate combinations that are derived from rankings of more than three candidates are qualitatively different from rankings of elections with exactly three

[16]The fact that less than 12% of the available elections had enough voters to be useful highlights the paucity of data as well as the value of identifying a statistical model that can be used to simulate elections.

[17] Dependence among the elections requires that the likelihood function be calculated from the conditional, rather than the marginal, distributions of the six vote-share vectors over all elections. However, because we use the likelihood functions to compare the accuracy of each model with that of the other models, ignoring the dependence is unlikely to affect these relative assessments.

candidates – for example, because it is often simpler to rank three candidates than a larger number of candidates. However, we have individual ballot data for only eight genuine three-candidate elections, which is not sufficient to draw reliable statistical inference about the appropriate model of vote-casting.

Because no reasonable voting method is immune to strategic voting, it is possible that the rankings in our data set reflect voters' strategic considerations. We therefore examine a second ranking data set that is derived from survey data rather than election data, because the strategic considerations of survey respondents are likely to differ from those of voters. We assemble our second data set from the "thermometer" scores that are part of 18 surveys conducted by the American National Election Studies (ANES) between 1970 and 2004. These surveys are conducted every two years, and participants are asked to rate politicians on a scale from 0 to 100 (the thermometer). We refer to these persons as "candidates."

The number of respondents in a survey ranges from 1,212 in 2004 to 2,705 in 1974, and the number of candidates included in the surveys ranges from 3 in 1986 and 1990 to 12 in 1976. As before, we construct all possible combinations of three candidates within a year, for a total of 913 three-candidate combinations from all 18 surveys, with between 759 and 2,521 responses and a mean of 1,566.7 responses. For simplicity, we will refer to the survey respondents as "voters" and to the three-candidate combinations as "elections." For each response, we rank the three candidates according to their thermometer scores, thereby eliminating any information about the intensity of the voter's preferences. If a response yields a strict ranking of candidates, then we count it as one vote for this ranking. Voters are allowed to assign equal scores to different candidates, and we adopt the following intuitive rule of accommodating ties: If all candidates are tied, then we count the response as 1/6 vote for each ranking, and if two candidates are tied, then we count the response as half a vote for each of the two possible strict rankings that break the tie. Thus our adjusted data set consists of the total number of votes for each of the six strict rankings in each of the 913 three-candidate elections. (Had all voters in every election ranked all available candidates, this procedure would have yielded 913 three-candidate elections with a mean of 1,989 voters.)

Table 9.2 shows, for both data sets, the number of elections with different numbers of candidates and the average number of voters in the original data sets for elections with 350 or more voters, the number of three-candidate elections we could have extracted from each of these elections, the number of three-candidate elections that had 350 or more voters that we did extract, the number of original elections whose ballots we used to construct these three-candidate elections with 350 or more voters, and the average number of voters in each of these three-candidate elections with 350 or more voters.

It is notable that both data sets have voting cycles – the 913 ANES surveys have four cycles (0.44%), and the 20,087 ERS elections have 476 cycles (2.37%). However, there are only 101 voting cycles (1.45%) among the 6,794 ERS elections with 21 or more voters, and only six voting cycles (0.68%) among the 883 ERS elections with 350 and more voters. Thus the frequency of voting cycles falls fairly quickly as the number of voters increases.

Table 9.2 Properties of the original ERS and ANES data sets and of the extracted three-candidate elections

	Number of candidates in the original elections	Number of elections with 350 or more voters	Average number of voters per election	Potential number of three-candidate elections (with any number of voters)	Number of three-candidate elections with 350 and more voters	Number of original elections that lead to three-candidate elections with 350 and more voters	Average number of voters per three-candidate election
	(1)	(2)	(3)	(4)	(5)	(6)	(7)
ERS data:	3	0					
	4	6	614	24	4	1	1,005
	5	3	1,238	30	10	1	1,859
	6	2	770	40	19	1	414
	7	3	714	105	37	2	390
	8	3	541	168			
	9	2	450	168			
	10	5	679	600	120	1	665
	11	2	913	330	108	2	404
	12–20	9	986	3,883	585	2	531
	Sum	35		5,348	883	10	
ANES data:	3	2	2,078	2	2	2	1,599
	4	0					
	5	2	1,396	20	20	2	1,248
	6	2	1,607	40	40	2	1,403
	7	5	1,858	175	175	5	1,540
	8	3	2,184	168	168	3	1,699
	9	2	1,561	168	168	2	1,228
	10	1	2,257	120	120	1	1,826
	11	0					
	12	1	2,248	220	220	1	1,663
	Sum	18		913	913	18	

9.4.2 Assessment of the Eight Models with Fewer than Five Degrees of Freedom per Election

SPP predicts that, in every election, one of the candidates will never be ranked last. The fact that our two data sets contain predominantly elections in which every candidate is ranked last by some voters is conclusive evidence against the empirical relevance of this model, at least for our two data sets. We therefore do not consider SPP in our further analysis. Table 9.3 reports the log-likelihood values and the AIC and BIC as well as our estimates of the Dirichlet δs for the remaining seven models whose support is a proper subset of the five-simplex. For both data sets, the three measures of accuracy agree about the relative ranking of these models. IC, UUP, and DC have the smallest log-likelihoods, the largest values of AIC and BIC, the smallest values of δ, and thus the lowest accuracy. The Borda model consistently has the fourth highest accuracy, while EPSF and the Condorcet model are the third and second most accurate.

All our measures of accuracy indicate that the spatial model describes the observed data much better than any of the other seven models. Likelihood ratio tests of the spatial model and the nested IC, DC, and Borda models indicate that the improvement in the likelihood justifies the spatial model's additional degrees of freedom. For all models, AIC and BIC also suggest that, by a wide margin, the spatial model provides the best description of the ERS elections and the ANES surveys, despite its much larger use of degrees of freedom. The estimate of δ of the spatial model is close to being infinite, which indicates that the strict spatial model provides a very good explanation of almost all of the variation among the observed vote shares, and that perturbing the predicted vote-shares in the manner described by the Dirichlet process provides no significant improvement in the spatial model's fit. In contrast, the estimated δs of the other models are very small in comparison, meaning that adding variation through the Dirichlet process beyond the variation explained by the respective model improves the fit of these models considerably.

9.4.3 Assessment of the Four Models with Five Degrees of Freedom per Election

We analyze next whether the election data are consistent with any of the four models with five degrees of freedom per election, IAC, $\text{IAC}_b(k_b)$, $\text{IAC}_t(k_t)$, and $\text{IAC}_c(k_c)$. We report only the results from the ANES data because the results from the ERS data are qualitatively the same. We calibrated the three parameters – the shares of last, first, and middle ranks of the candidates with the fewest last, first, and middle ranks – as $k_b = 0.218$, $k_t = 0.201$, and $k_c = 0.226$. The fact that these values differ notably from the value of 1/3 predicted by IAC is some evidence against the hypothesis that IAC has generated the observed ANES data.

Table 9.3 Assessment of seven models whose support is a subset of the five-simplex

		883 ERS elections					913 ANES surveys				
Mean number of voters:		716.4					1,566.7				
	DF for ERS; ANES	LLH Multinomial	LLH Pólya	δ	AIC	BIC	LLH Multinomial	LLH Pólya	δ	AIC	BIC
Model:	(1)	(2)	(3)	(4)	(5)	(6)	(7)	(8)	(9)	(10)	(11)
IC	0; 0	−113, 551	−23, 084	15.93	227,102	227,102	−195, 310	−27, 811	19.29	390,621	390,621
UUP	5; 5	−72, 581	−22, 133	25.51	145,172	145,204	−171, 991	−27, 586	21.42	343,992	344,019
DC	2,649; 2,739	−93, 902	−23, 028	17.22	193,102	210,036	−127, 510	−26, 915	29.70	260,498	278,099
EPSF	2,649; 2,739	−51, 295	−21, 091	43.92	107,888	124,822	−91, 278	−26, 000	45.23	188,034	205,635
Borda model	883; 913	−54, 255	−21, 215	40.16	110,276	115,921	−122, 264	−26, 679	32.53	246,354	251,218
Condorcet model	883; 913	−32, 614	−19, 590	93.49	66,994	72,639	−93, 229	−25, 986	44.96	188,284	193,148
Spatial model	3,532; 3,652	−14, 267	−14, 267	> 1.E08	35,598	58,177	−16, 428	−16, 428	> 1.E08	40,160	59,614

(1) We calculated the AIC and BIC in columns 5, 6, 10, and 11 using the multinomial log-likelihood functions
(2) For the spatial model, the differences between the multinomial and Pólya log-likelihood function values are less than 0.5

As discussed in Sect. 9.3.2, the accuracy of these four models cannot be evaluated by calculating a likelihood of observed outcomes given p, because a model with five degrees of freedom can match the six observed shares of any three-candidate election. We therefore drew, for each of these four models, 1,000 samples of probability vectors from the distribution on the unit five-simplex specified by the model, and evaluated these samples in terms of σ_{ij}, $j = 5, 6, 8, \ldots, 12$. If any of the four models with five degrees of freedom per election has generated the observed data, then we would expect that, for data simulated with that model, σ_{ij}, will be similar to σ_{Oj}, $j = 5, 6, 8, \ldots, 12$.

Column 1 of Table 9.4 shows the σ_{Oj}s that we calculated from the observed ANES data, and columns 2–5 show the corresponding values of σ_{ij} that we calculated from our simulations. The differences are highly significant. We obtained the same result with other measures of distance.[18] Our results mean that the vectors of vote shares in the ANES data are much more clustered than what is permitted by any of the four models that assign equal probabilities to either all possible vectors (in case of IAC) or vectors in which two probabilities sum to no more than a value less than 1/3 (in case of $\text{IAC}_b(k_b)$, $\text{IAC}_t(k_t)$, and $\text{IAC}_c(k_c)$). Specifically, our estimate of an effectively infinite δ for the spatial model implies that the vote-share vectors that generate elections are clustered extremely closely around the vote share vectors predicted by the strict spatial model. We obtained the same results when we calibrated the parameters of models 2–4 from the ERS data and compared the σ_{ij}s with the ERS σ_{Oj}s. Thus the hypothesis that our two data sets were generated by IAC, $\text{IAC}_b(k_b)$, $\text{IAC}_t(k_t)$, or $\text{IAC}_c(k_c)$ cannot be sustained.

In columns 2–8 of Table 9.5 we repeat this exercise with data that we simulated with the seven remaining models 5, 6, 8, …, 12 whose support is a proper subset of the five-simplex. With the exception of the data simulated under the spatial model, the σ_{ij}s are significantly different from the σ_{Oj}s, which is yet further evidence that neither of our two data sets is likely to have been generated by any of these models. In contrast, the σ_{12j}s that we calculated from the data simulated under the spatial model are not significantly different from the $\sigma_{\text{O}j}$s. The one exception is $\sigma_{12,12}$, which is significantly smaller than $\sigma_{\text{O},12}$. This suggests that, even though the generating mechanism of the ANES data is likely to be very close to the spatial model, it is not identical to it. One possible explanation is that our simulation framework assumes that voters know all candidates, even though not all voters in the ANES data ranked every candidate. We found that it is possible to improve upon the spatial model by relating candidate recognition to the δ of the multivariate Pólya distribution, but we leave a more thorough investigation of this issue for future research.

[18]We weighted the σs by their standard deviations and we also evaluated the likelihood ratios that we determined from the simulated and the observed data.

Table 9.4 Mean Euclidian distances σ calculated from observed and simulated data

	Observed data	Simulated data			
Data source:	ANES	IAC	$IAC_b(k_b)$	$IAC_t(k_t)$	$IAC_c(k_c)$
			$k_b = 0.218$	$k_t = 0.201$	$k_c = 0.226$
Mean number of voters:	1,566.7	1,567	1,567	1,567	1,567
Model:	(1)	(2)	(3)	(4)	(5)
IC	0.197 (0.0021)	0.335 (0.0031)	0.277 (0.0022)	0.291 (0.0023)	0.289 (0.0023)
UUP	0.153 (0.0017)	0.331 (0.0032)	0.268 (0.0030)	0.251 (0.0030)	0.248 (0.0027)
DC	0.148 (0.0023)	0.253 (0.0030)	0.197 (0.0023)	0.214 (0.0026)	0.238 (0.0027)
EPSF	0.126 (0.0020)	0.252 (0.0030)	0.202 (0.0022)	0.240 (0.0026)	0.210 (0.0024)
Borda model	0.149 (0.0018)	0.272 (0.0029)	0.236 (0.0022)	0.250 (0.0026)	0.228 (0.0021)
Condorcet model	0.121 (0.0015)	0.211 (0.0024)	0.211 (0.0020)	0.211 (0.0022)	0.190 (0.0018)
Spatial model	0.008 (0.0002)	0.061 (0.0018)	0.086 (0.0021)	0.083 (0.0021)	0.080 (0.0020)

Standard errors of estimate are in parentheses

9.5 Cycles

As a final comparison of the contenders for the model of P, we examine how accurately each model predicts the observed occurrence of voting cycles in our two data sets. A common definition of a voting cycle is the absence of a strict pairwise majority rule winner (SPMRW), because there is a cycle (or at least a "semi-cycle" with one or more ties) if no candidate beats all other candidates in pairwise comparisons. We simulate data under different models of P to assess whether the frequencies of cycles in these simulated data correspond to the frequencies of cycles that we observe in the ANES and ERS data.

To estimate the frequency of SPMRWs for the eleven models of P, we simulated one million elections for each of the eleven models, and we recorded the number of SPMRWs in these simulated elections. To be able to compare our estimates with the observed frequency of SPMRWs in our two data sets, we had to account for the variations in the number of voters in the observed elections. To do so we recorded the numbers of voters in each of the 883 ERS elections and 913 ANES elections, and then simulated either 11,325 or 11,326 elections for each number of voters in the ERS data set (for a mean of 716.4 voters) and either 10,952 or 10,953 elections for each number of voters in the ANES data set (for a mean of 1566.7 voters). To

Table 9.5 Mean Euclidian distances σ calculated from observed and simulated data

	Observed data	Simulated data						
Data source:	ANES	IC	UUP	DC	EPSF	Borda	Condorcet	Spatial model
Mean number of voters:	1,566.7	1,567	1,567	1,567	1,567	1,567	1,567	1,567
Model:	(1)	(2)	(3)	(4)	(5)	(6)	(7)	(8)
IC	0.197 (0.0021)	0.022 (0.0002)	0.079 (0.0004)	0.267 (0.0035)	0.260 (0.035)	0.116 (0.0019)	0.138 (0.0023)	0.193 (0.0022)
UUP	0.153 (0.0017)	0.059 (0.0003)	0.022 (0.0002)	0.261 (0.0034)	0.250 (0.0034)	0.098 (0.0016)	0.097 (0.0019)	0.156 (0.0020)
DC	0.148 (0.0023)	0.016 (0.0002)	0.067 (0.0003)	0.016 (0.0003)	0.225 (0.0003)	0.114 (0.0019)	0.133 (0.0021)	0.145 (0.0023)
EPSF	0.126 (0.0020)	0.016 (0.0002)	0.061 (0.0003)	0.231 (0.0030)	0.016 (0.0003)	0.047 (0.0006)	0.076 (0.0014)	0.127 (0.0020)
Borda model	0.149 (0.0018)	0.018 (0.0002)	0.068 (0.0003)	0.267 (0.0035)	0.111 (0.0016)	0.019 (0.0002)	0.077 (0.0014)	0.142 (0.0019)
Condorcet model	0.121 (0.0015)	0.017 (0.0002)	0.048 (0.0003)	0.265 (0.0034)	0.158 0.0022)	0.058 (0.0010)	0.018 (0.0002)	0.120 (0.0016)
Spatial model	0.008 (0.0002)	0.008 (0.0002)	0.020 (0.0003)	0.006 (0.0008)	0.006 (0.0003)	0.007 (0.00021	0.018 (0.0003)	0.006 (0.0002)

Standard errors of estimate are in parentheses

assess the differences among the models as the number of voters becomes either small or large, we also simulated ten sets of 1,000,000 elections each, with 10, 20, 40, 100, 500, 1,000, 5,000, 10,000, 100,000, and 1,000,000 voters per election.[19]

Column 2 of Tables 9.6 and 9.7 shows the share of elections with an SPMRW in the ERS and ANES data. Columns 3–7 of Table 9.6 show the shares of elections with an SPMRW among the elections that we simulated with the spatial model, the Borda model, the Condorcet model, EPSF, and UUP, while Columns 3–8 of Table 9.7 show these shares for IC, DC, IAC, $IAC_b(k_b)$, $IAC_t(k_t)$, and $IAC_c(k_c)$. Consider first the two shaded rows with elections whose mean numbers of voters correspond to those of the ERS and ANES elections. We arranged the models of P so that models whose simulated shares are closer to the observed shares are further to the left. The predictions of the spatial model for the shares of elections with an SPMRW come closest to the observed shares, followed by the predictions of the Borda model, the Condorcet mode, EPSF, and UUP.[20] (Although UUP predicts the share of SPMRWs for elections with a mean of 1566.7 voters slightly better than the spatial model, UUP's prediction for elections with a mean of 716.4 voters is much worse.) The shares predicted by the five models in Table 9.6 are reasonably accurate, while the predictions of the six models in Table 9.7 are much worse. This result is not surprising because the five models in Table 9.6 all say that all elections will have SPMRWs as the number of voters approaches infinity, while the six models in Table 9.7 all say that the limit of the share of elections with SPMRWs is less than one. That is, the models in Table 9.6 predict that voting cycles will occur very rarely, if at all, in elections with many voters, while the models in Table 9.7 predict that voting cycles will occur in more than 6% of all elections, even if the number of voters is very large. Thus the small number of voting cycles in our two data sets is further evidence that IAC, $IAC_b(k_b)$, $IAC_t(k_t)$, $IAC_c(k_c)$, DC, and IC should not be used to make predictions about what happens in actual elections. [21]

For IC and IAC, Gehrlein (2002) determined analytically the expected frequency of SPMRWs for elections with different numbers of voters; for $IAC_b(k_b)$ and $IAC_c(k_c)$, Gehrlein and Lepelley (2009) determined analytically the expected frequency of SPMRWs for different k as the number of voters goes to infinity. The fact that our simulations of IC and IAC for 10, 20, 40 and 100 voters as well as our simulations of $IAC_b(k_b)$ and $IAC_c(k_c)$ for 1,000,000 voters yield frequencies of

[19]When the number of voters is small, the percentages for elections with odd numbers of voters differ from those of elections with even numbers. Out of space considerations we report only percentages for elections with even numbers of voters.

[20]For each model, we undertook the simulations using the parameters that we calibrated from the ANES elections. Each model's prediction of the frequency of SPMRWs in elections with an average of 716.4 voters can therefore be interpreted as an out-of-sample prediction of the observed frequency of ERS elections with an SPMRW.

[21]Gehrlein (2002, p.189) reports analytic results for a model of P that we do not analyze here, the "maximum culture." He finds that only 90.83% of all elections generated under the maximum culture have an SPMRW as the number of voters approaches infinity. His result is strong evidence that the maximum culture also does not describe vote-casting in actual elections.

Table 9.6 Shares of strict pairwise majority rule winners (SPMRWs) in observed and simulated elections

Mean number of voters	Observed elections	Simulated elections: Spatial model	Borda model	Condorcet model	EPSF (6)	UUP
(1)	(2)	(3)	(4)	(5)	(6)	(7)
10		69.68%	67.39%	67.82%	81.44%	60.01%
20		80.83%	79.07%	79.36%	89.67%	69.42%
40		88.79%	88.01%	88.15%	94.52%	77.62%
100		94.98%	95.27%	95.39%	97.73%	86.51%
500		98.92%	99.41%	99.46%	99.54%	97.43%
716.4	*99.32%*	*99.36%*	*99.55%*	*99.62%*	*99.73%*	*98.40%*
1,000		99.47%	99.77%	99.82%	99.77%	99.14%
1,566.7	*99.56%*	*99.78%*	*99.89%*	*99.92%*	*99.91%*	*99.77%*
5,000		99.89%	99.98%	99.99%	99.95%	99.99%
10,000		99.94%	99.99%	100.00%	99.98%	100.00%
100,000		100.00%	100.00%	100.00%	100.00%	100.00%
1,000,000		100.00%	100.00%	100.00%	100.00%	100.00%

All shares are determined from the results of one million simulated elections. To simulate one million elections with a mean number of 716.4 voters, we simulated either 11,325 or 11,326 elections with the number of voters from each of the 883 ERS elections. To simulate one million elections with a mean number of 1,566.7 voters, we simulated either 10,952 or 10,953 elections with the number of voters from each of the 913 ANES elections. For all other rows, we simulated one million elections, each with the number of voters shown in the first column

Table 9.7 Shares of strict pairwise majority rule winners (SPMRWs) in observed and simulated elections

	Observed elections	Simulated elections:					
Mean number of voters		IAC	$IAC_t(k_t)$	DC	$IAC_c(k_c)$	$IAC_b(k_b)$	IC
(1)	(2)	(3)	(4)	(5)	(6)	(7)	(8)
10		73.46%	65.65%	61.94%	70.69%	68.12%	58.44%
20		81.98%	76.03%	69.99%	77.66%	77.34%	66.85%
40		87.31%	83.41%	76.34%	83.94%	80.54%	73.46%
100		91.07%	89.00%	82.32%	87.79%	86.41%	79.69%
500		93.20%	92.10%	88.53%	90.36%	90.07%	85.90%
716.4	*99.32%*	*93.46%*	*92.57%*	*91.18%*	*90.65%*	*90.43%*	*88.67%*
1,000		93.45%	92.64%	90.00%	90.69%	90.54%	87.55%
1,566.7	*99.56%*	*93.65%*	*92.90%*	*92.12%*	*90.86%*	*90.77%*	*89.59%*
5,000		93.69%	92.99%	92.06%	90.91%	90.86%	89.51%
10,000		93.75%	93.02%	92.57%	90.96%	90.87%	90.06%
100,000		93.73%	93.00%	93.34%	91.08%	90.98%	90.82%
1,000,000		93.71%	93.05%	93.60%	90.99%	90.91%	91.10%

All shares are determined from the results of one million simulated elections. To simulate one million elections with a mean number of 716.4 voters, we simulated either 11,325 or 11,326 elections with the number of voters from each of the 883 ERS elections. To simulate one million elections with a mean number of 1,566.7 voters, we simulated either 10,952 or 10,953 elections with the number of voters from each of the 913 ANES elections. For all other rows, we simulated one million elections, each with the number of voters shown in the first column

SPMRWs that correspond almost exactly to the analytical results suggests that our simulations are reliable.[22]

Although the predictions of EPSF and UUP for the ERS and ANES elections are comparable to the predictions of the first three models, their predictions become notably different as the number of voters becomes either very small or very large. For elections with 10 voters, the predicted shares of EPSF and UUP differ by about ten percentage points from those of the spatial model. In addition, UUP predicts that every election has an SPMRW when there are 10,000 and more voters, while our simulations with the spatial model yielded about 500 elections out of a million with 10,000 voters without an SPMRW.

The shares predicted by the Borda and the Condorcet models are most similar to those of the spatial model, but even those two models underpredict the share of SPMRWs for elections with few voters and overpredict this share for elections with many voters; our simulations with 10,000 voters yielded fewer than 100 elections without an SPMWR for the Borda model and fewer than 50 elections without an SPMRW for the Condorcet model; much less than what is predicted by the spatial model.

Because the eleven models make significantly different predictions about the number of voting cycles for elections with different numbers of voters, it is crucial to select the right model if one wants to make accurate predictions of the frequency of cycles in actual elections. The results of our analyses presented in this chapter suggest that the spatial model is likely to yield accurate predictions about the occurrence of voting events in actual elections, while all other models are inadequate for this purpose. [23]

---

[22]For IC, Gehrlein (2002) established that an SPMRW exists in 58.34% of elections with 10 voters, 66.86% of elections with 20 voters, 73.46% of elections with 40 voters, and 91.23% of elections in the limit as the number of voters approaches infinity. For IAC he established that one can expect an SPMRW in 73.43% of elections with 10 voters, 81.99% of elections with 20 voters, 87.35% of elections with 40 voters, 91.05% of elections with 100 voters, and 93.75% of elections in the limit as the number of voters approaches infinity. For elections with an infinite number of voters, Gehrlein and Lepelley (2009) established that an SPMRW exists in 89.05% of elections for $\text{IAC}_b(k_b)$ if $k_b = 0.23$ and in 92.03% of elections for $\text{IAC}_b(k_b)$ if $k_b = 0.21$ (we use the calibrated value $k_b = 0.218$ in our simulations). For $\text{IAC}_c(k_c)$, they established that an SPMRW exists in 90.83% of elections if $k_c = 0.23$ and in 91.74% of elections if $k_c = 0.21$ (we use the calibrated value $k_c = 0.226$ in our simulations).

[23]The main purpose of this section is to establish that the spatial model predicts the occurrence of cycles more accurately than any of the other models. Because it might be possible to further improve upon the predictions of the spatial model (for example, by considering finite values of the δ of the multivariate Pólya distribution), the frequencies reported in column 3 of Table 9.6a might not be the most accurate predictions possible.

9.6 Conclusion

The starting point of our inquiry is the observation that no theoretical analysis of probability structures can tell us anything about the probability of observing vectors of rankings in actual elections. This is clearly not a new discovery. It parallels the mundane observation that no analysis in theoretical econometrics provides any information about the particular estimates that one obtains from analyzing specific data. The theoretical econometrics literature investigates the properties of different models that can describe data, but to analyze a specific data set, one needs to identify the model or the set of models that is most likely to have generated these data.

The relationship between our work and the theoretical literature on voting is like that between theoretical and empirical econometrics. The existing literature on voting is largely theoretical, in the sense that it does not seek to identify systematic patterns in ranking data from actual elections. The large literature on the probabilities of finding Condorcet cycles and the Condorcet efficiency of different voting rules has established that the results vary greatly across different models of P. But which of these models best describes the distribution of rankings in actual elections? The theoretical literature on voting cannot answer this question.

Our results suggest that a spatial model describes the statistical structure of P in actual elections much better than any other model that has been proposed so far, and so well that it may be difficult to find a model whose accuracy is significantly higher. We consider our result to be very encouraging, but more work needs to be done. For example, we draw our current conclusions on the basis of two data sets, one compiled from elections and the other from surveys. Our analyses suggest that the two data sets have somewhat different properties, but it is not clear whether these differences stem from their different sources or from the fact that the average ANES "election" has almost twice the number of voters than the average ERS election. Analyses of additional election data are necessary to answer this question and to determine the robustness of our results. We also focus exclusively on three-candidate elections, partly because this simplifies the exposition and makes it easier to relate our analysis to the previous literature, and partly because we are currently only able to evaluate the spatial model for three candidates. Extending our analysis to elections with more than three candidates will provide important insights into the general relative accuracy of the different models.

Our analysis applies to inquiries into the frequency of both common and rare voting events, for example, the possibility that strategic voting by a single voter could alter the outcome of an election, the existence of dominant candidates, or the frequencies of voting paradoxes. Our framework makes it possible to develop realistic models of vote-casting for such analyses and thereby to improve significantly the accuracy of their predictions for actual elections. Such new inquiries into old questions are likely to yield interesting new insights.

Appendix A: Model parameterization, parameter estimates, and simulations:

Two of our 12 models, IC and IAC, do not have any parameters that need to be estimated to either fit the model to observed voting data or to undertake Monte Carlo simulations. We do not calibrate SPP because the model assigns zero probability to the possibility that every candidate is ranked last, and we observed numerous elections in both data sets in which every candidate is ranked last by some voters. This appendix describes our strategy for calibrating the parameters of the remaining nine models and for simulating elections with each of these models. We report the calibrated values for the ERS and ANES data sets in Table A.1.

A.1 $\text{IAC}_b(k_b)$, $\text{IAC}_t(k_t)$, and $\text{IAC}_c(k_c)$

The support of these models is the entire five-simplex and each model can be calibrated to describe any set of observed vote shares by setting p_r equal to the observed vote share q_r. We calibrate the parameters k_b, k_t, and k_c as the mean over all elections of the smallest shares by which a candidate is ranked either last (k_b), first (k_t), or second (k_c). To simulate elections we assume that each parameter follows a beta distribution over the interval [0, 1/3], whose mean and variance coincide with those that we observe in the actual elections. We draw a share k from this beta distribution, determine p_1 as a draw from a uniform distribution on $[0, k]$, set $p_2 = k - p_1$, draw p_3, p_4, p_5, and p_6 from the unit four-simplex, rescale these four values so that they sum to $1 - k$, and continue only with those shares that fulfill the minimum criterion of the respective model.[24] We also simulated elections under the assumption that the support of p is the lattice on which the elements of p yield integer values when multiplied by n (we obtained identical results). We then use these six shares to draw the six n_rs (the number of votes for each of the six rankings) from density function (9.5) or (9.6).

A.2 UUP

The five parameters of this model are five of the six vote shares (that sum to 1) that describe the expected ranking. These parameters are the same for all elections, and we calibrate these parameters by identifying the share vector p that maximizes

[24] That is, if we label the candidates so that candidate C fulfills the respective criterion, then we continue only with those sets of shares for which $p_2 + p_3 > k_b$ and $p_4 + p_5 > k_b$ when we simulate elections with $\text{IAC}_b(k_b)$ (which assumes $p_1 + p_6 \leq k_b$), $p_1 + p_2 > k_t$ and $p_5 + p_6 > k_t$ when we simulate elections with $\text{IAC}_t(k_t)$ (which assumes $p_3 + p_4 \leq k_t$), and $p_1 + p_4 > k_c$ and $p_3 + p_6 > k_c$ when we simulate elections with $\text{IAC}_c(k_c)$ (which assumes $p_2 + p_5 \leq k_c$).

Table A.1 Parameter values calibrated from ERS and ANES data

Model →	$\text{IAC}_b(k_b)$	$\text{IAC}_t(k_t)$	$\text{IAC}_c(k_c)$	Borda	Condorcet	
Data set ↓	k_b	k_t	k_c	β	γ	
ERS	0.182 (0.002)	0.187 (0.002)	0.248 (0.001)	0.517 (0.009)	0.513 (0.008)	
ANES	0.218 (0.002)	0.201 (0.002)	0.226 (0.002)	0.358 (0.006)	0.360 (0.006)	
Model →	Spatial model					
		Angles of the perpendicular bisectors with the line $\overline{TO}$				
Data set ↓	$\overline{TO}$	β_1	β_2	β_3		
ERS	0.597 (0.009)	0.548 (0.011)	1.549 (0.011)	2.560 (0.011)		
ANES	0.445 (0.008)	0.556 (0.012)	1.550 (0.015)	2.592 (0.012)		
		Corresponding angles between pairs of bisectors:				
		A_1	A_2	A_3		
ERS		1.130 (0.006)	1.000 (0.006)	1.011 (0.006)		
ANES		1.105 (0.010)	0.994 (0.011)	1.042 (0.010)		
Model →	UUP					
Data set ↓	p_1	p_2	p_3	p_4	p_5	p_6
ERS	0.294	0.167	0.127	0.108	0.120	0.184
ANES	0.233	0.150	0.140	0.148	0.160	0.169

(1) The values in parentheses are standard errors
(2) We calibrated all values from the multinomial model (the values calibrated from the multivariate Pólya model are very similar)
(3) The entries for UUP are the six shares, calibrated over all elections in the respective data set, which minimize the multinomial likelihood function. The entries for the other six models are the means of the values that we calibrated for each election in the respective data set
(4) For DC and EPSF, aggregation across elections is meaningless because of the arbitrariness of labeling a candidate A, B, or C, and we did not calibrate any values for these two models

the likelihood function (9.5) or (9.6) over all elections. The maximum value of the likelihood function also determines the fit of UUP. Note that the five parameters of UUP are constant across elections, while the parameters of the other eight models are calibrated separately for each election. To simulate elections with UUP, we use the calibrated shares to draw six vote shares from the unit six-simplex, and then use draws from density function (9.5) or (9.6) to determine the six n_rs.

A.3 DC and EPSF

Both models have two parameters per election – two of the three pairs of probabilities of opposite (for DC) and neighboring (for EPSF) rankings. We fit these models to the data by using, for each election, the average value of the two observed vote shares of each pair of rankings as the predicted vote shares for that pair. To simulate elections with these models, we assume that the three pairs of shares follow a Dirichlet distribution whose mean and variance coincides with those of the shares that we observe in the actual elections. We draw three shares from the Dirichlet distribution, which we use as input to draw six vote-shares from the unit six-simplex, and then use the multinomial or multivariate Pólya distributions to draw the six n_rs.

A.4 The Borda model and the Condorcet model

Each model has one parameter per election. For the Borda model the parameter β is the increase in log probability associated with an increase by one in the rank that a voter assigns to the "best" candidate. For the Condorcet model the parameter γ is the increase in log probability associated with a reduction of one adjacent-pair permutation in the difference between the "best" ranking and the one that a voter reports. We fit these models to the data by calibrating, for each election, the values of β and γ in (9.1) and (9.3), respectively, so that the resulting vector p maximizes the likelihood for the observed vote shares. To simulate elections with these models, we assume that β and γ follow a gamma distribution whose mean and variance coincides with those of the values of β and γ that we calibrated in the actual elections. We then draw a value of β or γ from the gamma distribution that we use to determine six shares from either (9.1) and (9.2) or (9.3) and (9.4); we proceed as above to draw six vote shares from the unit six-simplex and six n_rs from the multinomial or multivariate Pólya distribution.

A.5 The spatial model

This model has four parameters per election, and Fig. A.1 shows one way of using the four degrees of freedom (as in Good and Tideman 1976). The intersection of the perpendicular bisectors T is placed at the origin of a Cartesian coordinate system.

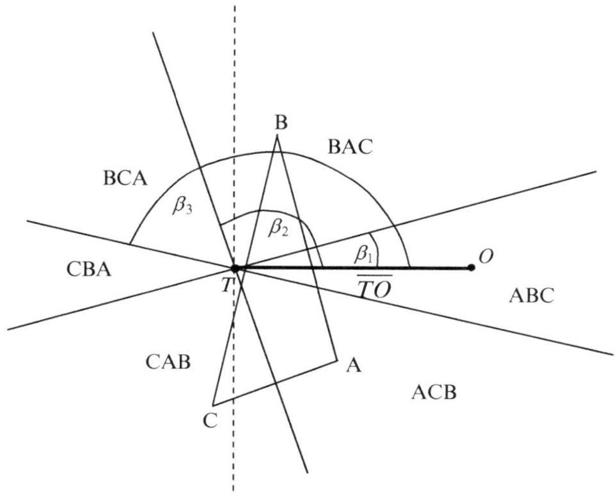

Fig. A.1 The four parameters $\overline{TO}$, β_1, β_2, and β_3 that define a spatial model observation

The fact that the vote shares are independent of rotations around the mode of the distribution of voters' ideal points, O, permits us to rotate the coordinate system so that O is located on its horizontal axis. The first degree of freedom then specifies the distance between T and O. The remaining degrees of freedom specify the angles β_1, β_2, and β_3 formed by the line $\overline{TO}$ and the three perpendicular bisectors. Thus any feasible set of values of the four degrees of freedom corresponds to a set of p_r. We calibrate the spatial model by placing, for each election, the borders between pairs of adjacent rankings and the distance in such a way as to create sectors that match the six probabilities p_r (the integrals over the triangular-shaped slices under the bivariate normal distribution) as closely as possible to the six observed vote shares, q_r.

To simulate elections with the spatial model, we assume that the three vote shares follow a Dirichlet distribution and that the distance between T and O follows a Weibull distribution; the means and variances of these distributions coincide with those of the parameters that we calibrated in each of the actual elections. We then use draws from the Dirichlet and the Weibull distributions to construct six shares, and proceed as above to draw six vote shares from the unit six-simplex and six n_rs from the multinomial or multivariate Pólya distribution.

Acknowledgements We express our gratitude to an anonymous referee, who read our paper with unusual care and pointed out several errors and omissions, as well as to Dan Felsenthal and Moshé Machover for their patience and persistence during the editing process. All remaining errors are ours.

References

Brams, S.J., & Fishburn, P. (2001). A nail-biting election. *Social Choice and Welfare*, *18*, 409–414.
Campbell, C., & Tullock, G. (1965). A measure of the importance of cyclical majorities. *Economic Journal*, *75*, 300 & 853–857.
Cervone, D., Gehrlein, W.V., Zwicker, W. (2005). Which scoring rule maximizes Condorcet efficiency under IAC. *Theory and Decision*, *58* (2), 145–185.
Chamberlin, J.R., & Featherston, F. (1986). Selecting a voting system. *Journal of Politics*, *48* (2), 347–369.
Condorcet, M.JAN., Marquis de (1785). *Essai sur l'application de l'analyse à la probabilité des décisions rendues à la pluralité des voix* (Paris).
Conitzer, V., & Sandholm, T. (2005). Common voting rules as maximum likelihood estimators. *Proceedings of the 21st Annual Conference on Uncertainty in Artificial Intelligence (UAI-05)* (pp. 145–152). UK: Edinburgh.
Davis, O.A., Hinich, M.J., Ordeshook, P.C. (1970). An expository development of a mathematical model of the electoral process. *American Political Science Review*, *44*, 426–448.
DiDonato, A.R., & Hageman, R.K. (1980). Computation of the integral of the bivariate normal distribution over arbitrary polygons. Naval Surface Weapons Center, Government Accession Number ADA102466.
Enelow, J.M., & Hinich, M.J. (1984). *The spatial theory of voting: An introduction*. Cambridge University Press.
Enelow, J.M., & Hinich, M.J. (1990). *Advances in the spatial theory of voting*. Cambridge University Press.
Feld, S., & Grofman, B. (1990). Collectivities as actors. *Rationality and Society*, *2* (4), 429–448.
Feld, S., & Grofman, B. (1992). Who's afraid of the big bad cycle? Evidence from 36 elections. *Journal of Theoretical Politics*, *4* (2), 231–237.
Felsenthal, D.S., Maoz, Z., Rapoport, A. (1993). An empirical evaluation of six voting procedures: do they really make any difference? *British Journal of Political Science*, *23* (1), 1–27.
Felsenthal, D.S., & Machover, M. (1995). Who ought to be elected and who is actually elected? An empirical investigation of 92 Elections under three procedures. *Electoral Studies*, *14* (2), 143–169.
Gehrlein, W.V., & Fishburn, P.C. (1976). Condorcet's paradox and anonymous preference profiles. *Public Choice*, *26*, 1–18.
Gehrlein, W.V. (1978). Condorcet winners in dual cultures. Presented at the National Meeting of the Public Choice Society, New Orleans.
Gehrlein, W.V. (2002). Condorcet's paradox and the likelihood of its occurrence: Different perspectives on balanced preferences. *Theory and Decision*, *52* (2), 171–199.
Gehrlein, W.V. (2004). Consistency in measures of social homogeneity: A connection with proximity to single-peaked preferences. *Quality and Quantity*, *38*, 147–171.
Gehrlein, W.V. (2006). *Condorcet's paradox*. Berlin and Heidelberg: Springer.
Good, I.J., & Tideman, T.N. (1976). From individual to collective ordering through multidimensional attribute space. *Proceedings of the Royal Society of London (Series A)*, *347*, 371–385.
Kemeny, J. (1959). Mathematics without numbers. *Daedalus*, *88*, 571–91.
Kendall, M., & Gibbons, J.D. (1990). *Rank correlation methods*. New York: Oxford University Press.
Kotz, S., Balakrishnan, N., Johnson, N.L. (2000). *Continuous multivariate distributions*. New York: Wiley.
Kuga, K., & Nagatani, H. (1974). Voter antagonism and the paradox of voting. *Econometrica*, *42*, 1045–1067.
Lepelley, D. (1995). Condorcet efficiency of positional voting rules with single-peaked preferences. *Economic Design*, *1*, 289–299.
Merrill, S. (1984). A comparison of efficiency of multicandidate electoral systems. *American Journal of Political Science*, *28*, 23–48.

Mosimann, J.E. (1962). On the compound multinomial distribution, the multivariate beta distribution, and correlations among proportions. *Biometrika*, *49*, 65–82.
Regenwetter, M., Grofman, B., Marley, A.A.J. (2002). On the model dependence of majority preference relations reconstructed from ballot or survey data. *Mathematical Social Sciences*, *43*, 451–466.
Regenwetter, M., Marley, A.A.J., Grofman, B. (2003). General concepts of value restriction and preference majority. *Social Choice Welfare*, *21*, 149–173.
Regenwetter, M., Grofman, B., Marley, A.A.J., Tsetlin, I.M. (2006). *Behavioral social choice: Probabilistic models, statistical inference, and applications*. Cambridge University Press.
Saari, D.G. (1990). Susceptibility to manipulation. *Public Choice*, *64*, 21–41.
Saari, D.G. (2001). Analyzing a nail-biting election. *Social Choice and Welfare*, *18*, 415–430.
Tideman, T.N., & Richardson, D. (2000). Better voting methods through technology: the refinement-manageability trade-off in the single transferable vote. *Public Choice*, *103* (1–2), 13–34.
Tideman, T.N. (2006). *Collective decisions and voting*. Ashgate: Burlington VT.
Truchon, M. (2006). Borda and the maximum likelihood approach to vote aggregation. Mimeo.
Young, H.P. (1988). Condorcet's theory of voting. *American Political Science Review*, *82* (4), 1231–1244.

Part III
Theory and Practice: Additional Considerations in Selecting a Voting Procedure

Chapter 10
On the Relevance of Theoretical Results to Voting System Choice

Hannu Nurmi

10.1 Introduction

The first systematic comparisons of voting procedures appeared in the 1970's. The journal *Behavioral Science* became a major forum for these early publications. Especially notable are the article by Fishburn (1971) and a series of works by Jeffrey T. Richelson. This series culminated in a summary (Richelson 1979) that is perhaps the most extensive of its kind in terms of both the number of systems and the number of criteria. These were followed by a book-length treatise by Straffin (1980) and perhaps most notably by William H. Riker's (1982) *magnum opus*. While these texts explicitly dealt with voting systems, they were preceded and inspired by several path-breaking works in the more general field of social choice functions, e.g. (Fishburn 1973, 1977) and (Young 1975). The history of voting procedures had also been discussed in Black (1958) and Riker (1961). The wider public was first made aware of the theory of comparative voting systems by an article in *Scientific American* written jointly by Niemi and Riker (1976).

From those early years on there has been a relatively clear distinction between theoretical and applied works. Fishburn, Richelson and Young are obviously theoretical scholars, while Riker and Straffin had a more applied focus. Indeed, Riker (1982) can be seen as an attempt to justify a specific theory of democracy by invoking theoretical results achieved in social choice theory. More specifically, Riker argues that since

H. Nurmi (✉)
Public Choice Research Centre and Department of Political Science, University of Turku, 20014 Turku, Finland
e-mail: hnurmi@utu.fi

D.S. Felsenthal and M. Machover, *Electoral Systems*, Studies in Choice and Welfare, DOI 10.1007/978-3-642-20441-8_10, © Springer-Verlag Berlin Heidelberg 2012

- all known voting procedures have at least one serious flaw,
- voting equilibria are extremely rare in multidimensional spatial voting models, and
- strategic manipulation opportunities are ubiquitous,

it is erroneous to equate voting results with the "will of the people" or expressions of collective opinion for the reason that the latter is a meaningless notion. Hence, defining democracy as a system ruled in accordance with the will of the people is indefensible. His favorite – liberal – view of democracy, on the other hand, is immune to the negative results of social choice theory because it does not require more of a voting – or, more generally, ruling – system than that it enables the voters to get rid of undesired rulers. For this purpose, continues the argument, the plurality rule is a particularly apt instrument.

Riker's view is thought-provoking. Many authors, while accepting its premises based on social choice theory, have questioned the conclusions (Lagerspetz 2004; Mackie 2003; Nurmi 1984, 1987). This paper dwells on the premises and their significance for voting system design. We shall first outline the standard view which looks at various voting systems and evaluates them in terms of criteria of performance. This approach in essence deems all systems satisfying a given criterion of performance as equivalent and those which don't also equivalent. Starting from the 1960s a rich literature on probability and simulation modeling of voting systems performance has emerged to give a somewhat more nuanced picture (Klahr 1966; Niemi and Weisberg 1972). We shall discuss the nature and relevance of these results. We then deal with the intuitive difficulty of devising examples of various criterion violations and discuss whether this should play a role in voting system evaluations. Finally, we shall scrutinize the "givens" of the theory used in the evaluation.

10.2 The Standard Approach

The motivation for introducing a new voting system or criticizing an old one is often a counterintuitive or unexpected voting outcome. A case in point is Borda's memoir where he criticized the plurality voting and suggested his own method of marks (McLean and Urken 1995). With time this approach focusing on a specific flaw of a system has given way to studies dealing with a multitude of systems and their properties. An example of such studies (e.g. (Nurmi 2002, 36), (Nurmi 2006, 136–137)) is summarized in Table 10.1.

Here criterion a denotes the Condorcet winner criterion, b the Condorcet loser one, c strong Condorcet criterion, d monotonicity, e Pareto, f consistency, g independence of irrelevant alternatives and h invulnerability to the no-show paradox. A "1" ("0", respectively) in the table means that the system represented by the row satisfies (violates) the criterion represented by the column.

The systems are viewed as choice rather than preference functions. This distinction makes a difference especially in the case of the Kemeny rule. As a preference

Table 10.1 A comparison of voting procedures

Voting system	Criterion							
	a	*b*	*c*	*d*	*e*	*f*	*g*	*h*
Amendment	1	1	1	1	0	0	0	0
Copeland	1	1	1	1	1	0	0	0
Dodgson	1	0	1	0	1	0	0	0
Max-min	1	0	1	1	1	0	0	0
Kemeny	1	1	1	1	1	0	0	0
Plurality	0	0	1	1	1	1	0	1
Borda	0	1	0	1	1	1	0	1
Approval	0	0	0	1	0	1	0	1
Black	1	1	1	1	1	0	0	0
Pl. runoff	0	1	1	0	1	0	0	0
Nanson	1	1	1	0	1	0	0	0
Alternative vote	0	1	1	0	1	0	0	0

function it is consistent (Young and Levenglick 1978), but as a choice rule it isn't.[1] It will be recalled that choice functions map preference profiles into subsets of alternatives. Denoting by Φ the set of all preference profiles and by A the set of alternatives, we thus have

$$f : \Phi \rightarrow 2^{A}$$

for social choice functions.

Preference functions, in contradistinction, map preference profiles into rankings over alternatives (cf. social welfare functions). I.e.

$$F : \Phi \rightarrow \mathcal{R}$$

where $\mathcal{R}$ denotes the set of all preference rankings over A.

Consider now a partition of a set N of individuals with preference profile ϕ into two separate sets of individuals N_1 and N_2 with corresponding profiles ϕ_1 and ϕ_2 over A and assume that $f(\phi_1) \cap f(\phi_2) \neq \emptyset$. The social choice function f is consistent iff $f(\phi_1) \cap f(\phi_2) = f(\phi)$, for all partitionings of the set of individuals.

The same definition can be applied to social preference functions. F is consistent iff whenever $F(\phi_1) \cap F(\phi_2) \neq \emptyset$ implies that $F(\phi_1) \cap F(\phi_2) = F(\phi)$.

It turns out that, like all Condorcet extensions, Kemeny's rule is an inconsistent social choice function. An example is provided by Fishburn (1977, 484). However, as a preference function it is consistent, i.e. whenever two distinct subsets of individuals come up with some common preference rankings, these common rankings must also be the result when the sub-profiles are put together. Young's result that all Condorcet extensions are inconsistent is visible in Table 10.1 where

[1] I am grateful to Dan Felsenthal for calling my attention to the apparent discrepancy between Young and Levenglick's claim that the Kemeny rule satisfies both the Condorcet winner criterion and consistency, and Fishburn's demonstration that the rule is not consistent.

Table 10.2 Monotonicity and vulnerability to no-show paradox among anonymous and neutral systems: examples

	Monotonic	Non-monotonic
Vulnerable	Copeland	Plurality runoff
Invulnerable	Borda	Empty

all those systems with a 1 in column *a* have a 0 in column *f*. The satisfaction of the Condorcet winner criterion is, however, just a sufficient, not necessary, condition for inconsistency: plurality runoff and alternative vote system fail on both the Condorcet winner criterion and on consistency.

Of particular interest in Table 10.1 is column *h*, the invulnerability to the no-show paradox. One of the main motivations for elections is to get an idea of voter preferences. Systems that are vulnerable to the no-show paradox are at least prima facie incompatible with this motivation. It has been shown by Moulin (1988) and Pérez (1995) that all Condorcet extensions are vulnerable to the no-show paradox and, indeed, as shown by Pérez (2001), most of them to the strong version thereof whereby by abstaining a group of voters may get their first-ranked alternative elected, while some other alternatives would be elected if they would vote according to their preferences.

At first sight, monotonicity is closely related to invulnerability to the no-show paradox. On closer scrutiny the situation gets more nuanced. Firstly, among monotonic systems there are both systems that are vulnerable to the no-show paradox and those that are not (see e.g. Nurmi (2002, 103)). In other words, monotonicity does not imply invulnerability to the no-show paradox. By Moulin's result all monotonic Condorcet extensions – e.g. Copeland's and Kemeny's methods – are vulnerable to the no-show paradox. More obviously, monotonicity does not imply vulnerability either since e.g. the plurality rule is both monotonic and invulnerable to the paradox. The same is true of the Borda count. But what about non-monotonicity? Does it imply vulnerability? Again Moulin's result instructs us that non-monotonic Condorcet extensions – e.g. Dodgson's and Nanson's methods – are vulnerable. So are plurality runoff and alternative vote. Indeed, in Table 10.1 all non-monotonic systems are vulnerable to the no-show paradox. Campbell and Kelly (2002) have shown, however, that this is not the case in general, i.e. there are non-monotonic systems that are invulnerable to the no-show paradox. These are, however, either non-anonymous or non-neutral (or both). Hence, within the class of anonymous and neutral procedures we get the following table (Table 10.2).

10.3 Standard Approach and System Choice

Table 10.1 gives a summary information of some criteria and systems. To justify a "1" in the table one has to show that the criterion represented by the column is under no profile violated by the system represented by the row. To justify a "0" requires no more than an example where the system violates the criterion. This information may be useful in choosing a voting system. Suppose that one is primarily interested in only one criterion, say Condorcet winning. Then one's favorite systems are those with a "1" in column a. This in itself sensible way of proceeding leaves, however, one with many systems. So, we need additional considerations to narrow the choice down.

A more "graded" approach to comparing two systems with respect to one criterion has also been suggested (Nurmi (1991); see also Lagerspetz (2004)). The superiority of system A with respect to system B takes on degrees from strongest to weakest as follows:

1. A satisfies the criterion, while B doesn't, i.e. there are profiles where B violates the criterion, but such profiles do not exist for A.
2. in every profile where A violates the criterion, also B does, but not vice versa.
3. in *practically all profiles* where A violates the criterion, also B does, but not vice versa ("A dominates B almost everywhere").
4. in a plausible probability model B violates the criterion with higher probability than A.
5. in those political cultures that we are interested in, B violates the criterion with higher frequency than A.

We shall return to items 4 and 5 in the next section. Comparing systems with respect to just one criterion is, however, not plausible since criteria tend to be contested not only among the practitioners devising voting systems, but also within the scholarly community. Suppose instead that one takes a more holistic view of Table 10.1 and gives some consideration to all criteria. A binary relation of dominance could then be defined as follows:

Definition 1. A system A (strictly) dominates system B in terms of a set of criteria, if and only if whenever B satisfies a criterion, so does A, but not the other way around.[2]

In Table 10.1, e.g. Kemeny's rule dominates all other systems except Copeland, Black, plurality, Borda and approval voting. Regardless of what relative weights one assigns to various criteria, it seems natural to focus on the undominated systems. Thus in Table 10.1 one is left with the six systems just mentioned.

But all criteria are not of equal importance. Nor are they unrelated. To wit, if a system always ends up with the Condorcet winner, i.e. satisfies criterion a, it also elects the strong Condorcet winner, that is, satisfies criterion c. It is also known that the Condorcet winner criterion is incompatible with consistency (Young 1974a,b). Some criteria seem to be context-related in the sense that they lose their practical relevance in some specific contexts. E.g. one could argue that consistency has no

[2] A referee suggests a more general version: "To whatever degree B satisfies a criterion, A satisfies it to at least the same degree, but not the other way around." Since we are primarily dealing with dichotomous criteria, we shall use the less general version.

practical bearing on committee decisions since the results are always determined by counting the entire set of ballots. This observation notwithstanding, there is a more subtle argument one can build against the standard approach: the finding that a criterion is not satisfied by a system tells us very little – in fact nothing – about the likelihood of violation. For that we need to focus on the likelihood of "problematic" profiles since these – together with choice rules – determine the outcomes. It is when we compare the outcomes with the profiles that we find out whether a criterion violation has occurred.

10.4 How Often are Criteria Violated?

To find out how often a given system violates a criterion – say, elects a Condorcet loser – one has to know how often various preference profiles occur and how these are mapped into voting strategies by voters. Once we know these two things, we can apply the system to the voting strategy n-tuples (if the number of voters is n), determine the outcomes, and, finally, compare these with the preference profile to find out whether the choices dictated by the criterion contradict those resulting from the profile, e.g. if an eventual Condorcet loser was chosen. Traditionally, two methods have been resorted to in estimating the frequency of criterion violations: (1) probability modeling, and (2) computer simulations. Both are based on generating artificial electorates and calculating how frequently the criterion is violated or some other incompatibility encountered in these electorates.

The literature on probability and computer simulations is vast (see e.g. Gehrlein (1997, 2002, 2006); Gehrlein and Lepelley (2004); Lepelley (1993); Merlin et al. (2000); Saari and Tataru (1999)). Of particular interest has been the occurrence of cyclic majorities. The early models were based on the impartial culture (IC) assumption. Under it each voter is randomly and independently assigned to a preference ranking over alternatives. So, the voters are treated as random samples – with replacement – from a uniform distribution over all preference rankings. The method devised by Gehrlein and Fishburn (1978) is useful in deriving limit probabilities when the number of voters increases. IC is a variation of the principle of insufficient reason: since we cannot know which preference profiles will emerge in the future, we assume that all individual preference relations are purely random in the sense that each individual's preference relation is independently drawn from a uniform distribution of preferences over all possible preference rankings. Like all versions of the principle of insufficient reason, IC is based on untenable epistemology: it is not possible to derive knowledge about probabilities of rankings from complete ignorance regarding those probabilities. Despite its implausibility, this assumption could still be made because of its technical expediency if one could point out that the results based on IC do not deviate very much from those obtained under other more plausible assumptions. But, alas, this is not the case: the IC simulations results often differ dramatically from other simulation results (see e.g. Nurmi (1999)).

Regenwetter et al. (2006) strongly criticize the IC assumption by arguing that it in fact maximizes the probability of majority cycles. Their criticism aims at playing down the empirical significance of the results that – under the IC assumption – suggest that the probability of majority cycles is reasonably high even in the case of just three alternatives. Surely, the fact that the probability of cycles is estimated at 0.09 in IC's when the number of voters approaches infinity and the number of alternatives is 3, does not imply that the probability of cycles would be of the same order in current, past or future electorates. What those results literally state is that if the opinions of the voters resemble IC, then the probability of encountering majority cycles is as specified. The interest of these estimates is not in their predictive success in real world, but in their ability to provide information about variables and parameters that increase or decrease the likelihood of cycles. Probability models are in general more useful in providing this kind of information. Often the interest is not so much in the probability estimates themselves but on their variability under various transformations in the models.

Consider the studies on Condorcet efficiency of various voting procedures, i.e. on the probability that the Condorcet winner is chosen by a procedure under various cultures. Those studies that focus on Condorcet efficiency are typically reporting the probability of the Condorcet winner being chosen, provided that such a winner exists in the profile. In other words, these studies (e.g. Merrill (1984)) do not aim at predicting how often Condorcet winners are elected, but, by focusing instead on just those profiles where a Condorcet winner exists, help to identify the propensity of various procedures to elect the Condorcet winner (see also Merrill (1988)). Similarly studies reporting the probability of various systems to come up with Condorcet losers are not predicting the relative frequency of Condorcet losers being elected in current elections, but are aiming at disclosing factors, variables or parameters that increase or decrease such choices under profiles where a Condorcet loser exists. Yet, the argument of Regenwetter et al. is supported by simulations where IC assumption is slightly perturbed by assuming that a small minority of the electorate – say, 5 or 10% of the total – forms a homogeneous sub-culture of voters with identical preferences while the rest of the electorate remains an IC. It then turns out that the Condorcet efficiencies of various systems change quite significantly. More importantly, even the ranking of systems in terms of Condorcet efficiency can change for some combinations of alternatives and voters (Nurmi 1992). Similar observation can be made about differences in choice sets of various systems under IC and small perturbations thereof. IC seems to be associated with larger discrepancies of systems than systems where a minuscule group representing identical preferences is immersed in IC (Nurmi 1988a,b, 1992).

Despite its tendency to exaggerate Condorcet cycles and dampen Condorcet efficiencies of systems that are not Condorcet extensions, IC may be a useful construct in illuminating the differences of voting rules. By estimating the likelihood that two rules make overlapping choices in IC's we get a profile-neutral view of how far apart they are as choice intuitions. For example, IC simulations suggest – unsurprisingly – that two Condorcet extensions, Copeland's rule and max-min method (also known as Simpson's method), are relatively close to each other in the sense of resulting rarely in distinct choice sets. More interesting is the finding that the Borda count is nearly as close to Copeland's rule as the max-min method is (Nurmi 1988a,b). This is consistent with the relatively high Condorcet efficiency of the Borda count reported in several studies (e.g. Merrill (1984); Nurmi (1988a,b)). As is well-known, the Borda scores of alternatives can be computed from the outranking matrix by taking row sums. This binary implementation of the Borda count already hints that, despite its positional nature, the method is reasonably close to the idea that the winners be detected through binary comparisons.

The criticism of IC has so far not produced many alternative culture assumptions. Perhaps the most widespread among the alternative assumptions is that of impartial anonymous culture (IAC). Consider an electorate of n voters considering the set of k alternatives. The number of rankings of alternatives is then $k!$. Let n_i denote the number of voters with ith preference ranking ($i = 1, \ldots, k!$). Each anonymous profile can be represented by listing the n_i's. The profile satisfies anonymity since transferring j voters from n_s to n_t when accompanied with transferring j voters from n_t to n_s leaves the distribution of voters over preference ranking unchanged. In IAC's every distribution of voters over preference rankings is assumed to be equally probable. This changes the Condorcet efficiency as well as Borda paradox estimates by increasing the former and decreasing the latter (Gehrlein 1997, 2002).

Is IAC then more realistic than IC? Both IC and IAC are poor proxies of political electorates. Given any election result it is inconceivable that the profile emerging in the next election would, with equal probability, be any distribution of voters over preference rankings. The same is true of committees and other bodies making several consecutive collective choices. There is in general far more interdependence between voters than suggested by IAC. Indeed, it can argued (Nurmi 1988a,b) that in reconstructing the profile transformation over time, one should distinguish two mechanisms: (1) one that determines the initial profile, and (2) one which determines the changes from one time instant (ballot) to the next. Both IC and IAC collapse these two into one mechanism that generates each voting situation de novo. This is certainly not the way in which everyday experience suggests that opinion distributions are formed. If it were, the electoral campaigns would take on heretofore unknown forms: the distinctions between core constituencies and moving voters would vanish as would that between government and opposition etc. So, it seems that everyday observations fly in the face of IC, IAC and many other models used in generating voter profiles. This does not play down the importance of those models as theoretical tools, i.e. in enhancing our understanding of the mechanisms increasing or decreasing the occurrence of various paradoxes, incompatibilities or discrepancies related to voting systems. Nevertheless, to render choice theoretic results more relevant for the evaluation of voting rules, one should bring the incompatibility results closer to practice by finding out what the problematic profiles look like, i.e. what kinds of opinion distributions underly them. If it is very difficult to envision how those profiles would emerge in practice, then arguably the corresponding incompatibility results do not have much practical importance.

10.5 Counterexamples Are Sometimes Difficult to Come By

Summaries like Table 10.1 provide information that is of somewhat asymmetric nature. To prove that a system is incompatible with a criterion one needs to find a profile where – under the assumed mechanism concerning voting behavior – the system leads to a choice that is not consistent with the range of choices allowed for by the criterion. To find such a profile when, theoretically, one should exist, is, however, not always easy. At the behest of and in cooperation with Dan S. Felsenthal the present author embarked upon looking for examples illustrating the incompatibility of the Condorcet winning criterion and invulnerability to the no-show, truncation and twin paradoxes. The background of this search is the result proven by Moulin (1988) and subsequently strengthened by Pérez (2001) saying that all Condorcet extensions are vulnerable to the no-show paradox. In the subsections that follow these incompatibilities are illustrated for some well-known Condorcet extensions (for fuller discussion, see Felsenthal (2010)).

10.5.1 Black's Procedure

Black' procedure is vulnerable to the no-show paradox, indeed, to the strong version thereof. This is illustrated in Table 10.3.

Here D is the Condorcet winner and, hence, is elected by Black.

Suppose now that the right-most voter abstains. Then the Condorcet winner disappears and E emerges as the Borda winner. It is thus elected by Black. E is the first-ranked alternative of the abstainer. Hence we have a strong version of the paradox.

Truncation paradox is closely related to the no-show one. It occurs whenever a group of individuals gets a better outcome by revealing only part of their preference ranking rather than their full ranking. Obviously, not voting at all is an extreme version of truncation and thus the above can be used to show that Black is also vulnerable to truncation. If more specific demonstration is needed, then one might consider the modification of the above example whereby the right-most voter truncates his preference after A, i.e. does not express any view regarding C and D. Then, the Condorcet winner again disappears and the Borda winner E emerges as the Black winner. Again the strong version of the truncation paradox emerges.

Table 10.3 Black's system is vulnerable to strong no-show paradox

1 voter	1 voter	1 voter	1 voter	1 voter
D	E	C	D	E
E	A	D	E	B
A	C	E	B	A
B	B	A	C	D
C	D	B	A	C

Table 10.4 Nanson's method is vulnerable to strong no-show paradox

5 voters	5 voters	6 voters	1 voter	2 voters
A	B	C	C	C
B	C	A	B	B
D	D	D	A	D
C	A	B	D	A

Table 10.5 Dodgson's method is vulnerable to no-show and twin paradoxes

42 voters	26 voters	21 voters	11 voters
B	A	E	E
A	E	D	A
C	C	B	B
D	B	A	D
E	D	C	C

10.5.2 *Nanson's Method*

Nanson's Borda-elimination procedure is vulnerable to the strong version of no-show paradox as well as Table 10.4 illustrates.[3] Here Nanson's method results in B.

If one of the right-most two voters abstains, C – their favorite – wins. Again the strong version of the no-show paradox appears.

The twin paradox occurs whenever a voter is better off if one or several individuals, with identical preferences to those of the voter, abstain. In Table 10.4 we have an instance of the twin paradox as well: if there is only one CBDA voter, C wins. If he is joined by another, B wins.

Nanson is also vulnerable to truncation: if the 2 right-most voters indicated only their first rank, C would win (not B).

10.5.3 *Dodgson's Method*

In Table 10.5, A is closest to becoming the Condorcet winner, i.e. it is the Dodgson winner.[4]

Now take 20 out the 21 voter group out. Then B becomes the Condorcet and, thus, Dodgson winner. B is preferred to A by the abstainers, demonstrating Dodgson's vulnerability to the no-show paradox. Adding those 20 "twins" back to retrieve the original profile shows that Dodgson is also vulnerable to the twin paradox.

---

[3]This subsection is partly based on the author's correspondence with Dan S. Felsenthal on May 25, 2001.

[4]This example is an adaptation of one given by Fishburn (1977, 478).

Table 10.6 Schwartz's method violates the Pareto condition

1 voter	1 voter	1 voter
A	D	B
B	C	D
D	A	C
C	B	A

Table 10.7 Schwartz's method is vulnerable to no-show and twin paradoxes if voters are risk-averse

23 voters	28 voters	49 voters
A	B	C
B	C	A
C	A	B

10.5.4 Pareto Violations, No-Show and Twin Paradoxes of Schwartz

As will be recalled, the Pareto condition states: if everybody strictly prefers x to y, then y is not chosen. Schwartz's method violates this condition as shown in Table 10.6.

Table 10.6 exhibits a top cycle: $A \succ B \succ D \succ C \succ A$. Hence this is the choice set of Schwartz. Yet, C is Pareto dominated by D.

To find out whether Schwartz is vulnerable to the no-show paradox we have to make assumptions regarding the risk-posture of voters. If they are assumed to be risk-averse, then the following example demonstrates the vulnerability of Schwartz to both no-show and twin paradoxes.

In Table 10.7 the Schwartz choice set is A, B, C. With four voters from the BCA voters abstaining, C becomes the Condorcet – and thus Schwartz – winner. Starting from the 96-voter profile and adding BCA voters one by one, we can demonstrate the twin paradox.

In case of risk-neutral voters, we can demonstrate these paradoxes through the profile of Table 10.8:

Here the Schwartz (GOCHA) choice set is A, B, C, D. With four voters of the BCDA group abstaining, C again becomes the Condorcet winner and is thus elected. This shows the no-show paradox. The twin paradox emerges when one starts with the 96-voter profile and adds BCDA voters one by one as above.

10.5.5 Max-Min Rule

The max-min rule is also vulnerable to no show, truncation and twin paradoxes. Table 10.9 illustrating this is an adaptation of Pérez (1995).

The outranking matrix of Table 10.9 profile is in Table 10.10.

Table 10.8 Schwartz's method and risk-neutral voters

23 voters	28 voters	49 voters
A	B	C
B	C	D
D	D	A
C	A	B

Table 10.9 Max-min method is vulnerable to no show, truncation and twin paradoxes

5 voters	4 voters	3 voters	3 voters	4 voters
D	B	A	A	C
B	C	D	D	A
C	A	C	B	B
A	D	B	C	D

Table 10.10 Outranking matrix of Table 10.9

	A	B	C	D	Row min
A	–	10	6	14	6
B	9	–	12	8	8
C	13	7	–	8	7
D	5	11	11	–	5

Thus, B is elected. However, with the 4 CABD voters abstaining, the outcome would be A. With only 1 CABD voter added to the 15-voter profile, A is still elected. If one then adds 3 "twins" of the CABD voter, one ends up with B being elected. Hence twins are not welcome. If those 4 voters reveal their first preference only, the minimum entry in B's row drops to 4 and C emerges as the winner. Hence the truncation paradox. This outcome assumes that winners are determined on the basis of minimum support in pairwise comparisons. If a voter does not reveal his preference between x and y, he gives no votes to either one in the corresponding pairwise comparison. This is in line with Brams (1982) who first introduced the notion of preference truncation. Of course, other interpretations can be envisaged.

10.5.6 Young Fails On No Show and Twin Paradoxes

Young's method is a Condorcet extension that looks for the largest subset of voters which contains a Condorcet winner and elects the Condorcet winner of that subset of voters. Being a Condorcet extension, Young's rule is also vulnerable to the no-show and twin paradoxes as illustrated by Table 10.11. The illustration is again inspired by and adapted from Pérez (2001) and Moulin (1988).

In this profile E is elected (needs only 12 removals). Add now 10 voters with ranking EDABC. This makes D the Condorcet winner. Hence, the 10 added voters are better off abstaining. Indeed we have an instance of the strong version of the no-show paradox. Obviously, twins are not always welcome here.

Table 10.11 Young's method is vulnerable to no show and twin paradoxes

11	10	10	2	2	2	1	1
B	E	A	E	E	C	D	A
A	C	C	C	D	B	C	B
D	B	D	D	C	A	B	D
E	D	B	B	B	D	A	E
C	A	E	A	A	E	E	C

10.5.7 Kemeny Fails On No-Show and Twin Paradoxes

The example of Sect. 10.5.5 is applicable here. In the 15-voter profile (the four left-most groups of voters), the Kemeny-ranking is DBCA. Now add 4 voters with DABC ranking. A now becomes the Condorcet and Kemeny winner. Hence these four voters are better off not voting.

The twin paradox occurs when we start with the 15-voter profile adding voters one by one until the winner changes from D to A. The last added voter is the unwelcome twin.

Counterexamples are, indeed, important in proving incompatibilities of systems and criteria. However, they vary a great deal in terms of the underlying difficulty of constructing them. The above counterexamples dealing with the no show paradox and Condorcet extension methods show that even though a general result – here due to Moulin and Pérez – is known, it is not necessarily straight-forward to find examples to illustrate the incompatibility. This suggests that perhaps the compatibility should be viewed as a matter of degree rather than a dichotomy. In fact, we are here encountering the same problem as when discussing the relevance of simulation models: how often are problematic profiles likely to emerge? We just don't know, but if the difficulty of finding examples of some incompatibilities – e.g. between Young's method and invulnerability to the no-show paradox – is anything to go by, some of the problematic profiles occur only in very specific circumstances. Hence their practical relevance is limited.

In addition to the empirical frequency of problematic profiles, the relevance of choice theoretic results also hinges upon the acceptability of the assumptions made in the theory. This is an issue we now turn to.

10.6 Another Look at Behavioral Assumptions

The bulk of social choice theory is based on the assumption that the individuals are endowed with complete and transitive preference relations over the alternatives. While there are good grounds for making this assumption, it is not difficult to construct examples where a reasonable individual might not satisfy it. Consider Table 10.12.

Table 10.12 Performance of three universities on three criteria

Criterion (i)	Criterion (ii)	Criterion (iii)
A	B	C
B	C	A
C	A	B

The Dictator of Universities (DU, so far a purely fictitious figure) ponders upon the evaluation of three universities A, B and C in terms of three criteria: (i) research output (scholarly publications), (ii) teaching output (degrees), (iii) external impact (expert assignments, media visibility, R & D projects, etc.). DU deems these criteria of roughly equal importance in determining the future funding of the universities. His observations are summarized in Table 10.12.

Since the criteria are of roughly equal importance DU comes up with the following list of binary preferences: $A \succ B \succ C \succ A \succ \ldots$. There is nothing unreasonable in this obviously intransitive preference relation. So, perhaps we should give some thought on alternative foundations of choice theory. There are basically two ways to proceed in searching for those foundations: (1) assume something less demanding, or (2) something more demanding than preference rankings.

10.6.1 Asking for Less Than Rankings

It is well-known that Arrow's focus on social welfare functions was eventually replaced by apparently less demanding concept of social choice function. In similar vein, one could replace the notion of complete and transitive individual preference relation with that of a choice function, i.e. a rule indicating for each subset of alternatives the set of best alternatives. In Arrovian spirit one could then look for plausible conditions on methods of aggregating the individual choice functions into collective ones.

The following would seem plausible conditions on collective choices based on individual choice functions:

- Citizen sovereignty: For any alternative x, there exists a set of individual choice function values so that x will be elected,
- Choice-set monotonicity: If x is elected under some profile of individual choices, then x should also be elected if more individuals include x in their individual choices
- Neutrality
- Anonymity
- Choice-set Pareto: If all individuals include x in their individual choice sets, then the aggregation rule includes x as well , and if no voter includes y in their individual choice set, then y is not included in the collective choice.

Table 10.13 Two choice function aggregation rules

Alt. set	Ind. choice sets			Rule 1	Rule 2
	Ind. 1	Ind. 2	Ind. 3		
$\{x, y, z\}$	$\{x\}$	$\{z\}$	$\{y\}$	$\emptyset$	$\{x, y, z\}$
$\{x, y\}$	$\{x\}$	$\{x\}$	$\{y\}$	$\{x\}$	$\{x\}$
$\{x, z\}$	$\{x\}$	$\{z\}$	$\{x\}$	$\{x\}$	$\{x\}$
$\{y, z\}$	$\{y\}$	$\{z\}$	$\{y\}$	$\{y\}$	$\{y\}$

- Chernoff's condition: If an alternative is among winners in a large set of alternatives, it should also be among the winners in every subset it belongs to (Chernoff's postulate 4 (Chernoff 1954, 429)).
- Concordance: Suppose that the winners in two subsets of alternatives have some common alternatives. Then the rule is concordant if these common alternatives are also among the winners in the union of the two subsets (Chernoff's postulate 10 (Chernoff 1954, 432; Aizerman and Aleskerov 1995, 19–20)).

Incompatibilities can also be encountered in this less demanding setting. To wit, consider two rules for making collective choices. Rule 1: whenever an alternative is included in the choice sets of a majority of voters, it will be elected. Rule 2 (plurality): whichever alternative is included in more numerous choice sets than any other alternative, is elected. Table 10.13 presents an example of a three-member voting body pondering upon the choice from $\{x, y, z\}$. The individual choice sets as well as those resulting from the application of Rule 1 and Rule 2 are indicated (Aizerman and Aleskerov 1995, 237).

Concordance is not satisfied by Rule 1, since x is chosen from $\{x, y\}$ and $\{x, z\}$, but not from $\{x, y, z\}$. Rule 2 fails on Chernoff since z is in the choice set from $\{x, y, z\}$, but from $\{x, z\}$. It is also worth noticing that plurality (Rule 2), but not majority (Rule 1) fails on choice-set monotonicity.

Aggregating choice profiles instead of preference ones is in a way natural when one is dealing with collective choices rather than rankings. Yet, as we just saw, incompatibilities between various desiderata can be encountered here as well. Individual choice functions are less demanding than preference rankings. All one needs to assume regarding the underlying preference relations is completeness. A step towards more demanding ways of expressing preferences is individual preference tournament. Tournaments – it will be recalled – are complete and asymmetric relations. One could argue that when the individuals take different properties or aspects of choice options into account when forming their preference between different pairs of options, the satisfaction of completeness and asymmetry comes naturally. Yet, transitivity is less obvious. Tversky's (1969) experiments with choices involving pairs of risky prospects illustrate this.

Now, if tournaments instead of rankings or choice functions are taken as proper descriptions of individual opinions, we have readily at hand several solution concepts, to wit, the uncovered set, top cycle set, Copeland winners, the Banks set (Banks 1985; Miller 1995; Moulin 1986). Typically these specify large subsets of alternatives as winners and are, thus, relatively unhelpful in settings where single winners are sought. There are basically two ways of utilizing individual tournament matrices in making collective decisions:

1. Given the individual $k \times k$ tournaments, construct the corresponding collective one of the same dimension by inserting 1 to position (i, j) if more than $n/2$ individuals have 1 in the (i, j) position. Otherwise, insert 0 to this position. The row sums then indicate the Copeland scores. Rows with sum equal to zero correspond to the Condorcet losers, those with sums equal to $k - 1$ to the Condorcet winners. Uncovered and Banks' sets can be computed as well (the latter, though, is computationally hard). Also Dodgson scores can determined.
2. Construct the collective opinion matrix as an outranking matrix where the entry in the (i, j) position equals the number of individuals with 1 in the (i, j) position. The row sums then indicate the "Borda scores". Max-min scores can also be determined.

So, the concepts of preference aggregation can be re-invoked in tournament aggregation.

10.6.2 Asking for More Than Rankings

Another way of responding to social choice incompatibilities is to start from more, rather than less, demanding notions than individual preference rankings. In fact, this response has a firm foundation in the classic utility theory. Over the past decade it has been reiterated by several authors. To quote one of them (Hillinger 2005):

> ... a new 'paradox of voting': It is theorists' fixation on a context dependent and ordinal preference scale; the most primitive scale imaginable and the mother of all paradoxes.

The step from complete and transitive preference relations to utility functions representing these functions is short, in fact, in the finite alternative sets nonexistent. Given the preference relations one can eo ipso construct the corresponding utility functions. These might then be used in preference aggregation. Since the cardinal utilities thereby obtained are unique up to affine transformations, one can transform all utility functions into the same scale by restricting the range of values assigned to each alternative. The utility values can, then, be used in defining social choice functions in many ways. Hillinger (2005) suggests the following. Let P_i be a strict preference relation of voter i and let P_i assign the set of candidates into disjoint subsets $A_1, \ldots, A_K, K \geq 1$ such that the voter is indifferent between candidates in the same subset and strictly prefers $a_i \in A_i$ to candidate $a_j \in A_j$ iff $i > j$. K is given independently of the number of candidates. For a given K, the voter is asked to assign to each candidate one of the numbers $x_0, x_0 + 1, \ldots, x_0 + K - 1$. The utilitarian voting winner is the alternative with the largest arithmetic mean or sum of assigned numbers.

This method simply sums up the scores – or utilities expressed in the $[x_0, x_0 + K - 1]$ interval – to determine the winning candidate or ranking of the candidates. Now this method has many names. Riker (1982) calls it Bentham's method, Hillinger the utilitarian or evaluative voting and Warren D. Smith the range voting. It is worth pointing out that the cumulative voting method whereby each voter can freely allocate a fixed stock of votes to various candidates, is not equivalent to utilitarian voting, although somewhat similar in spirit to the latter.

The just mentioned methods invoke a new criterion of performance: the maximization of collective utility. What is then maximized is the sum of utilities assigned to an alternative by all voters. Summation is, of course, just one possible way of handling the utilities. In addition to various non-anonymous (weighted) methods of summation, one could also maximize the product of the utility values. Riker calls this Nash's method with an obvious reference to the Nash product in bargaining theory.

The most recent entrant in the class of systems dealt with in this subsection is the majoritarian judgment introduced and elaborated by Balinski and Laraki (2007). It works as follows:

1. Each voter gives each candidate an ordinal grade (e.g. poor, medium, good, excellent)
2. The median grade of each candidate is determined
3. The winner is the candidate with the highest median grade
4. A specific tie-breaking rule is defined

Felsenthal and Machover (2008) have given an evaluation of the majoritarian judgment in terms of criteria applied in the ordinal social choice framework. The result is a typical mixture of good and bad showings. To summarize their evaluation: the majoritarian judgment does satisfy the Chernoff property, it is monotonic and is immune to cloning. These are undoubtedly desirable properties. In contrast to these, the system is inconsistent, vulnerable to the no-show paradox and may result in a Condorcet loser.

The evaluation shows that the ordinal choice theory criteria can be applied to voting systems that utilize richer information about voter opinions than just the ranking of candidates. However, one could ask whether the evaluation based solely on criteria borrowed from the ranking environment misses something relevant, viz. the fact that these systems are devised to attain goals (such as maximizing social welfare) that cannot be expressed in terms of ordinal concepts only. If this is the case, then at least some of the evaluation criteria should be specific to systems based on aggregating cardinal utilities. For example a person resorting to utilitarian voting might not be at all worried if the method fails on Condorcet winner criterion as long as it maximizes the sum of expressed utilities. Much work remains to be done in devising non-trivial criteria for such more specific evaluations. Until they have been invented, the best we can do is to proceed in the manner suggested by Felsenthal and Machover (2008).

10.7 Concluding Remarks

The most significant results of social choice theory pertain to compatibilities of various choice desiderata. Some of these take the form of proving the incompatibility of various properties of choice rules, others do the same for specific choice rules and voting procedures. The choice of the best rule is complicated by the sheer number of desiderata that one intuitively would like to see fulfilled, but even within relatively small subsets of important choice criteria one typically finds no procedure that would satisfy them all. Even dominance relations between procedures are uncommon. Since the procedures are intended for use in future collective decision making contexts, their success in avoiding anomalies or paradoxes is highly contingent upon encountering problematic preference profiles. Probability models and simulations have often been resorted to in order to obtain estimates about the theoretical frequencies of problematic profiles. This approach can be complemented by another one focusing on the difficulty of finding counterexamples showing various incompatibilities. Arguably it is only by looking at the structure or details of the problematic profiles that one can obtain information about their likelihood in practice. In the preceding we have also briefly touched upon alternative foundations of choice theory. Some of them require more information from the individuals, others less than the ordinal ranking approach. Setting up useful criteria for analyzing systems aggregating this new type of information is still largely to be done.

Acknowledgements The author wishes to thank Dan S. Felsenthal for inspiration, criticism and guidance far beyond the call of duty of an editor. Numerous constructive comments of an anonymous referee are gratefully acknowledged as well.

References

Aizerman, M. & Aleskerov, F. (1995). *Theory of choice*. Amsterdam: North-Holland.
Balinski, M. & Laraki, R. (2007). A theory of measuring, electing and ranking. *Proceedings of the National Academy of Sciences, 104*, 8720–8725.
Banks, J. (1985). Sophisticated voting outcomes and agenda control. *Social Choice and Welfare, 1*, 295–306.
Black, D. (1958). *The theory of committees and elections*. Cambridge: Cambridge University Press.
Brams, S. J. (1982). The AMS nomination procedure is vulnerable to 'truncation of preferences'. *Notices of the American Mathematical Society, 29*, 136–138.
Campbell, D. & Kelly, J. (2002). Nonmonotonicity does not imply the no-show paradox. *Social Choice and Welfare, 19*, 513–515.
Chernoff, H. (1954). Rational selection of decision functions. *Econometrica, 22*, 422–443.
Felsenthal, D. S. (2010). Review of paradoxes afflicting various voting procedures where one out of m candidates ($m \geq 2$) must be elected. Prepared for presentation in a symposium and workshop on *Assessing Alternative Voting Procedures*. Chateau du Baffy, Normandy, France, 30 July–2 August, 2010.
Felsenthal, D. & Machover, M. (2008). The majority judgment voting procedure: A critical evaluation. *Homo Oeconomicus, 25*, 319–334.

Fishburn, P. (1971). Comparative analysis of group decision methods. *Behavioral Science, 16*, 538–544.
Fishburn, P. (1973). *The theory of social choice*. Princeton: Princeton University Press.
Fishburn, P. C. (1977). Condorcet social choice functions. *SIAM Journal on Applied Mathematics, 33*, 469–489.
Gehrlein, W. V. (1997). Condorcet's paradox and the Condorcet efficiency of voting rules. *Mathematica Japonica, 45*, 173–199.
Gehrlein, W. V. (2002). Condorcet's paradox and the likelihood of it occurrence: Different perspectives on balanced preferences. *Theory and Decision, 52*, 171–199.
Gehrlein, W. V. (2006). *Condorcet's paradox*. Berlin: Springer.
Gehrlein, V. W. & Fishburn, P. C. (1978). Probabilities of election outcomes for large electorates. *Journal of Economic Theory, 19*, 38–49.
Gehrlein, V. W. & Lepelley, D. (2004). Probability calculations in voting theory. In M. Wiberg (Ed.), *Reasoned choices*. Helsinki: The Finnish Political Science Association.
Hillinger, C. (2005). The case for utilitarian voting. *Homo Oeconomicus, 22*, 295–321.
Klahr, D. (1966). A computer simulation of the paradox of voting. *American Political Science Review, 60*, 384–390.
Lagerspetz, E. (2004). Evaluating voting systems. In M. Wiberg (Ed.), *Reasoned choices*. Helsinki: The Finnish Political Science Association.
Lepelley, D. (1993). On the probability of electing the Condorcet loser. *Mathematical Social Sciences, 25*, 105–116.
Mackie, G. (2003). *Democracy defended*. Cambridge: Cambridge University Press.
McLean, I. & Urken, A. B. (Eds.). (1995). *Classics of social choice*. Ann Arbor: The University of Michigan Press.
Merlin, V., Tataru, M., & Valognes, F. (2000). On the probability that all decision rules select the same winner. *Journal of Mathematical Economics, 33*, 183–207.
Merrill, S. (1984). A comparison of efficiency of multicandidate electoral systems. *American Journal of Political Science, 28*, 23–48.
Merrill, S. (1988). *Making multicandidate elections more democratic*. Princeton: Princeton University Press.
Miller, N. (1995). *Committees, agendas, and voting*. Chur: Harwood.
Moulin, H. (1986). Choosing from a tournament. *Social Choice and Welfare, 3*, 271–291.
Moulin, H. (1988). Condorcet's principle implies the no show paradox. *Journal of Economic Theory, 45*, 53–64.
Niemi, R. & Riker, W. (1976). The choice of voting systems. *Scientific American, 234*, 21–27.
Niemi, R. G. & Weisberg, H. F. (Eds.). (1972). *Probability models of collective decision making*. Columbus: Charles E. Merrill.
Nurmi, H. (1984). Social choice theory and democracy. *European Journal of Political Research, 12*, 325–333.
Nurmi, H. (1987). *Comparing voting systems*. Dordrecht: D. Reidel.
Nurmi, H. (1988a). Discrepancies in the outcomes resulting from different voting schemes. *Theory and Decision, 25*, 193–208.
Nurmi, H. (1988b). Inferential modes in applying social choice theory. In B. Munier & M. Shakun (Eds.), *Compromise, negotiation and group decision*. Dordrecht: D. Reidel.
Nurmi, H. (1991). Att jämföra röstningsförfaranden. In H. Nurmi, B. Höglund, J.-E. Lane, & S. Berg (Eds.), *Gruppbeslut och rangordning*. Stockholm: Forskningsrådsnämden.
Nurmi, H. (1992). An assesment of voting system simulations. *Public Choice, 73*, 459–487.
Nurmi, H. (1999). *Voting paradoxes and how to deal with them*. Berlin: Springer.
Nurmi, H. (2002). *Voting procedures under uncertainty*. Berlin: Springer.
Nurmi, H. (2006). *Models of political economy*. London: Routledge.
Pérez, J. (1995). Incidence of no show paradoxes in Condorcet choice functions. *Investigaciones Economicas, XIX*, 139–154.
Pérez, J. (2001). The strong no show paradoxes are a common flaw in Condorcet voting correspondences. *Social Choice and Welfare, 18*, 601–616.

Regenwetter, M., Grofman, B., Marley, A., & Tsetlin, I. (2006). *Behavioral social choice: Probabilistic models, statistical inference, and applications*. Cambridge: Cambridge University Press.

Richelson, J. T. (1979). A comparative analysis of social choice functions: I, II and III: A summary. *Behavioral Science, 24*, 355.

Riker, W. (1961). Voting and the summation of preferences. *American Political Science Review, 55*, 900–911.

Riker, W. (1982). *Liberalism against populism: A confrontation between the theory of democracy and the theory of social choice*. San Francisco: W. H. Freeman.

Saari, D. & Tataru, M. (1999). The likelihood of dubious election outcomes. *Economic Theory, 13*, 345–363.

Straffin, Ph. (1980). *Topics in the theory of voting*. Boston: Birkhäuser.

Tversky, A. (1969). Intransitivity of preferences. *Psychological Review, 76*, 31–48.

Young, H. P. (1974a). An axiomatization of Borda's Rule. *Journal of Economic Theory, 9*, 43–52.

Young, H. P. (1974b). A note on preference aggregation. *Econometrica, 42*, 1129–1131.

Young, H. P. (1975). Social choice scoring functions. *SIAM Journal on Applied Mathematics, 28*, 824–838.

Young, H. P. & Levenglick, A. (1978). A consistent extension of Condorcet's election principle. *SIAM Journal of Applied Mathematics, 35*, 285–300.

Chapter 11
Putting Paradoxes into Perspective: in Defence of the Alternative Vote

Ken Ritchie and Alessandro Gardini

11.1 Introduction

On 5 May 2011 electors in Britain voted in a referendum on changing the system for electing Members of Parliament from first-past-the-post (FPTP, as single-member plurality is called in the UK) to the alternative vote[1] (AV – also known as "STV for single-member constituencies" and, in the US, instant run-off voting). On a 42% turnout, 32% voted in favour of change while 68% voted to retain FPTP. This paper is not concerned with the referendum campaigns or the outcome, but whether AV was the right option for change to offer.

While there has been dissatisfaction with FPTP, principally over its failure to deliver a parliament that broadly reflects the distribution of political support for the parties, the options for change were restricted by the desire of most politicians to retain a system based on single-member constituencies. Nevertheless, although there were other systems that might have been chosen, AV was the only alternative to FPTP that was even considered.

Much of the debate over whether and how Britain should change its electoral system has been driven by parties' calculations of how a change might affect their electoral prospects, even if the arguments have been presented in terms of how reform might improve British democracy. However, the democratic arguments for change are strong – the reform lobby has a case when it describes FPTP as "a broken

[1] In this paper we use the abbreviation AV for the alternative vote following practice in the UK. When referring to Approval Voting we do not abbreviate.

K. Ritchie (✉)
Chief Executive of the Electoral Reform Society (UK)
e-mail: ken.ritchie@lineone.net

A. Gardini
Research Officer for the 'Yes to AV' campaign
e-mail: a.g.gardini@gmail.com

D.S. Felsenthal and M. Machover, *Electoral Systems*, Studies in Choice and Welfare,
DOI 10.1007/978-3-642-20441-8_11, © Springer-Verlag Berlin Heidelberg 2012

system" – even if there is no clear consensus on what change should be made. Few, if any, maintain that AV is a complete solution but, as we will argue, it is a compromise that would at least overcome some of the defects of FPTP.

Most of these arise from FPTP's inability to ensure a broadly proportional outcome. AV, however, is not a proportional system, and in some circumstances could produce even more distorted results than FPTP.[2] No system that elects only one candidate can guarantee a proportional result. For most electoral reformers, the election of a national parliament or local authority requires at least a degree of proportionality, otherwise there is the risk of electing a body that will not be representative of the diversity of views of the electorate, that may not give any voice to significant but minority opinions, and that consequently may not be effective in providing a body that can hold broad debates and hold executives to account. Thus democracy campaigners tend to view AV as an inadequate reform, but the only change that at present is politically feasible and one that could open the door to further more radical changes.

However, choosing the "best" system for the election of a single candidate remains an important issue. Until the opposition of most MPs to any change which would undermine what they regard as their right to be the unique representative of their constituents can be overcome, the British parliament will continue to be elected using single-member constituencies. Moreover, there are, of course, situations in which an election must be held for only a single position: in London and a number of local authorities in Britain, mayors and many local councillors are elected in single-member districts (unfortunately British democracy has not advanced as far as allowing us elections for our head of state) and it is clearly important that the most suitable system is used for such elections.

In this paper we therefore confine our attention to elections for single positions and do not consider the range of proportional or semi-proportional systems that we regard as more appropriate choices for British elections (although we recognise that some of the systems we consider have proportional variants, such as the single transferable vote, STV, a proportional version of AV, which is our preferred choice of system).

We will argue that AV, for whatever reasons it emerged as the leading opponent of FPTP, is a better choice for public elections than any of its single-member-election rivals.

There are, however, criticisms of AV of a logical and mathematical nature, the most serious being that AV does not necessarily elect a Condorcet winner and can be non-monotonic. It has been described as "highly paradoxical" (see Chap. 3). Although these criticisms did not feature in Britain's debate on its voting system in any significant way, it does not mean they are unimportant. In this paper we will examine these criticisms, comparing the performance of AV with competing options and in doing so we will attempt to put these paradoxes into perspective.

[2] In 1997, for example, there was a large swing to Labour which would have given it not just more first preference votes but also more second preferences enabling it to win even more seats under AV that it would under FPTP.

We will conclude that, for whatever reasons AV might have been chosen as Britain's preferred alternative to FPTP, in the prevailing political circumstances AV was the right choice.

11.2 The Background to the UK's Referendum on the Voting System

Parliamentary debates on how Members of Parliament (MPs) are elected are not new in the UK but, prior to 1997 when Labour was elected after 18 years of Conservative government, they have been rare. In 1918 and in 1931, the Commons (the lower house) voted in favour of AV, but in each case the Lords (the upper house) wanted STV and the result was no progress (Hart 1992). Until the middle of the last century a small number of MPs were elected in two-member constituencies and 12 were elected by STV to "university seats", but since 1951 all MPs have been elected by FPTP.

Not even in 1951 and 1974 when FPTP resulted in a "wrong winner" (a party winning an absolute majority of seats in spite of another party receiving more votes) was there any serious challenge to FPTP. Britain had an essentially two-party system – not until 1997 did Labour and the Conservatives together win less than 90% of the seats – and neither of the main parties were willing to consider any reform that could change the nature of the "two horse race" of British politics.

Prior to the 1997 election, however, Labour, fearful of the consequences of yet another defeat, was anxious to secure Liberal Democrat backing should they fail to win an absolute majority of seats. They agreed with the Liberal Democrats that a new Labour government would set up a commission to recommend an alternative voting system which would be put to the electorate in a referendum.

Following Labour's 1997 victory, the Jenkins Commission was established and it recommended a new system, the alternative vote plus (AV+), a form of mixed member proportional whereby 80–85% of MPs would be elected in constituencies using AV and the remainder would be elected from regional party lists to compensate for the disproportionality that might arise from AV (Independent Commission on the Voting System, 1998). But, finding that 43% of the votes had given them 63% of the seats, Labour had little appetite for changing the system that had rewarded them so generously and consequently the referendum was never held.

Labour must nevertheless be given credit for the reforms it introduced in its first term (1997–2001), even if the reforms were driven by reasons of political expediency rather than concerns for the quality of democracy. The Scottish Parliament and the Welsh Assembly, created with powers devolved from central government, were elected by mixed member proportional systems (referred to in the UK as AMS – the additional member system) and the Northern Ireland Assembly was established and elected by STV in six-member constituencies. A regional closed-list system was introduced for European parliament elections, other than in Northern Ireland where STV was used. Mayors in London and those local authorities that decided to have one (other than mayors with only ceremonial functions) were elected by the supplementary vote (see Sect. 11.4.4.1 below). In Scotland, the new Scottish parliament used its powers to introduce STV in three- and three-member wards for local government.

However, not until the second half of 2009, when it was clear that FPTP would not deliver a Labour victory, did Labour return to plans for a referendum on the voting system for the election of MPs. But even at that stage Labour was not prepared to contemplate a change to a proportional system which could have made it difficult for them ever again to hold power on their own, and hence they proposed a referendum on AV (with the hope that Liberal Democrat supporters would use their second preference votes to help Labour defeat Conservatives in marginal seats).

Although Labour lost in May 2010, the result was a hung parliament and the Liberal Democrats extracted from the Conservatives a commitment to a referendum as the price of support in a coalition. The Liberal Democrats were strongly committed to a proportional system, but AV was as much as the Conservative leadership was prepared to offer.

Thus the debate on voting systems has, unsurprisingly, been dominated by the interests of the major parties.

However, there has long been a popular campaign for electoral reform in the UK – the Electoral Reform Society was formed in 1884. But while cross-party/non-party campaigns for reform have no doubt had influence, it seems that change was never likely to happen without political circumstances that made change attractive to a majority of politicians. Such circumstances were there in the run up to the 2010 general election when Labour faced defeat under FPTP and in the immediate aftermath of the election, although the Conservatives did not want change, they recognised they would need to offer at least a referendum to win Liberal Democrat support in forming a government.

During the debates of 2009 and 2010, however, the possibility of changing to a single-member constituency system other than AV never seems to have been considered, either by the politicians or the electoral reform lobby. We expect that very few were aware of any alternatives. Not only did AV have an historical place in Britain's debates on electoral reform, but proponents of STV saw AV as a step towards what they wanted and other reformers may have hoped for future moves to the Jenkins Commission's AV+ recommendation of 1998. Moreover, all of the major parties were already using AV for the selection of their candidates.

Although AV was defeated in the referendum, we doubt that it was defeated on its merits. Campaigners for change presented AV as a first step towards a new political culture, but the campaigns were fought on the old: debates were often acrimonious with both sides accusing each other of spreading misinformation, perhaps the most serious being the "no" campaign's claim that the extra costs of AV would lead to the closure of schools and hospitals. In the end, the "yes" campaigners were unable to overcome a general public resistance to change.

11.3 How AV Improves on FPTP

Those arguing for a change from FPTP in the UK have some powerful arguments. Many of them stem from FPTP's disproportionality:

- **FPTP produces distorted outcomes.**
  While FPTP has produced "wrong winners", more often it produces exaggerated and undeserved majorities. In 2005, for example, Labour won 55% of the seats on 35% of the votes. That Labour had a majority sufficient to push through parliament almost all of its proposals when nearly two out of every three voters voted for a candidate of a party other than Labour raised questions about the legitimacy of government.
  At regional level, the distortions are even more pronounced. In 2010 the Conservatives, although the dominant coalition partner, won only one seat in Scotland in spite of receiving 16.7% of the votes, while Labour won only 2 of the 58 seats in Eastern region with 19.6% of the votes.
- **FPTP makes it difficult for significant minorities to gain representation.**
  Although the Green Party enjoys considerable popularity, it did not win a seat until the 2010 election. In that election the UK Independence Party received 919,546 votes but won no seats.[3]
- **FPTP is a barrier to increased representation of women and other under-represented groups.**
  Parties select one candidate for each constituency. With proportional systems that use multi-member constituencies where parties elect several candidates simultaneously, parties have an incentive to select a diverse group of candidates to maximise their support in all parts of the electorate.
  However, AV, not being a proportional system, would do little to overcome these problems (although in most elections it would be less disproportional than FPTP).[4] Nevertheless, with some of the major problems of FPTP, AV would make a difference:
- **FPTP encourages tactical voting, unnecessary under AV.**
  What we mean by a tactical vote in this paper is an "insincere" vote cast for the voter's preferred candidate of those the voter believes can win, rather than for the candidate the voter would like to win (described by Franklin et al. (1994) as an "instrumental" tactical vote rather than an "expressive" one cast in hope that it will influence the terms of the debate or the policies of the winning candidate).
  For example, in a Labour-Conservative marginal, a Liberal Democrat supporter may decide to vote, say, Labour to prevent a Conservative victory rather than "wasting" their vote on the candidate of their own party. It has been

[3] Small parties will highly localised support may, however, do better under FPTP (and AV) than under some proportional systems that use multi-member districts (Reynolds 2006).

[4] If, however, an index of proportionality that took account not just of first preferences were used, then AV might score reasonably well on proportionality.

estimated that as many as 9% of voters vote tactically in Britain's recent general election (Fisher and Curtice 2006).

While it is not strictly accurate to claim that AV makes tactical voting impossible, we will argue below that the opportunities for tactical voting under AV are so limited that it is effectively eliminated. With AV voters can vote sincerely. For example, in the above example a voter could vote for the Liberal Democrat candidate in the knowledge that through a lower preference they can still influence the outcome.

In British election campaigns it is common for candidates to claim (not always truthfully) that they are in a "two-horse race", arguing "I know you support another candidate, but in this particular contest it makes more sense to vote for me and prevent someone you really don't want from winning". With AV this form of argument holds no water – voting for the candidates they really want will, other than in very exceptional circumstances (which we consider in section 6 below), be the best strategy for voters.

A consequent advantage of AV is that elections should give a better picture of the support that parties enjoy. It is likely, for example, that at present election results grossly understate support for the Green Party as many Green supporters will vote for someone with a better chance of winning (the Green Party received just 1.0% of the vote in the 2010 general election, but 8.6% in the 2009 European parliament election which was fought on a party list system).

- **FPTP allows candidates to win with less than majority support but AV, at least in a qualified sense, requires winners to have majority support**.

  In 2010 only 35% of MPs elected had over 50% of the votes in their constituencies. Some have been elected on much less: in the 2010 general election the winner in the Norwich South seat received only 29.4% of the votes, and in the 1992 general election a Liberal Democrat was elected with just 26.0% in a four-way contest. It is quite possible that in some cases the most unpopular candidate won: in FPTP local elections, the extremist British National Party has won seats with less than 30% of the votes[5] when polling evidence suggests that a large majority of the voters would have preferred any candidate to that of the BNP.

  An AV winner must have an absolute majority of the votes at the final stage of the count. Of course this does not mean that winners must have a majority of first preference votes, which would be an impossible requirement, and it does not even mean that the winner will have support through some preference from a majority of voters as some may not express preferences for all candidates. Nevertheless, with AV winners will have a greater sense of legitimacy and more voters will feel that they have an MP of their choice.

- **FPTP elections are fought only in marginal seats, of which there will be more under AV**.

  Most seats are perfectly "safe" for one party or another in UK elections, some having been held by candidates of the same party for more than a century.

[5] E.g., the BNP won South Oxhey, Hertfordshire, in 2009 with 29.2%,

Even in the 2010 election, in which there were more uncertainties than usual, the Electoral Reform Society was confidently able to send victory congratulations to candidates in a majority of seats before the election was even held. In safe seats supporters of parties other than the dominant party vote knowing that their vote is only a gesture of defiance. Parties focus their campaign resources onto the seats that matter – those they need to fight to hold or where there is a realistic chance of winning from an opponent. In 2005 it was estimated (in "The Times", London, 6 April 2005) that the two major parties targeted only 2% of the electorate – the swing voters in swing seats.

AV will not eliminate "safe seats", but it will reduce their number. With FPTP many seats are safe although the candidate of the dominant party has significantly less than 50% of the votes because the remainder of the votes are distributed between the other candidates. With AV, transfers of votes between these candidates may present the front-runner with a more serious challenge. Fewer seats will be safe and in more constituencies there will be competitive elections, and where MPs must work harder to secure their election their sense of accountability to their electors is likely to be higher.

- **FPTP elections result in many "wasted" votes, but fewer would be "wasted" under AV.**

  By "wasted" votes we mean votes that do not contribute to the election of a candidate, either because they are cast for losing candidates (52.8% in 2010 election) or because they are votes that only add to superfluous majorities (18.3% in 2010). The term is one that can be debated, but for many voters is nevertheless meaningful. For example, in most UK constituencies a vote for the Green Party candidate is "wasted" in that the Green candidate's chances of winning are extremely remote.

  If we regard a vote for a loser as a wasted vote, then no system will eliminate wasted votes. But with AV, through transfers, more voters will have voted for winners and fewer votes will be wasted in this sense.

- **FPTP encourages negative campaigning and a form of adversarial politics that is unattractive to voters: AV has the potential to change this.**

  With FPTP it is only necessary to get more votes than any other candidate. It can be easier to do that by attacking opponents in a disparaging way than by promoting one's own ideas positively. Negative campaigning is a major feature of British general election campaigns (a problem that is not, of course, unique to Britain).

  AV's effect on the nature of election campaigns is harder to predict. Political debate will always be robust but, with AV, a winner may need transfers of votes from supporters of other candidates, giving candidates an incentive to emphasise where they agree with others as well as where they differ. Unreasonable attacks on other candidates may not be the best strategy for gaining transfers from their supporters. A study of San Francisco's first use of IRV (reported in FairVote 2005) found that:

> "several of the races were marked by less mudsling and more coalition-building and issue-based campaigning than in previous San Francisco elections, because with ranked choice voting candidates have incentive to build coalitions rather than attacking opponents as a successful winning strategy, since winners may need to attract the second or third rankings from the supporters of other candidates."

Given these advantages of AV, one might ask why change should be controversial. Much opposition to change is, we suspect, based on political calculations, but the major arguments are that FPTP is more likely to deliver majority government (whether or not the majority is justified by electoral support) and FPTP's simplicity and transparency. However, there is nothing complex about an AV count, and it can be assumed that UK general elections will continue to be counted manually without counts taking much more time than with FPTP. Moreover, with AV candidates and their supporters should have little difficulty in understanding the procedures and the manner in which the winner is determined – particularly as AV is widely used in the UK by political parties, trade unions, professional associations and voluntary organisations.

11.4 The Defects of AV's Competitors

11.4.1 Alternative Non-ranked Voting Methods and Their Problems

If FPTP were to be replaced, are there other systems not involving the ranking of candidates which should be considered? The main contenders are plurality with run-off, successive elimination, and approval voting, but here we argue that none of them is suitable.

11.4.1.1 Plurality with Run Off

Plurality with Run Off (i.e. holding a second election between the front-runners when no candidate gets at least 50%) is simply not a practical choice. In the UK, a system that requires a second public election in most constituencies would be a non-starter because of costs to the state and to the political parties, and it is doubtful that electors would have an appetite for returning to the polling stations for a second round. Experience in the US, where the system is used in many local elections, shows that turn outs drop dramatically on the second round (an average drop of 35% in elections from 1994 to 2008 (FairVote 2009). It should be noted that the use of AV is gaining ground in the US under the name instant run off voting (IRV) partly because it saves the expense of a second round.

However, even leaving the practicalities aside, it is an unsatisfactory system.[6] If it is accepted that FPTP does not necessarily produce the right winner, why should it be assumed that one of the first two past the post should be the right winner?[7] Unlike AV, the first round still encourages a certain amount of tactical voting because of risk of a compromise choice not reaching the second round. If a compromise candidate does not reach the second round it can lead to surprising outcomes. To illustrate, Jean-Marie Le Pen of the French National Front qualified for the second round in the French Presidential election in 2002 with only 16.9% of the vote, defeating Jospin who had only 16.2% because the left vote was fragmented amongst a number of minor candidates. This ultimately gave Chirac one of the biggest landslides in French history.

11.4.1.2 Successive Elimination

If plurality with run-off is unsatisfactory because of the need for two ballots, this system with its need for a whole series of eliminating ballots, quite apart from its other defects, makes it a non-starter.

11.4.1.3 Approval Voting

With approval voting, voters can vote for all candidates whom they would tolerate, or "approve", but cannot express preferences between those they approve. Approval voting therefore seriously restricts voters' ability to express their views. If, with five candidates, A, B, C, D and E, a voter votes for ABC, we know that the voter rates each of A, B and C above either D or E, but we do not know the extent of their approval of A, B and C. It may mean the voter regards A as by far the best candidate, B as someone who would be "not bad" and C as on the margins of acceptability, or it could mean the voter's preferences for the three candidates are quite the reverse. As a result, approval voting can give strange results which are very dependent on voter tactics rather than voter preferences, as voters can harm candidates they strongly favour by voting for others they merely think will not eat their children.

Consider, for example, an election in which the ordered preferences for the five candidates are:

ABCDE	50
BACDE	20
DECAB	30

Candidate A would win under AV, is a Condorcet winner and would win using a Borda count.

[6] Plurality with Run Off would nevertheless be more satisfactory than FPTP in that it does not allow a Condorcet loser to win.

[7] We are grateful to David Hill for pointing out this absurdity.

However, if all voters "approve" their first three choices, then with approval voting C is the winner (with 100 votes compared to 70 for each of A and B). Approval voting can therefore lead to the election of candidates disliked by few but not necessarily strongly liked by many.

Suppose, however, that those preferring ABCDE also "approve" of D and that those for whom D is their first preference "plump" for D (i.e. voting only for D rather than DEC), then we would have (with upper case indicating an approval):

ABCDe	50
BACde	20
Decab	30

Now D would win with approval voting (with 80 votes compared to 70 for each of A, B and C). But if supporters of D voted sincerely, expressing support for E and C, then C would again be the winner. That supporters of D could contribute to D's defeat by C by expressing support, even if less strongly felt, for other candidates is surely a near-fatal flaw.

Approval voting is thus very prone to tactical/strategic voting: parties have an interest in ensuring that their supporters vote only for their candidate and for no others. If in the above example, A could persuade all their supporters to vote for A and only A, then A could not be overtaken. Consequently approval voting is a system in which it can be difficult for voters to decide how best to vote to achieve an outcome they want – by voting for more than one candidate they can cause their favoured candidate to lose, but by only voting for one they may lose the opportunity to influence the result. A public election with approval voting would therefore be a tactical battle, with sincere voters likely to lose out.

Thus, while there may be forms of collective decision-making in which approval voting might have a use, as a system for public elections it is deeply flawed.[8]

11.4.2 Condorcet Systems and Pair-wise Problems

Here we consider systems based on pair-wise comparisons which are popular amongst many social-choice theorists. Where there is a Condorcet winner, such systems will elect that candidate. Where there is not, there are a number of variants for determining the winner, including:

- Copeland's method, in which the winner is the candidate who wins the greatest number of pair-wise contests (although there is a high risk of ties)
- Kemeny's method, which seeks the most favoured ranking of candidates (but suffers from the complexity of the count)

[8] Range voting and majority judgement systems are susceptible to tactical voting in the same manner as approval voting and we therefore do not consider them.

- Black's method, which resorts to a Borda count where there is no Condorcet winner

In spite of the apparent sophistication of these methods, they suffer from a number of flaws.

A Condorcet Winner may not Always be the "best" Winner

If a candidate would defeat any other in a pair-wise contest, the logic of declaring that candidate the winner may seem compelling, but we question whether it is always the "right" decision. To take a simple case, suppose the votes in a three-candidate election were as follows:

ACB	48
BCA	47
CAB	5

C is the Condorcet winner. The result, however, suggests that C is a candidate with little personal support. Supporters of the main rivals, A and B, would prefer to see C win rather than their candidate's principal opponent, but that does not necessarily mean they are giving C positive support as a person with the qualities needed to be an effective political representative. Indeed, the very low first-preference support for C might suggest that few regard C as someone of sufficient calibre and expertise to be a good representative.

It is even possible for a candidate to be a Condorcet winner without any first preferences whatsoever. Suppose D were to enter the above contest and the votes were:

ACDB	22
ADCB	26
BDCA	25
BCDA	22
CDAB	5

While A would still win under AV, D now replaces C as the Condorcet winner. We cannot imagine the electorate understanding the election of a candidate with no first-preference support[9], and we would argue that such a candidate is unlikely to be a good choice.

[9] David Hill has pointed out to us that results would make more sense to the ordinary elector if reported as "D beats A by 60–40, D beats B by 40–38, D beats C by 43–35", but it would be surprising if the fact that D had no first-preference votes did not emerge from the reporting of the count.

Clearly it is better for a winner to be a Condorcet winner, but whether a candidate wins under Condorcet should not be the only consideration. Elections can be regarded as about choosing the most representative candidate, but they can also be seen to be about choosing the person best able to represent the interests of constituents.

In the above example, A would win under AV. While it can be argued that AV's disregard for some lower preferences may prevent a candidate with the lowest level of unpopularity but few higher preferences from winning, there is a risk that methods based on pair-wise comparisons may result in victories for candidates who are least objectionable rather than for candidates with stronger merits. In electing members of a national parliament, that does not seem desirable.

Not all Preferences Should be Regarded as Having Equal Value

A second concern is that these systems attach an equal weighting to all of a voter's rankings. We question whether this can be justified. Most voters will know who they want to win, and many will have a clear idea which candidate is their second choice, and even their third. But with lower preferences there is less guarantee that voters will be as discriminating in their choices. Having completed long STV ballot papers ourselves, we are aware that once we have ranked our favoured candidates there is an element of arbitrariness about how we rank the remaining candidates. If a voter ranks candidates A, B,... H (say) in that order, can it be right that we regard A being placed above B as being no more important than G being above H? We think not.

Systems that require voters to rank all candidates will be particularly susceptible to these risks. We should not expect all voters to have opinions on all candidates, and to compel them to express preferences where none exist can only give a false picture of voters' views and introduce an element of chance. Where voters are not required to rank all candidates, few will do so: in the Scottish STV elections of 2007, although voters were electing either three or four candidates with an average of 7.4 candidates in each electoral district, the median voter only expressed three preferences and many (perhaps this being the first election with preferential voting) only marked the ballot paper with a cross (taken to mean a first preference with no other preferences expressed) (Baston 2007, p. 69).

When AV was being debated in the British parliament in 1931, Winston Churchill, later to become Britain's Prime Minister, criticised AV for taking account of "the most worthless votes for the most worthless candidates" (Rogaly 1976, p. 84)) and went on to complain that the system contained an element of blind chance and accident. His concern was that who wins might be decided by quite low preferences. Although his criticism of AV is misplaced as AV only considers a lower preference when all of the voter's higher preferences have been eliminated (and in the great majority of cases the outcome will be determined by higher preferences), his concern has validity when we consider Condorcet systems.

Voting by Ballot Paper Order ("Donkey Voting")

There is the added risk of "donkey voting" – having decided on their first few preferences, some voters might complete the ballot paper by numbering other candidates in the order shown on the ballot paper. (This may also happen with AV, but with AV it is less important as AV only considers lower preferences if higher preferences are eliminated.)

Evidence from the 2007 Scottish local elections – an STV election using multi-member districts – demonstrates that voting by ballot paper order can be a serious problem (Denver et al. 2009). Although voters were not required to rank all candidates, where a party had two candidates it can be assumed that most supporters of that party would want to vote for both. However, where a party's support was only sufficient to secure the election of one candidate, the successful candidate was around six times more likely to be the candidate placed higher on the ballot paper. It appears that most voters did not discriminate between the candidates but merely followed the ballot-paper ordering. Walsh and Robson (1973) have reported on similar problems in Irish and Australian elections.

This problem could, of course, be resolved by using Robson rotation as is done in parts of Australia (whereby candidates' names appear on ballot papers in a different, pre-determined order) or by randomising the order on the ballot paper at the point of printing, as is done in some local elections in New Zealand. However, this introduces further complexity into the electoral process.

Lower Preferences may Disadvantage Higher Preferences

With AV, voters can safely rank candidates knowing that showing lower preferences cannot disadvantage the candidates they really want. Condorcet systems do not have this important property and that we consider as serious a flaw as any other paradox of voting systems.

If A is a Condorcet winner, then voting AB rather than A will not change the outcome, but if there is not a Condorcet winner, expressing a lower preference could lead to the defeat of a higher one. Consider the following example:

A	28
AE	8
B	12
BC	4
CBDA	18
DC	3
DCE	4
DE	8
E	11
EB	4

Here there is no Condorcet winner, but A wins under both Copeland's and Black's methods (as well as under AV). This set of votes could have arisen if A had an astute campaign manager who was able to persuade most of A's supporters to vote for A and only for A.

Suppose, however, that just five of those who voted just for A had decided to give a second preference for B (i.e. A's supporters voting A 23; AB 5; AE 8), then B would become the Condorcet winner! (B would also win a Borda count, but A would remain the winner with AV.)

This contrasts with AV with which a lower preference cannot disadvantage a higher preference. A vote for ABC will remain with A until such time as it is determined that A cannot win, and only then will B be considered. C will not be considered until both A and B have been eliminated. This, we believe, is of crucial importance in a public election, and that Condorcet systems do not have this property is, for us, a major reason for rejecting them as viable options.

If by voting ABC the preference for B could result in the elimination of A, a voter wanting A to win would be advised to vote for no candidate other than A. Parties would campaign for voters to vote for their candidates and not to express preferences for any others. Electoral outcomes could depend on which voters decided to risk expressing a list of preferences rather than on voters' sincere rankings of the candidates' merits. The whole rationale for preference voting would be lost. (Some might argue that this problem could be overcome by requiring voters to rank all candidates but that, as we have argued, would be a foolish move.)

11.4.3 Why Borda won't do

The Borda count does not use pair-wise comparisons but it suffers from similar problems. Consider the following election:

A	41
BCA	39
CAB	20

Here A wins under AV and with Borda (there is no Condorcet winner). But with Borda:

- if just eight of those who voted A had voted AB then B would have won;
- if just seven of those who voted A had voted AC then C would have won;[10]

[10] Here where voters give truncated preference lists we assume unranked candidates receive the average of remaining scores (i.e. a vote for A alone gives scores 3, 1.5 and 1.5 for A, B and C respectively) but, if we use the version of Borda in which unranked candidates are given the lowest score, similar results can be obtained.

thus illustrating how with Borda a lower preference can upset a higher one. In most circumstances a party would advise its supporters only to vote for that party's candidate and to show no other preferences, thereby defeating the purpose of the system.

Borda can also be criticised for the way it assigns values to preferences. With AV and systems based on pair-wise comparisons, we are only concerned with voter's preferences for one candidate over another. But there is no reason for assuming that the strength of voters' preferences should be valued as 4, 3, 2, 1 rather than, say, 4, 2, 1, 0.5. In our example, with this scoring system B would have defeated A.

11.4.4 Variants of AV

11.4.4.1 The Supplementary Vote

The Supplementary Vote (SV), a truncated form of AV that has been used in England for the election of mayors. The position of Mayor of London was introduced in 2000 and later local authorities were allowed to elect mayors, subject to the consent of the electorate in a referendum. However, in only 37 of England's 152 local authorities have referendums been held, and in only 12 of these have they been successful.

With SV, voters are only allowed to express a first and second preference. If no candidate wins by receiving over 50% of first preferences, then all but the leading two candidates are eliminated and any second preferences for either of these candidates transferred from those eliminated.

Where the count goes to a second stage, winners have a majority of the votes counted at that stage, but not necessarily a majority of all votes. Indeed, in most SV elections the winners have had less than 50% of the votes: in Torbay in 2005 the winner had less than 30%. Thus, while SV might be a slight improvement over FPTP in that voters can express at least a second preference and winners may need more votes than with FPTP, it does not provide the full advantages of AV and we see no reason for considering it further.

11.4.4.2 Coombs Method

Coombs method is similar to AV, but where no candidate has a majority the candidate with most last preferences is eliminated. This creates a problem when voters do not give full preference lists and, as we have argued in Sect. 11.4.2, it would be foolish to require them to do so. However, we also have reservations about using a system based as much on who voters do not want as on who they support. In the UK, elections suffer from much negative campaigning: with Coombs method there is a risk that this problem would be accentuated. We do not therefore favour it.

11.5 Would AV's Paradoxes be a Problem in UK Elections?

11.5.1 AV and Paradoxes

A system in which voters rank their preferences for candidates cannot meet all the requirements of a voting system that might be regarded as reasonable, as Arrow (1951) has demonstrated. AV is no exception:

1. AV may not always elect a Condorcet winner.
2. AV can be non-monotonic, creating the potential for forms of tactical voting.
3. Voters may obtain a more favourable result by not listing all preferences (the "truncation paradox").
4. Voters may obtain a more favourable result by not voting (the "no show" paradox).
5. A candidate who wins in every sub-district may not win in the constituency as a whole (the "reinforcement paradox").

A first question that arises is how serious are these paradoxes. An outcome that some may consider paradoxical may not be problematic: a voting system sets the rules of the game to which the players – the candidates and the voters – accept beforehand. As Tideman (2006, p. 140) puts it:

> "it is not necessary that the members of a collectivity believe that voting identifies the available option that is indisputably best; it is only necessary that they believe the overall pattern of outcomes that will result from their agreeing that a prescribed set of collective decisions will be made by voting will be better... than if they do not agree to make such decisions by voting."

However, if a voting system were to produce outcomes so perverse as to undermine the legitimacy of elections, then the paradox would indeed be a problem.

Secondly, we need to consider how frequently paradoxes occur. The rare odd result might, in our view, be a price worth paying for a voting system that behaves in an apparently rational manner in the great majority of contests.

Thirdly, we must consider the extent to which paradoxes offer opportunities for tactical voting. The Gibbard–Satterthwaite theorem tells us AV cannot escape from motivating voters on occasions to vote tactically (Gibbard 1973), but we must consider whether and how often in the circumstances of a real public election tactical voting is ever a viable strategy.

11.5.2 Testing for Paradoxes using the 2010 and 2005 UK General Elections

We have examined whether paradoxes would be problems in UK general elections by looking at 2010 and 2005 election results. As these elections were held under FPTP, we have assumed that FPFP votes represent first preferences (we know that this is not strictly true because of tactical voting, but it should provide a reasonable basis for measuring AV's behaviour).[11] We have used opinion polls conducted around the time of these elections for data on voter's subsequent preferences. These polls, however, do not give us all the information we need for modelling how the elections might have looked under AV – they asked only for second preferences and for the minor parties the sample sizes do not provide adequate samples.

We have therefore confined our attention to England where contests are, with very rare exceptions, between the three major parties – the Conservatives, Labour and the Liberal Democrats (Northern Ireland has its separate parties and Scotland and Wales both have significant fourth parties). However, English constituencies comprise 82% of the UK total, and in them the major parties received 92% of the votes in 2010 and 94% in 2005: we therefore consider that looking at votes for the major parties in England give us reasonable data sets for assessing the performance of AV.

In testing for paradoxes, we have only considered vote transfers, as measured by the opinion polls, from one major party to another. By doing so, however, we may have over-estimated the number of voters giving only one preference and the numbers whose second and subsequent preferences are non-transferable (i.e. preference lists do not include a second major party). For 2010 we have therefore also tested with the numbers in these categories halved and transfers between the major parties increased proportionately. We have thus used three tables (Table 11.1) for vote transfers:

Test 1: 2010: general election results for 2010 and 2010 opinion polls;
Test 2: 2010: general election results for 2010 and revised opinion polls for 2010;
Test 3: 2005: general election results for 2005 and 2005 opinion polls.

The appendix contains the figures we have used.

11.5.3 AV and Condorcet Winners

In our analysis we found a small number of cases in which AV would not have elected a Condorcet winner:

Table 11.1 AV and Condorcet winners in English elections

	Number of constituencies		
Test	1 (2010)	2 (2010 rev)	3 (2005)
No Condorcet winner	6	3	1
AV elects Condorcet winner	516	494	525
AV does not elect Condorcet winner	10	34	2
AV does not elect Condorcet winner (%age)	1.9%	6.4%	0.3%

[11] Moreover, the opinion poll questions we have used for second preferences analyse voters in terms of their voting intentions under FPTP rather than what their first preferences would be under AV.

However, whether AV's "failure" to always elect a Condorcet winner, where one exists, should be regarded as a defect of the system depends on the importance we attach to Condorcet. We have argued (in Sect. 11.4.2 above) that a Condorcet winner might not be the right winner if that candidate has little first-preference support: in analysing winners other than Condorcet winners in test 2 (2010 rev) we found that 12 had less than 20% of first preferences, a further 15 had between 20% and 25%, and none of the remaining 7 had more than 30%.

Thus, we conclude that cases in which AV does not elect a Condorcet winner would be rare and rarely serious.

11.5.4 AV and Non-monotonicity

Non-monotonicity can arise with AV where voters, by changing their preference rankings, can affect the order in which candidates are eliminated and in turn can affect the eventual outcome.

If in an AV election involving candidates of Labour (L), the Conservatives (C) and Liberal Democrats (D), with three candidates, the votes cast were:

L	38
DL	17
DC	15
CL	5
CD	25

then the Conservative candidate is eliminated and the Liberal Democrat wins. But if three of the Labour supporters were to lower Labour in their preferences by voting CL, the Liberal Democrat would be eliminated and Labour wins.

However, in our three tests on 2005 and 2010 data, the highest number of constituencies where such non-monotonicity could have occurred was 3 out of a total of 531, and even in these constituencies voting tactically to take advantage of the non-monotonicity would have been a risky strategy (see "Case 1" of Sect. 11.6 below). Non-monotonicity would not, therefore, in our view, be a problem if AV were to be used in UK general elections.

11.5.5 AV and the Truncation Paradox

Voters may achieve a more desirable outcome by not listing all their preferences. With AV, however, we do not regard this paradox as a serious one.

Suppose a voter ranks the candidates ABCD. With AV, the prospects of A winning are the same whether the voter votes A, AB, ABC or ABCD, and similarly B's chances do not depend on whether the voter votes AB, ABC or ABCD. This is because with AV no preference is considered until all higher preferences have been eliminated.

Circumstances could arise, however, in which a voter gets a more desirable outcome by suppressing a preference, effectively voting for a candidate not so highly ranked. However, these circumstances are similar to those in which tactical voting is possible, and in Sect. 6 below we have found them to be rare.

11.5.6 AV and "No Show"

The "no show" paradox arises when a voter obtains a more favourable result by not voting. With our three-candidate example, if a, b and c are the sincere first-preference votes of A, B and C and $a > b > c$, it may be possible for some supporters of B, by not voting and securing the elimination of B, to allow C to win. The requirements for this to happen are similar to those of "case 2" considered under tactical voting, but the inequalities to be tested become:

$$a < (b + c) \tag{11.1}$$

$$a + c_a > b + c_b \tag{11.2}$$

$$(b - c) < b_c \tag{11.3}$$

$$c + b_c - (b - c) > a + b_a \tag{11.4}$$

The numbers of constituencies satisfying these inequalities are shown in Table 11.2 (see following page).

Thus the "no show" paradox would occur only very infrequently. However, if "no show" were a deliberate tactic to influence an electoral outcome and we reject any constituencies in which a party would need to persuade more than 10% of its supporters not to vote, these numbers are reduced to:

Test 1:	3
Test 2	5
Test 3:	0

Of the five constituencies identified under Test 2 (2010 rev), however, two are such close three-way contests that a "no show" strategy could not be contemplated.

11.5.7 AV and the Reinforcement Paradox

With AV it is possible for a candidate A to win in a constituency but, if votes were counted in all sub-districts of the constituency, for another candidate, B, to win in each sub-district.

Table 11.2 Number of constituencies where "no show" would have been possible

	Number of constituencies		
Test	1 (2010)	2 (2010 rev)	3 (2005)
"No show" possible	4	7	0
"No show" not possible	527	524	530

Table 11.3 Example of reinforcement paradox.

	District 1	District 2	Constituency
AB	8	6	14
AC	10	8	18
B	22	20	42
CA	6	14	20
CB	4	2	6

Such a result may appear paradoxical, but it need not be regarded as problematic. Consider a two-district constituency and candidates A, B, and C and the following results in Table 11.3 below:

Here B wins narrowly in each district, defeating A in one and C in the other (it is perhaps more paradoxical that the paradox could not occur if A came second in both districts). When combined, however, C's supporters' strong support for A over B gives A victory (support that does not come into play in district 2 because of A's elimination). The result may be surprising, but there is nothing unreasonable about it. However, as UK elections are not counted at sub-constituency level, it is not possible for us to assess how often this paradox would occur.[12]

11.6 AV and Tactical Voting

In Sect. 11.3 above we discussed tactical voting under FPTP. With FPTP, if it is anticipated by a voter who supports C that $a > b > c$ (where a, b, c are the votes of A, B and C), the voter may choose to vote for B in hope that B defeats A. This form of tactical voting, which is common under FPTP, is unnecessary under AV as by voting CB the voter achieves the same result.

The nature of tactical voting under AV is quite different and much more sophisticated. In three-candidate elections it is possible for supporters of one of the leading two candidates to vote for the third in order to change which candidate is eliminated at the first stage of the count.

There are three cases we need to consider:

1. If $a > b > c$ and B wins after the elimination of C, some ACB voters may vote CAB in hope that A will win after the elimination of B.

[12] With AV, any change to counting at sub-constituency level is unlikely as counting centres would need to determine not just a single total for each candidate but a total for each preference list.

2. If $a > b > c$ and B wins after the elimination of C, some ACB voters may vote CAB allowing C to win rather than B.
3. If $a > b > c$ and A wins after the elimination of C, some BCA voters may vote CBA in hope that C will defeat A after the elimination of B (their own candidate).

11.6.1 Case 1: Supporting Your Candidate by Voting for an Opponent

Suppose that in an election with candidates A, B and C, A has more first preference votes than B but, following the elimination of C, B wins. Non-monotonicity can arise, allowing enough of A's supporters to give a first-preference to C to eliminate B resulting in victory for A.

If a, b and c are the sincere first preference votes for candidates A, B and C, then $(b-c)$ supporters of A (or, more precisely, $(b-c+1)$) could ensure the elimination of B rather than C by switching their votes to C. This form of tactical voting would be successful if:

$$\mathrm{a} < (\mathrm{b} + \mathrm{c}) \tag{11.5}$$

$$\mathrm{a} + \mathrm{c_a} < \mathrm{b} + \mathrm{c_b} \tag{11.6}$$

$$[\mathrm{a} - (\mathrm{b} - \mathrm{c})] + \mathrm{b_a} > [\mathrm{c} + (\mathrm{b} - \mathrm{c})] + \mathrm{b_c} \tag{11.7}$$

$$[\mathrm{a} - (\mathrm{b} - \mathrm{c})] > \mathrm{b} \tag{11.8}$$

where n_m is the number who vote NM.

Unless (11.5) is true A wins at the first stage of the count. Unless (11.6) is true, A would win with the elimination of C and tactical voting would not be necessary. To ensure that A would win after tactical voting, (11.6) must be true, and if (11.7) were not true then the switch in votes would result in A's elimination.

Table 11.4 shows the outcome of our search for these inequalities occurring in our data on English elections.

Results for the main parties in the three constituencies we found with Test 2 are shown in Table 11.5.

In Bristol South, 12.7% (i.e. (28.1 – 22.9)/41.1) of Labour supporters (2363 voters) would have needed to switch to the Conservatives. It is difficult to believe that a party could engineer, or attempt to engineer, such a massive transfer of its support to its principal rival. In Colne Valley and Reading East tactical voting might appear more feasible but, when in 2005 the Conservatives were in second place in Colne Valley and less than 2% ahead of Labour in Reading East, tactical voting would have been a surprising strategy.

Table 11.4 Potential for "case 1" tactical voting in English elections

	Number of constituencies		
Test	1 (2010)	2 (2010 rev)	3 (2005)
Possible	1	3	1
Not possible	530	528	529

Table 11.5 Test 2 constituencies with the potential for "case 1" tactical voting

	Cons	Lab	Lib Dem
Bristol South	22.9%	41.1%	28.1%
Colne Valley	37.0%	26.4%	28.2%
Reading East	42.6%	25.5%	27.3%

In the only constituency we found for 2005, Walsall North, tactical voting would have been even more politically unrealistic. Votes for the major parties were:

Conservative	29.6%
Labour	33.6%
Liberal democrat	31.2%

Here 1.6% of voters would have needed to switch from Labour to the Conservatives, reducing the Labour vote to 32.0%, only 0.8% above its nearest rival. Here it is unimaginable that some Labour voters would have taken the risk in voting Conservative.

11.6.2 Case 2: Supporting your Second Choice (I)

As in case 1 (above), suppose A has more first preference votes than B but on the elimination of C, B wins. Again, it may be possible for enough ACB supporters to switch to CAB to prevent C's elimination and allow C to win. The number of switched votes must be at least $(b - c)$. There are two situations to consider:

1. $a - (b - c) > \mathrm{b}$ in which case B is eliminated;
2. $a - (b - c) < \mathrm{b}$ in which case A is eliminated.

What we must test for here is:

$$\mathrm{a} < (\mathrm{b} + \mathrm{c}) \tag{11.9}$$

$$\mathrm{a} + \mathrm{c_a} < \mathrm{b} + \mathrm{c_b} \tag{11.10}$$

$$\mathrm{a_c} > \mathrm{b} - \mathrm{c} \tag{11.11}$$

and either

$$\mathrm{c} + \mathrm{b_c} > \mathrm{a} + \mathrm{b_a} \tag{12i}$$

$$\mathrm{c} + \mathrm{a_c} > \mathrm{b} + \mathrm{a_b} \tag{12ii}$$

Table 11.6 Potential for "case 2" tactical voting in English elections

	Number of constituencies		
Test	1 (2010)	2 (2010 rev)	3 (2005)
Possible	0	6	3
Not possible	531	525	525

Table 11.7 Test 2 constituencies in 2010 with the potential for "case 2" tactical voting

	Cons	Labour	Lib Dem
Sherwood	39.2%	38.8%	14.9%
Broxtowe	39.0%	38.3%	17.0%
Stockton South	38.9%	38.3%	15.1%
Amber Valley	38.6%	37.4%	14.4%
Lancaster and Fleetwood	36.1%	35.3%	19.1%
Warrington South	35.8%	33.0%	27.5%

Table 11.8 Test 2 constituencies in 2005 with the potential for "case 2" tactical voting

	Cons	Labour	Lib Dem
Coventry North East	35.4%	34.3%	24.2%
Doncaster Central	37.3%	34.3%	25.4%
Durham North	37.7%	34.1%	22.8%

Using our three tests on 2010 and 2005 data, the number of English constituencies in which this form of tactical voting would be possible are shown in Table 11.6 above.

However, although the numbers of constituencies in which this form of tactical voting is theoretically possible are very small, on looking at the votes cast in Tables 11.7 and 11.8 above the chances of voters deciding to vote tactically are even smaller.

In all of these constituencies, the Conservatives have only a slender lead over Labour, and Liberal Democrat transfers to Labour (roughly a net 12% of the Liberal Democrat vote in 2010 and a net 28% in 2005) would only give Labour a small majority. As voters cannot know with any precision what percentage of first preferences the candidates will receive, tactical voting could risk losing a seat which might otherwise be won. Thus in none of these constituencies is it likely many voters would choose to vote tactically, and in all of these constituencies a very high proportion of conservative voters (in many cases more than half) would need to switch to the Liberal Democrats for tactical voting to be successful. A change in result through tactical voting therefore seems highly improbable.

11.6.3 Case 3: Supporting Your Second Choice (II)

If $a > b > c$ and, on the elimination of C, A would win, circumstances could arise in which some BCA supporters could secure the election of C by voting for C to eliminate their preferred candidate, B (Table 11.9).

Table 11.9 Numbers of constituencies with the potential for "case 3" tactical voting

	Number of constituencies		
Test	1 (2010)	2 (2010 rev)	3 (2005)
Possible	16	35	0
Not possible	515	496	530

Table 11.10 Numbers of constituencies with the potential for "case 3" tactical voting if not more than 10% of a party's supporters vote tactically

	Number of constituencies		
Test	1 (2010)	2 (2010 rev)	3 (2005)
Possible	7	7	0
Not possible	524	524	530

For this to happen,

$$a < (b + c) \tag{11.13}$$

$$a + c_a > b + c_b \tag{11.14}$$

$$(b - c)/2 < b_c \tag{11.15}$$

$$c + b_c > a + b_a \tag{11.16}$$

Unless (11.13) is true, no amount of tactical voting will prevent A from winning. If (11.14) is not true, B will win when C is eliminated and tactical voting by supporters of B is unnecessary and counter-productive. At least $(b - c)/2$ supporters of B must switch to C to eliminate B rather than C, and they are unlikely to do this if C is not their second preference: hence (11.15). Finally, (11.16) is needed to ensure C wins after the elimination of B.

Testing these inequalities with our transfer assumptions we found:

This form of tactical voting was theoretically possible in 2010 because of the higher proportion of Labour votes that would transfer to the Liberal Democrats, particularly with Test 2 where it was assumed 78% of Labour voters would give the Liberal Democrats as their second preference.

However, parties cannot control how all of their supporters vote – many supporters will have little contact with their preferred parties' campaigns. If, stretching the bounds of what is credible, we were to assume that parties could get up to 10% (but not more) of their supporters to vote for a rival, the number of constituencies where tactical voting would be possible is considerably reduced. The numbers are shown in Table 11.10 and results in the constituencies identified in Table 11.11.

Of the constituencies shown in table 11.11, three were tight three-way contests – Derby North, Hampstead and Kilburn and Northampton North. With the Conservatives fighting hard to displace Labour in Derby North and Labour struggling, unsuccessfully, to hold the other two, it is almost inconceivable that supporters of these parties would have been prepared to contemplate voting for their principal opponents.

Table 11.11 Constituencies with potential for "case 2" tactical voting in 2010 if not more than 10% of a party's supporters vote tactically

	Con	Lab	LD
Bristol East	28.5%	36.9%	24.6%
Calder Valley	39.4%	27.0%	25.2%
Derby North	31.7%	33.0%	28.0%
Ealing Central and Acton	38.0%	30.1%	27.6%
Filton & Bradley Stoke	40.8%	26.4%	25.3%
Hampstead and Kilburn	32.7%	32.8%	31.2%
Northampton North	34.1%	29.3%	27.9%

11.6.4 Is Tactical Voting a Real Possibility?

In all of the above cases we have identified opportunities for tactical voting with the knowledge of how people voted. But before elections parties will rarely have a sufficiently accurate picture of how people will vote – not only do national opinion polls fluctuate on a daily basis, but they have a margin of error around 2% (and error margins will be much higher at constituency level). Moreover, parties' own canvas records are likely to be even less reliable. However, even if parties were confident in their predictions, they are not in a position to manage approximately the right number of vote switches to achieve the desired result – an inaccurate prediction or an excessive number of tactical votes could unnecessarily lose a seat.

Thus we conclude that the occasions in which voters could vote tactically are rare, but only in very exceptional cases would a voter know enough about how others might vote to be able to make an unintuitive tactical vote to influence the outcome successfully. As Tideman asserts (Tideman, p. 194), tactical voting with AV would be "generally a daunting task".

11.6.5 Tactical Voting with FPTP Compared with Tactical Voting Under AV

With FPTP voters cannot vote tactically to help their preferred candidates to win, but they can vote to help one of their lower preferences to win (and they frequently do). Theoretically, in any constituency in which the winner did not receive more than 50% of the votes, successful tactical voting would have been possible – that means 345 (65%) of the 531 English constituencies considered in 2010. However, if as above we assume that tactical voting requiring more than 10% of a party's vote will not be possible, we are still left with 40 constituencies, compared with 7 under AV.

However, here we are not comparing like with like. With FPTP, if it is assumed that, with sincere voting, $a > b > c$, then a supporter of C may vote for B, the second-placed candidate in hope that B will win. With AV, a supporter of B may vote for C, the third-placed candidate in hope that C will win. With FPTP tactical voting is intuitive – with AV it is far from it.

11.7 Conclusion

The UK's current electoral system, FPTP, is deeply flawed. In this paper we have compared AV with other possible alternatives to FPTP and have found no other system for single-member elections better suited for UK general elections. We have not argued that AV is perfect – no system is – but, in our view, its competitors all suffer from more serious defects.

The choice of electoral system is not just a matter of mathematics and logic. Other characteristics of voting systems (including issues relating to the conduct and costs of elections, transparency and the extent to which results will be understandable and acceptable) are also important, and here AV also scores over many of the alternatives.

We accept that AV can display paradoxical behaviour. However, our analysis suggests that the occurrence of these paradoxes would have been relatively rare in recent UK elections and our assessment of the extent to which they could have undermined the legitimacy of the outcomes suggest that they should not be a major concern. Our analysis has only used recent UK general election data and we accept in other situations AV's performance might be more problematic, but here our concern has been to find the best system for the UK.

However, even if we had formed a preference for another system, unless that preference had been a strong one we would not have argued for opposition to AV in the UK's referendum. Electoral reform is a political – not a mathematical – process and politics is the art of the possible. If the UK was going to move to a better electoral system than FPTP, then AV appeared to be the only possible first step. Unfortunately, in our view, it was a step the electorate was not prepared to take.

A Appendix: Second Preference Assumptions

As general elections in the UK are conducted using FPTP, election results do not tell us how voters might have used their second (or subsequent) preferences if a preferential system such as AV might have been used. In modelling how AV might have behaved in the 2005 and 2010 general elections we have therefore needed to rely on opinion polls which asked for second preferences.

We have tested for paradoxes using two assumptions about transfers in 2010 and one in 2005. The three "transfer tables" we have used are given below.

Table 11.12 Test 1 second-preference assumptions

	FPTP voting intention		
Second preference (below)	Con	Lab	Lib dem
Con	–	9.3	27.0
Lab	7.5	–	34.9
Lib dem	38.9	68.0	–
Other	22.1	8.1	14.4
None/don't know	31.4	14.5	23.7

A.1 Test 1: 2010 General Election

The election was held on 5 May 2010. A poll conducted on 25/26 April[13] provided the following information:

(We have calculated these percentages from the reported polling data by excluding respondents who gave the SNP or Plaid, which stand candidates only in Scotland and Wales, as their second preference.)

Although we have modelled only results in England, these results are from polling across the whole of Britain (excluding Northern Ireland). However, as 84% of British constituencies are in England and as the intentions of most voters would have been determined by national rather than local campaigns, we consider the figures a sufficiently good indicator of preferences for our purposes.

We have ignored responses from people whose first preference was for a minor party. In total the minor parties received only 8% of the vote and, if we exclude the exceptional cases of Brighton Pavilion (won by the Greens) and Buckingham (the seat of the Speaker which was not contested by the major parties), only in five constituencies did a minor party gain more than 10% of the votes. Consequently the number of opinion poll respondents intending to vote for any minor party was too small to give reliable results.

In our first test of AV we assumed that voters giving "Other", "None" or "Don't know" as their second preference either not giving a second preference or not giving a transferable one (Table 11.12).

A.2 Test 2: 2010 General Election with Revised Polling Data

The above assumption, however, may understate the number of voters who would give more than one preference. Those who gave their second preference to a minor party may well have given a subsequent preference for one of the major parties, and

[13] ComRes poll conducted for ITV News and The Independent, fieldwork 25 and 26 April 2010, page 20.

Table 11.13 Test 2 second-preference assumptions.

	FPTP voting intention		
Second preference (below)	Con	Lab	Lib dem
Con	–	10.7	35.3
Lab	11.8	–	45.6
Lib dem	61.3	78.0	–
No pref/non-transferable	26.9	11.3	19.1

Table 11.14 Test 3 second-preference assumptions.

	FPTP voting intention		
Second preference (below)	Con	Lab	Lib dem
Con	-	22	26
Lab	21	-	54
Lib dem	54	59	-
No pref/non-transferable	25	19	20

following an AV election campaign the number not voting for just one candidate might be reduced. (However, the assumption that a significant number might not give a second preference does not seem unreasonable when we consider that in Queensland, Australia, where voters are not required to give full preference lists, over 60% "plump" for a single candidate.)

We have therefore also tested how AV would perform if, for each party, the number not giving a useable second preference were reduced by half and preferences for the major parties increased proportionately. This leads Table 11.13 of transfers.

A.3 *Test 3: 2005 General Election*

Here we have used Table 11.14 with data from an opinion poll conducted on 30 April and 1 May 2005[14], just days before the election on 5 May.

Acknowledgements Dr Ken Ritchie was Chief Executive of the Electoral Reform Society (UK) from 1997 to 2010; Alessandro Gardini works with the "Yes to Fair Votes" campaign. The authors are grateful to Dan Felsenthal, David Hill, Geoffrey Goode, Colin Rosenstiel and Brian Wichmann for their comments on this paper.

[14] NOP/BBC poll of 30 April/1 May 2005, in Electoral Reform Society, 2005 p. 57.

References

Arrow, K. (1951). *Social Choice and Individual Values.*. New York: Wiley.

Baston, L. (2007). *Local Authority Elections in Scotland.* London: Electoral Reform Society.

Denver, D., Clark, A., Bennie, L. (2009). Voter reactions to a preferential ballot: the 2007 Scottish local elections. *Journal of Elections, Public Opinion & Parties, 19*(3), 265–282.

Electoral Reform Society (2005). *The UK General Election of 5 May 2005.* London: Electoral Reform Society.

FairVote (2005). Evaluation of San Francisco's First ranked choice election. _http://sfrcv.com/. Accessed26January2011.

FairVote (2009). Federal primary election runoffs and voter turnout decline, 1994–2008. http://archive.fairvote.org/?page=1489.Accessed26January2011.

Fisher, S., & Curtice, J. (2006). Tactical unwind: changes in party preference structure and tactical voting in Britain between 2001 and 2005. *Journal of Elections, Public Opinion and Parties, 16* (1), 55–76.

Franklin, M., Niemi, R., Whitten, G. (1994). The two faces of tactical voting. *British Journal of Political Science, 24*, 549–557.

Gibbard, A. (1973). Manipulation of voting schemes: a general result. *Econometrica, 41*, 587–601.

Hart, J. (1992). *Proportional representation: critics of the British electoral system 1820–1945.* UK: Clarendon Press.

Independent Commission on the Voting System (1998). *Report of the Independent Commission on the Voting System*, London: HMSO (Cm4090).

Reynolds, A. (2006). *Electoral systems and the protection and participation of minorities.* London: Minority Rights Group International.

Rogaly, J. (1976). *Parliament for the People.* London: Maurice Temple Smith.

Tideman, N. (2006). *Collective decisions and voting: the potential for public choice.* Aldershot: Ashgate.

Walsh, B., & Robson, C. (1973). *Alphabetical voting: a study of the 1973 general election in the Republic of Ireland.* Dublin: Economic and Social Research Institute (ESRI). General Research Series No 71.

Chapter 12
Approval Balloting for Fixed-Size Committees

D. Marc Kilgour and Erica Marshall

12.1 Introduction

Approval voting is a well-known voting procedure for single-winner elections. Voters approve as many candidates as they like; a candidate wins if and only if no other candidate receives more approvals (Brams and Fishburn 1978, 1983, 2005). But approval votes can be aggregated in different ways to serve different purposes, so it is reasonable to distinguish between *approval balloting*, in which each voter submits a ballot that identifies the voter's approved candidates, and *approval voting*, the single-winner procedure that selects the most-approved candidate(s) (Merrill and Nagel 1987).

Approval balloting is also appropriate for multi-winner elections, where the objective is to identify a "best" subset of candidates. There is a natural correspondence between an approval ballot and a winning subset – both are subsets of the set of candidates – so approval balloting may be particularly appropriate to multi-winner elections. Of course, one possible approval-balloting procedure is plurality, in which each approval ballot is interpreted as a vote for exactly the subset specified, and a subset wins if and only if no other subset receives more votes. But this procedure treats all subsets as unrelated, and makes no inferences about about the voter's preference for subsets similar but not identical to his or her ballot. Moreover, the similar subsets about which plurality extracts no information may be numerous, and presumably are often "almost as good" as the subset on the ballot. This observation motivates a search for procedures that exploit preference information that can be presumed to be implicit in the subset structure. In Kilgour (2010), 11 such procedures are identified and some of their basic properties described. But that study included only procedures and properties that arose in a very broad context discussed below.

D.M. Kilgour (✉) · E. Marshall
Department of Mathematics, Wilfrid Laurier University, Waterloo, ON, Canada N2L 3C5
e-mail: mkilgour@wlu.ca; mars5720@mylaurier.ca

D.S. Felsenthal and M. Machover, *Electoral Systems*, Studies in Choice and Welfare,
DOI 10.1007/978-3-642-20441-8_12, © Springer-Verlag Berlin Heidelberg 2012

The objective of this paper is to present and discuss procedures for determining the winning subset(s) that apply to one particular category of multi-winner elections, one for which approval balloting is particularly appropriate. In this category, (1) all *admissible* (or possibly winning) subsets are of exactly the same size (cardinality), and (2) whether a subset is admissible is easily computable. To be explicit, the procedures described here are particularly useful if the admissibility of a subset can be determined in polynomial time. If so, then the winning subset under a well-behaved voting procedure is easy to compute. More generally, condition (2) implies that the amount of computational effort to apply the procedure is determined by the procedure itself, and not by any admissibility criteria.

It is usual to study voting systems by identifying and comparing their properties (Arrow et al. 2002; Brams and Fishburn 2002), an approach followed in Ratliff (2003, 2006) in the investigation of procedures for multi-winner elections that are not based on approval balloting. After a general introduction (Sect. 12.2), the most common multi-winner elections will be described and discussed (Sect. 12.3), and a range of reasonable procedures for conducting them using approval balloting will be classified and studied in Sect. 12.4. Properties of these procedures are summarized in Sect. 12.5, and conclusions and directions for further research are summarized in Sect. 12.6. It should be noted that all procedures included here are anonymous (treat voters fairly) and neutral (treat candidates fairly), and share several other properties to be discussed later. The listing of properties is far from complete. Our intention is to make the procedures easy to compare, while postponing conclusions about which system is best for any particular purpose until more complete characterizations are available.

12.2 Multi-Winner Elections and Approval Balloting

We begin with basic terminology and notation for approval balloting and multi-winner elections. (Most of our notation and terminology is drawn from Kilgour (2010), where further references can be found.) Consider a multi-winner election with $n > 1$ voters and $m > 1$ candidates, and denote the set of all voters by $[n] = \{1, 2, \ldots, n\}$ and the set of all candidates by $[m] = \{1, 2, \ldots, m\}$. A voter's approval ballot specifies the subset of candidates the voter approves. Therefore, for $i \in [n]$, voter i's ballot can be represented as $V_i \subseteq [m]$. (Note that $V_i \in 2^{[m]}$, where $2^{[m]}$ is the "power set" of $[m]$, the set of all subsets of $[m]$. Voters may choose any subset in $2^{[m]}$, and in particular may approve no or all candidates, so $V_i = \emptyset$ and $V_i = [m]$ are both possible.) The ballot profile is $V = (V_1, V_2, \ldots, V_n)$, and the set of all possible ballot profiles is $\mathscr{V} = \left(2^{[m]}\right)^n$. We think of a voting procedure as taking a ballot profile V as input, and producing a winning subset as output.

A ballot profile is simply a list of the approval votes cast. Where no confusion is possible, we indicate ballots as subsets without punctuation. In Example 1, for instance, there are $n = 4$ voters and $m = 3$ candidates. Voter 1 votes for candidate 1 only, voter 2 votes for candidates 1 and 2 (but not 3), voter 3 votes for candidates

1 and 3 (but not 2), and voter 4's ballot is identical to voter 3's. Then we write the ballots as $V_1 = 1$, $V_2 = 12$, and $V_3 = V_4 = 13$, and the ballot profile as $V = (1, 12, 13, 13)$. A tabular presentation of a ballot profile is simpler, as follows:

Example 1.

Voter	1	2	3	4
Ballot	1	12	13	13

Later we will describe ballot profiles using only the second row of tables like the one above, i.e. without naming the voters.

In general, any of the 2^m different subsets in $2^{[m]}$ might win a multi-winner election. But in practice there are usually a priori restrictions. A subset of candidates is called *admissible* if it is eligible to win the election; the set (or class) of all admissible subsets is denoted $\mathscr{A}$. We assume that $|\mathscr{A}| > 1$ (since otherwise there is no point in conducting an election). To define the special case that is the subject of this study, suppose that k is a fixed integer satisfying $1 \leq k \leq m - 1$, and define $\mathscr{A}_k = \{S \subseteq [m] : |S| = k\}$. Thus, $\mathscr{A}_k$ is the class of subsets of $[m]$ containing exactly k candidates. All of the elections studied here satisfy $\mathscr{A} \subseteq \mathscr{A}_k$, and are called k-elections. In fact, many of the k-elections we study have $\mathscr{A} = \mathscr{A}_k$.

Consider, for example, the case $k = 1$. Then a 1-election (with $\mathscr{A} = \mathscr{A}_1$) is a single-winner election. It is clear that approval balloting can be used for such elections; the procedure is then called approval voting – see Brams (2008); Chap. 1 and 2 for recent references.

But if $k > 1$, a k-election is a multi-winner election in the sense that all k candidates in the subset selected win, and all other candidates lose. Nonetheless such an election could result in a tie of two or more admissible subsets. A voting procedure can be thought of as a function, possibly multi-valued, from the set of all possible ballot profiles, $\mathscr{V}$, to the class of admissible sets, $\mathscr{A}$. If the ballot profile is V, the winning subsets under a voting procedure $Proc$ can be denoted $Proc(V, \mathscr{A})$; note that $Proc(V, \mathscr{A}) \subseteq \mathscr{A}$ and $Proc(V, \mathscr{A}) \neq \emptyset$ provided $V_i \neq \emptyset$ for some voter $i \in [n]$. We will write $Proc(V, \mathscr{A}_k) = Proc_k(V)$.

Most of the procedures we discuss will select all admissible subsets that maximize (or minimize) some function, called a *score*. Formally, a score Sc is a function $Sc : \mathscr{V} \times \mathscr{A} \longrightarrow \mathbb{R}$. Note that, for any admissible subset $S \in \mathscr{A}$, the score $Sc(V, S) = Sc(S)$ depends only on the vote profile, $V \in \mathscr{V}$. Usually, $Sc(S)$ is a measure of the suitability of the admissible set S as winner of the election. The outcome of the voting procedure $Proc$ based on the score Sc is then

$$Proc(V, \mathscr{A}, Sc) = \arg\max_{S \in \mathscr{A}} Sc(V, S).$$

[For scores that measure unsuitability rather than suitability, the above maximization (arg max) is replaced by a minimization (arg min).] As indicated above, we also write $Proc(V, \mathscr{A}_k, Sc) = Proc_k(V, Sc)$.

Procedures based on maximization or minimization of a score share some important fundamental properties, in addition to the "fairness" properties of anonymity and neutrality. The relative scores of two admissible subsets S_1 and S_2, $Sc(V, S_1)$ and $Sc(V, S_2)$, depend only on V, and of course on S_1 and S_2, but not on any other subsets in the admissible class. Thus if $S_1, S_2 \in \mathscr{A}^0$, and S_1 wins and S_2 does not, then S_2 can never win in any admissible class $\mathscr{A} \supseteq \mathscr{A}^0$. It follows that such scoring procedures satisfy Independence of Irrelevant Alternatives and the Weak Axiom of Revealed Preference (Nurmi 2002).

Before discussing specific procedures, we turn to restrictions that are commonly applied in multi-winner elections conducted using approval balloting.

12.3 Common k-Elections

This paper concerns a special class of multi-winner elections distinguished by their admissible sets: k-elections, in which all admissible sets contain exactly k candidates. In practice, the most common k-elections are to determine the membership of a committee with k members. Further restrictions are common. For example, the committee may be required to include a minimum number of members, or a specific number, from one or more subsets of the set of candidates. A basketball all-star team, for instance, contains one center, two forwards, and two guards. Many committees must include at least one woman, or equal numbers of men and women, or members of each of several defined subgroups, such as departments, faculties, or schools within a university.

One approach to selecting subsets that meet such representational requirements is Constrained Approval Balloting, wherein all subsets in $\mathscr{A}_k$ are considered admissible (Brams 2008). Then subsets are rated not only by level of voter support but also by representativeness; usually, the winning subset(s) must be determined by trading off between these two measures. This method may be appropriate for complex representational problems in which, for example, individual candidates belong to several subsets so that one candidate may serve to represent more than one designated group. But if representational constraints are not too severe, then they can simply be built into the class of admissible sets (Fishburn and Pekeč 2004; Kilgour 2010). For instance, if a committee of b men and g women is required, then a $(b+g)$-election can be conducted in which the only admissible subsets contain exactly b men and exactly g women. The latter approach is taken here, as it is more consistent with the principle that the computational effort to determine whether a set is admissible should be low.

Ballot restrictions are often imposed in fixed-size committee elections, masking the fact that they are conducted under approval balloting, and imposing strategic constraints on the voters. For example, a ballot that names more than k candidates may be declared *spoiled*, and discarded. Analogous restrictions can be used to reflect representation conditions. The justification for such rules is uncertain, but their effects are clear; they preclude the common approval voting strategy of voting against one or a few candidates by supporting all others. Thus, restrictions on ballots make all ballots "positive" rather than "negative." We are not concerned with this "strategic" issues, however. Even if there are ballot restrictions, all ballots are approval ballots, and winners can be determined by any of the procedures discussed below.

12.4 k-Election Procedures and Their Properties

We now introduce procedures for k-elections, where the admissible sets satisfy $\mathscr{A} \subseteq \mathscr{A}_k$ for fixed k satisfying $1 \leq k < m$. In an important special case, $\mathscr{A} = \mathscr{A}_k$, i.e. every k-subset is admissible.

We begin with approval voting in single-winner elections. Given a vote profile V, the single-winner approval voting winners are the members of

$$AV(V) = \arg\max_{j \in [m]} |\{i : j \in V_i\}|.$$

Note that $AV(V)$ is a non-empty subset of $[m]$. It contains exactly the most approved candidates, who are considered to tie for winner of the Approval Voting election.

12.4.1 Generalized Approval Procedures

A natural generalization of approval voting is based on the idea of a *rep sequence*, which (in the context of k-elections) is a k-vector $r = (r(1), r(2) \ldots, r(k))$ with the properties that $r(1) \geq 0$ and $r(j) \geq r(j-1)$ for $j = 2, 3, \ldots, k$. Thus the rep coefficients $r(j)$ form a non-negative, non-decreasing sequence. By convention, $r(0) = 0$. To interpret the rep sequence, note that, from the point of view of voter i, the suitability of an admissible subset S to win the election is $r(j)$, where $j = |S \cap V_i|$. Thus the score of S equals $r(1)$ times the number of voters who approved of one candidate in S, plus $r(2)$ times the number of voters who approved of two candidates in S, plus $r(3)$ times ..., etc.

Definition 1. For fixed k and any $\mathscr{A} \subseteq \mathscr{A}_k$, the winning subset(s) under the *Generalized Approval* procedure based on the rep sequence $r = (r(1), r(2) \ldots, r(k))$ are

$$GA_k(r, V) = \arg\max_{S \in \mathscr{A}} f_r(S),$$

where $f_r(S) = \sum_i r(|S \cap V_i|)$ is the *rep score* of S.

Note that any Generalized Approval procedure is a scoring procedure.

The Generalized Approval winners are the admissible subsets that achieve maximum rep score. The rep score of an admissible subset $S \in \mathscr{A}$ is determined by adding contributions from each voter: for $j = 1, 2, \ldots, k$, S gains $r(j)$ for each voter who approved of exactly j members of S. Of course, a voter who approved of no members of S contributes nothing to the rep score of S.

Many approval balloting procedures identified by Kilgour (2010) can be expressed as Generalized Approval procedures using an appropriate rep sequence. The table below is easy to verify.

Procedure	Symbol	$r(j)$
Simple approval	AV	j
Proportional approval	PAV	$\sum_{\ell=1}^{j} \frac{1}{\ell}$
p representatives	REP_p	$\begin{cases} 0 \text{ if } j < p \\ 1 \text{ if } j \geq p \end{cases}$

In the REP_p procedures, p is assumed fixed, $1 \leq p < k$. The Simple Approval score of a subset S will be denoted $Sc_{AV}(S)$, the Proportional Approval score $Sc_{PAV}(S)$, and the p-Representative Approval score $Sc_{REP_p}(S)$. Of course, each of these scores is a sum of individual voters' scores. For an admissible subset S, the score contribution for voter i under Simple Approval equals the number of members of S approved by i. Under Proportional Approval, it equals 1 if i approves of one member of S, $1 + \frac{1}{2}$ if i approves of two members of S, $1 + \frac{1}{2} + \frac{1}{3}$ if i approves of three members of S, etc. Under p-Representative Approval, it equals 1 if i approves of at least p members of S, and 0 if i approves of fewer than p members of S.

To illustrate, consider Example 2, treating separately the k-elections with $k = 1, 2,$ and 3. In each cell of the results table below, the score of the winning subset(s) is given in parentheses. Note that "all" means that every admissible subset is a winner.

Example 2. $n = 6$ voters; $m = 5$ candidates

12	23	35	45	123	345

	AV	PAV	REP_1	REP_2
$\mathscr{A} = \mathscr{A}_1$	3 (4)	3 (4)	3 (4)	all (0)
$\mathscr{A} = \mathscr{A}_2$	23, 35 (7)	23, 25, 35 (6)	25 (6)	12, 23, 35, 45 (2)
$\mathscr{A} = \mathscr{A}_3$	235 (10)	235 (8)	125, 134, 135, 234, 235, 245 (6)	235 (4)

It is easy to check that if single-winner approval voting is applied to the vote profile in Example 2, the unique winner is candidate 3, with four approvals. In a 1-election, such as reported in the $\mathscr{A} = \mathscr{A}_1$ row of this table, the admissible sets consist of single candidates. Observe that the results of these elections match the approval voting results for all procedures except REP_2. Of the four procedures shown, REP_2 is the only one that fails the next property.

Definition 2. A procedure for k-elections is *based on approval voting* iff its winning subsets when $\mathscr{A} = \mathscr{A}_1$ are singletons containing exactly the candidates who win in single-winner approval voting.

In a Generalized Approval procedure applied to an election with $\mathscr{A} \subseteq \mathscr{A}_1$, a candidate's score increases by either $r(1)$ or 0 for each voter, according as the voter approved the candidate or not. From this fact, it is easy to show that any Generalized Approval procedure is based on approval voting iff $r(1) > 0$. A Generalized Approval procedure with a rep sequence satisfying this condition is called *proper*. Thus, REP_p procedures are based on approval voting iff they are proper.

Next we consider the relation of scores of subsets of different sizes for a fixed procedure and vote profile. For Simple Approval in Example 2, it is easy to check for subsets in $\mathscr{A}_1$ that $SC_{AV}(j) = 2, 3, 4, 2, 3$ for $j = 1, 2, 3, 4, 5$. Among subsets in $\mathscr{A}_2$, the maximum AV score, 7, is achieved by 23 and 35 only, and among subsets in $\mathscr{A}_3$, the maximum AV score, 10, is achieved uniquely by 235. Observe that the score of any subset of candidates is the sum of the scores of the candidates (treated as subsets of size 1).

Definition 3. For fixed $k \geq 1$ and $\mathscr{A} \subseteq \mathscr{A}_k$, a score is *candidate-wise* iff it satisfies $Sc(S) = \sum_{j \in S} Sc(j)$, where $Sc(j)$ is the score of j when $\mathscr{A} = \mathscr{A}_1$.

For instance, it is easy to verify, and confirmed by the calculations above, that the Simple Approval score is candidate-wise. We call a procedure based on a candidate-wise score a candidate-wise procedure. Proportional Approval is not a candidate-wise procedure, for observe that $Sc_{PAV}(j) = 2, 3, 4, 2, 3$ for $j = 1, 2, 3, 4, 5$ (since $Sc_{PAV}(S) = Sc_{AV}(S)$ when $|S| = 1$). Thus, $Sc_{PAV}(2) = 3$ and $Sc_{PAV}(3) = 4$, but $Sc_{PAV}(23) = 5 \neq Sc_{PAV}(2) + Sc_{PAV}(3)$. More generally, it can be shown that a Generalized Approval Voting procedure with rep sequence $r = (r(1), r(2), \ldots, r(k))$ is candidate-wise if and only if $r(j) = jr(1)$ for $j = 1, 2, \ldots, k$. Thus, Simple Approval is essentially the only Generalized Approval procedure that is candidate-wise.

In the context of k-elections, candidate-wise procedures are analogous to the additive procedures studied in a more general context in Kilgour (2010). They are important because they make it easy to determine all winning subsets.

Theorem 1. *In a candidate-wise procedure that maximizes {minimizes} a score $Sc(S)$, the winning subsets in the election with $\mathscr{A} = \mathscr{A}_k$ are exactly the subsets of candidates with the k largest {respectively, smallest} values of $Sc(j), j = 1, 2, \ldots, m$.*

Proof. In a candidate-wise procedure, the score of a set S is $Sc(S) = \sum_{j \in S} Sc(j)$. Thus any set of k candidates has maximum {minimum} score iff it contains the k candidates with the largest {smallest} individual scores. □

For example, for AV in Example 2, we noted that $Sc_{AV}(j) = 2, 3, 4, 2, 3$ for $j = 1, 2, 3, 4, 5$, and the unique winner when $k = 1$ is 3. Since $Sc_{AV}(j) = 3$ for $j = 2$ and 5 and $Sc_{AV}(j) < 3$ for $j = 1$ and 4, Theorem 1 implies that the winning subsets when $k = 2$ are 23 and 35. Similarly, the unique winning subset when $k = 3$ is 235, as 2, 3, and 5 are unequivocally the three highest scoring candidates.

The relationships suggested above among the AV winners of Example 2 are important. Every winning subset in the $\mathscr{A}_k$ election can be constructed by adjoining a candidate to some winning subset in the $\mathscr{A}_{k-1}$ election. For example, the winners of the $\mathscr{A}_2$ election are 23 and 35, obtained by adding either 2 or 5 to 3, the unique winner of the $k = 1$ election. Similarly, the winner of the $\mathscr{A}_3$ election is 235, constructed by adjoining 2 to 35 (which is one winner when $k = 2$) or by adding 5 to 23 (the other winner when $k = 2$).

Definition 4. A procedure is *upward-accretive* at $k \geq 1$ iff for each winning set $S \in \mathscr{A}_{k+1}$, there exists $j \in S$ such that $S - j$ is a winning set in $\mathscr{A}_k$.

Likewise, it is possible to ask whether the winners of the election in $\mathscr{A}_{k-1}$ can be constructed from the winners in $\mathscr{A}_k$.

Definition 5. A procedure is *downward-accretive* at $k \geq 1$ iff for each winning set $S \in \mathscr{A}_k$, there exists $j \notin S$ such that $S \cup j$ is a winning set in $\mathscr{A}_{k+1}$.

To summarize, a procedure is upward-accretive iff every winning subset in the (larger) $(k + 1)$-election can be obtained by adding a candidate to some winning subset in the (smaller) k-election. Consequently, *each* winning subset in the larger election differs by exactly one candidate from *some* winning subset in the smaller election. Similarly, a procedure is downward-accretive iff every winning subset in the smaller election can be obtained by deleting a member from some winning subset in the larger election. Consequently, *each* winning subset in the smaller election differs by exactly one candidate from *some* winning subset in the larger election. In particular, every candidate who belongs to some winning subset in the smaller election must be a member of some winning subset in the larger election. In general accretiveness is a sort of continuity; as the size of the winning set changes, the set of winners changes minimally.

Theorem 2. *A candidate-wise procedure is both upward- and downward-accretive.*

Proof. By Theorem 1, the winning subsets in a k-election conducted using a candidate-wise procedure consist of any k highest-scoring candidates, and the winning subset in a $k + 1$-election consist of any $k + 1$ highest-scoring candidates. Any subset containing the $k + 1$ highest-scoring candidates must include a subset containing k highest scoring candidates, so the procedure must be upward-accretive. Similarly, the procedure must be downward-accretive because every subset containing k highest-scoring candidates is contained in some subset containing $k + 1$ highest scoring candidates. □

We have already observed that AV is upward-accretive in Example 2, and can confirm directly that it is downward-accretive as well. Theorem 2 indicates that this is not a surprise. The results for Example 2 indicate that PAV is not upward-accretive, as 25 is a winning subset in the 2-election, but neither 2 nor 5 wins the 1-election. (Another example – see below – proves that PAV is not downward-accretive either.) Moreover, REP_1 is neither upward- nor downward-accretive, as 3 is the unique winner of the 1-election and 25 is the unique winner of the 2-election. Likewise, REP_2 is not downward-accretive, since 45 wins the 2-election, but 4 does not belong to any subset that wins the 3-election.

Theorem 3. *Let $1 \leq k_1 < k_2 \leq m$ and consider elections with vote profile V conducted according to a procedure $Proc$.*

(a) Suppose that $Proc$ is upward-accretive and that $S_2 \in Proc(V, \mathscr{A}_{k_2})$. Then S_2 has a subset that is a member of $Proc(V, \mathscr{A}_{k_1})$.
(b) Suppose that $Proc$ is downward-accretive and that $S_1 \in Proc(V, \mathscr{A}_{k_1})$. Then S_1 has a superset that is a member of $Proc(V, \mathscr{A}_{k_2})$.

Proof. Suppose that $Proc$ is upward-accretive and that $S_2 \in Proc(V, \mathscr{A}_{k_2})$. Then $|S_2| = k_2$, and S_2 must have a subset S_2^1 of size $k_2 - 1$ that belongs to $Proc(V, \mathscr{A}_{k_2-1})$. This completes the proof if $k_2 - 1 = k_1$. Otherwise, S_2^1 must have a subset of size k_2-2, S_2^2, that belongs to $Proc(V, \mathscr{A}_{k_2-2})$. Moreover, $S_2^2 \subseteq S_2^1 \subseteq S_2$. Continue this iteration for k_2-k_1 steps to obtain a subset $S_2^{k_2-k_1}$ of S_2, of cardinality k_1, that belongs to $Proc(V, \mathscr{A}_{k_2-(k_2-k_1)}) = Proc(V, \mathscr{A}_{k_1})$. This completes the proof of (a). The proof of (b) is similar. □

Recall that a proper Generalized Approval procedure has rep sequence satisfying $0 < r(1) \leq r(2) \leq \ldots \leq r(k)$. Our next example, in combination with Theorem 3, determines a sufficient condition for a proper Generalized Approval procedure to be neither upward- nor downward-accretive.

Example 3. $n = 6$ voters; $m = 4$ candidates

12	13	14	2	3	4

It is easy to verify that under any Generalized Approval procedure, the $\mathscr{A}_1$ election in Example 3 must be won by candidate 1, because $Sc_r(1) = 3r(1)$ whereas $Sc_r(j) = 2r(1)$ for $j = 2, 3$, and 4. For the $\mathscr{A}_3$ election, the score of any subset containing 1 must equal $Sc_r(123) = 2r(2) + 3r(1)$, whereas the score of the only subset excluding 1 is $Sc_r(234) = 6r(1)$. Clearly, the $\mathscr{A}_3$ election is won by 234 uniquely iff $6r(1) > 2r(2) + 3r(1)$, which is equivalent to $r(2) < \frac{3}{2}r(1)$. We conclude that, if a rep sequence satisfies this condition, then the associated Generalized Approval procedure is neither upward- nor downward-accretive. This conclusion can be generalized using other examples, but we note that, since AV is accretive, the greatest possible extent of the generalization is $r(2) < 2r(1)$.

Before turning to other procedures, we introduce one more property that a procedure might possess. A (b, g)-election is an election in which each candidate belongs to one of two disjoint classes, which for convenience we call "boys" and "girls." In a (b, g)-election, the objective is to select a subset consisting of exactly b boys and exactly g girls. There are two natural ways to conduct a (b, g)-election:

- Conduct a $(b + g)$-election in which the only admissible sets are sets containing exactly b boys and exactly g girls;
- Conduct a b-election among the boys and a g-election among the girls and then take the disjunction (union) of the winning subsets.

The next property specifies that these two processes produce exactly the same subset(s).

Definition 6. Assume that the set of candidates is $[m] = B \cup G$ where $B \cap G = \emptyset$, $0 < b < |B|$, and $0 < g < |G|$. Define $\mathscr{A}_B = \{S \subseteq B : |S| = b\}$, $\mathscr{A}_G = \{T \subseteq G : |T| = g\}$ and $\mathscr{A}_{BG} = \{R \subseteq [m] : R \cap B \in \mathscr{A}_B$ and $R \cap G \in \mathscr{A}_G\}$. Then a procedure $Proc$ satisfies *composition* iff, for any vote profile V on $B \cup G$,

$$\begin{aligned} &Proc_{b+g}(V, \mathscr{A}_{BG}) \\ &\quad = \{R \subseteq [m] : R = S \cup T, S \in Proc_b(V, \mathscr{A}_B) \text{ and } T \in Proc_g(V, \mathscr{A}_G)\}. \end{aligned}$$

Thus, the property of composition implements the idea that when a committee is to represent two distinct groups (in any proportion), it is immaterial whether the representation requirement is built into the class of admissible sets or whether it is guaranteed by means of separate elections.

Theorem 4. *A candidate-wise procedure satisfies composition.*

Proof. Assume $Proc$ is candidate-wise. By Theorem 1, $Proc_b(V, \mathscr{A}_B)$ contains all sets of b highest scoring candidates in B, and $Proc_g(V, \mathscr{A}_G)$ contains all sets of g highest scoring candidates in G. The union of two such sets must produce a member of $Proc_{b+g}(V, \mathscr{A}_{BG})$, since all of these subsets have the same score, and any $(b + g)$-subset of $[m]$ with a higher score could not possibly be admissible, for otherwise a higher-scoring subset could be found in either $\mathscr{A}_B$ or $\mathscr{A}_G$. □

12.4.2 Majority Threshold Procedures

Many threshold procedures have been proposed (Fishburn and Pekeč 2004); Majority Threshold (MT) and Strict Majority Threshold (SMT) have been singled out as particularly attractive (Kilgour 2010). A subset of candidates is defined to represent a voter if and only if a majority (or strict majority) of members of the subset are approved by the voter, and any admissible set that represents the most voters wins. In the context of k-elections, the representation condition is met for the MT procedure whenever the subset contains at least $p = \frac{k}{2}$ candidates approved by the voter; for the SMT procedure, this threshold is raised to $p = \frac{k+1}{2}$.

Thus, for fixed-size elections, both MT and SMT are equivalent to a REP_p procedures for the appropriate value of p. Moreover, k-elections held under MT and SMT are identical whenever k is odd, for then a subset contains a majority if and only if it contains a strict majority.

Because MT and SMT are REP_p procedures, they are also Generalized Approval procedures, and we have already identified some of their properties. For completeness, we mention that MT fails composition, as can be seen by setting $B = \{1, 2\}$ and $G = \{3, 4\}$ in Example 4 and considering a $(1, 1)$ election. The unique winner of the B-election is 1, the unique winner of the G-election is 3, but both 14 and 23 win the combined election.

Example 4. $n = 6, m = 4$

12	12	24	34	13	13

Another interesting example for MT and SMT elections is Example 5, below. Note that the $k = 1$ and $k = 3$ elections are identical (because k is odd), but the $k = 2$ elections are different (k is even).

Example 5. $n = 5, m = 4$

1	12	14	123	234

	MT	SMT
$\mathscr{A} = \mathscr{A}_1$	1 (4)	1 (4)
$\mathscr{A} = \mathscr{A}_2$	12, 13, 14 (5)	12, 23 (2)
$\mathscr{A} = \mathscr{A}_3$	124 (4)	124 (4)

Example 5 provides evidence that SMT is neither upward- nor downward-accretive. To show that SMT also fails composition, set $B = \{1, 2\}$ and $G = \{3, 4\}$, and consider a $(1, 1)$ election. The unique winner of the B-election is 1 (score 4), 3 and 4 tie in the G election, with score 2, and the unique winner of the combined $(1, 1)$ election is 23, with a score of 2.

12.4.3 Satisfaction-Related Procedures

Now we define three related procedures that are distinct from Generalized Approval. The first is included in Kilgour (2010), and discussed in detail in Brams and Kilgour (2011). We believe that the second and third are new.

Definition 7. For fixed k and any $\mathscr{A} \subseteq \mathscr{A}_k$, for any vote profile V, the following three *satisfaction-related* procedures choose as winning subset(s) $\arg\max_{S \in \mathscr{A}} f(S)$:

Procedure	Symbol	Score $f(S)$
Satisfaction approval	SAV	$f_{SAV}(S) = \sum_i \frac{\lvert S \cap V_i \rvert}{\lvert V_i \rvert}$
Capped satisfaction approval	CSA	$f_{CSA}(S) = \sum_i \frac{\lvert S \cap V_i \rvert}{\lvert S \rvert}$
Modified satisfaction approval	MSA	$f_{MSA}(S) = \sum_i \frac{\lvert S \cap V_i \rvert}{\min\{\lvert V_i \rvert, \lvert S \rvert\}}$

By convention, a fraction equals 0 if its denominator is 0.

SAV has been justified as a way to equalize voters, giving each the strategic choice of spreading support over several candidates, or concentrating it on a few. But, in the context of k-elections, SAV can be criticized for penalizing voters who adopt the standard approval strategy of voting against a few candidates by voting for all others, since the satisfaction score of such a voter cannot exceed $\frac{k}{\lvert V_i \rvert}$ for any subset. This upper bound can be very low if $k \ll m$. In contrast, voters who support no more than k candidates achieve satisfaction score 1 for some subsets, giving them greater influence when subsets are compared according to total satisfaction score. This observation motivates the introduction of CSA and MSA, which take into account that only subsets containing exactly k candidates are to be compared. Note that CSA offers a simple correction for the apparent bias of SAV, and MSA a correction that is more sophisticated, but more complex.

To illustrate these definitions, consider a new example, with results table below.

Example 6. $n = 9, m = 4$

12	12	2	13	3	3	14	4	4

	SAV	CSA	MSA
$\mathscr{A} = \mathscr{A}_1$	3, 4 (2.5)	1 (4)	1 (4)
$\mathscr{A} = \mathscr{A}_2$	34 (5)	12, 13, 14 (3.5)	34 (5)
$\mathscr{A} = \mathscr{A}_3$	134, 234 (7)	123, 124, 134 (3.33)	134, 234 (7)

The results for Example 6 are direct evidence that SAV is not based on approval voting, that CSA is not candidate-wise, and that MSA is not accretive. To complete the details, note that 1 is the unique approval winner; of the three satisfaction-related procedures, only SAV is not based on approval voting. Also $f_{CSA}(1) = 4$, $f_{CSA}(2) = 3$, and $f_{CSA}(12) = 3.5$, so $f_{CSA}(12) \neq f_{CSA}(1) + f_{CSA}(2)$, demonstrating that CSA is not candidate-wise. The facts that $MSA_1(V) = 1$ uniquely and $MSA_2(V) = 34$ uniquely imply that MSA is neither upward- nor downward-accretive. The results are also indirect evidence that MSA is not candidate-wise; if it were, it would be accretive by Theorem 2. Finally, we note that the calculations on which this table is based make it easy to show that the none of these procedures is a Generalized Approval procedure.

Satisfaction Approval Voting, though not based on approval voting, is otherwise a very well-behaved procedure.

Theorem 5. *SAV is a candidate-wise procedure.*

Proof. For any $S \in \mathscr{A}_k$ and for any voter i, observe that $|S \cap V_i| = \sum_{j \in S} |j \cap V_i|$. Therefore

$$f_{SAV}(S) = \sum_i \frac{|S \cap V_i|}{|V_i|} = \sum_i \frac{1}{|V_i|} \sum_{j \in S} |j \cap V_i|$$

$$= \sum_{j \in S} \sum_i \frac{1}{|V_i|} |j \cap V_i| = \sum_{j \in S} f_{SAV}(j),$$

as required. The reversal of the order of summation is justified because all sums are finite. □

As a corollary to Theorem 5, SAV is both upward- and downward-accretive, by Theorem 2.

Capped Satisfaction Approval, CSA, though not candidate-wise, does turn out to be accretive.

Theorem 6. *For all k such that $1 \leq k < m$ and all vote profiles V, $CSA_k(V) = AV_k(V)$. In particular, CSA is both upward- and downward-accretive, and satisfies composition.*

Proof. For fixed $k > 0$ and any $S \in \mathscr{A}_k$, $f_{CSA}(S) = \sum_i \frac{|S \cap V_i|}{|S|} = \frac{1}{k} \sum_i |S \cap V_i| = \frac{1}{k} f_{AV}(S)$. It follows that any subset $S \in \mathscr{A}_k$ that maximizes $f_{CSA}(S)$ also maximizes $f_{AV}(S)$, and vice versa. Since AV is both upward- and downward-accretive, so is CSA. □

By Theorem 4, candidate-wise procedures such as AV and SAV satisfy composition. So must CSA, since its winning sets are the same as those of AV. In particular, Theorem 6 shows that it is not necessary to be candidate-wise in order to be accretive. Of the satisfaction-related procedures, only MSA fails to satisfy composition, as shown by setting $B = \{1, 3\}$ and $G = \{2, 4\}$ in Example 6. It is then easy to verify that a separate 1-election in B, i.e. with admissible sets $\mathscr{A}_B = \{1, 3\}$ has unique winner 1, with a score of 4. (The score of 3 is 3.) Similarly, a separate 1-election in G, i.e. with admissible sets $\mathscr{A}_G = \{2, 4\}$ produces a tie between 2 and 4, both with score 3. But a 2-election with admissible sets $\mathscr{A}_{BG} = \{12, 14, 23, 34\}$ has unique winner 34, with a score of 5. (The scores of 12, 14, and 23 are 4, 4.5, and 4.5, respectively.) Thus, 1 is the unique winner in the separate election in B, but 1 is not a member of the unique winning subset in the combined election.

12.4.4 The Maximum Representation Procedure

We now turn to another procedure catalogued in Kilgour (2010) that applies to k-elections. "Representativeness," here termed $MAXREP$, originates in a general principle (Monroe 1995) that was subsequently adapted to approval balloting (Potthoff and Brams 1998) (See also (Brams 2008); Ch 6.). The definition below is equivalent.

First, for any $S \in \mathscr{A}_k$, let $X(S)$ be the set of 0–1 $n \times m$ matrices $x_{i,j}$ satisfying the following conditions:

(MR1) For each $i = 1, 2, \ldots, n, \sum_{j \in S} x_{i,j} = 1$,
(MR2) For each $j \notin S$, $x_{i,j} = 0$ for all $i = 1, 2, \ldots, n$,
(MR3) For each $j \in S, L \leq \sum_{i=1}^{n} x_{ij} \leq U$,

where $L = \lfloor \frac{n}{k} \rfloor$ and $U = \lceil \frac{n}{k} \rceil$. Note that if $\frac{n}{k}$ is an integer, then $L = U$, which implies that condition (MR3) is equivalent to

(MR3′) For each $j \in S, \sum_{i=1}^{n} x_{ij} = \frac{n}{k}$.

Thus, given any $S \in \mathscr{A}_k$, each $x \in X(S)$ is a 0–1 matrix with the properties that each row contains exactly one 1; if $j \notin S$, every entry in column j is 0; and if $j \in S$, column j contains either L or $U = L + 1$ 1's (unless n is a multiple of k, in which case there are exactly $\frac{n}{k}$ 1's in each column $j \in S$). Thus, any $x \in X(S)$ is a 0–1 matrix containing exactly n 1's, one in each row, all in columns corresponding to S; in these columns the numbers of 1's are as nearly equal as possible.

Definition 8. For fixed k, let $\mathscr{A} = \mathscr{A}_k$. For any vote profile V, a winning subset under the *Maximum Representation* procedure ($MAXREP$) is any member of

$$\arg\max_{S \in \mathscr{A}_k} \left\{ \max_{x \in X(S)} \sum_{j \in S} \sum_{i=1}^{n} x_{ij} \, |j \cap V_i| \right\}.$$

The definition of $X(S)$ effectively acts as a set of constraints on the maximization in Definition 8. To interpret these constraints, note that $x_{i,j} = 1$ indicates that candidate j is elected and is assigned to, or "represents" voter i. Each elected candidate represents at least L and at most U voters, and this representation is balanced, insofar as possible. (Specifically, each elected candidate represents exactly $\frac{n}{k}$ voters if $\frac{n}{k}$ is an integer; otherwise, each elected candidate represents at least L and at most $U = L + 1$ voters.) Every voter is represented, but not necessarily by a candidate the voter approved; the double summation represents the number of voters represented by a candidate they voted for. An integer program to determine the choice of x to achieve the maximum required for $MAXREP$ is available (Potthoff and Brams 1998).

It is easy to verify directly that the $MAXREP$ procedure is based on approval voting. However, calculations in Kilgour (2010) for Example 4 show that it is neither candidate-wise nor accretive (in any sense), as $MAXREP_1 = 1$ but $MAXREP_2 = 23$.

Moreover, $MAXREP$ fails composition, as can be seen from Example 4 by setting $B = \{1, 2\}$ and $G = \{3, 4\}$ and considering a $(1, 1)$ election. The unique winner of the election within B is 1 and the unique winner of the election within G is 3, but the unique winner of the combined election is 23.

12.4.5 Centralization Procedures

Centralization procedures for multi-winner elections with approval balloting are adaptations of an approach used in many problems: Since each voter's ballot can be considered to propose a committee, the most representative committee is the one that is "closest" to the ballots. These ideas have been adapted to voting in multi-winner elections (Kilgour et al. 2006; Brams et al. 2005). Here they are repeated for completeness, and then applied to k-elections.

The crucial concept is distance between subsets. For $S, T \subseteq [m]$, the *Hamming distance* between S and T, $d(S, T)$, is defined by

$$d(S, T) = |S \Delta T| = |S - T| \cup |T - S| = |\,(S \cap T^c) \cup (S^c \cap T)\,|.$$

Thus, the distance between two sets of candidates, S and T, equals the number of candidates in one of S and T but not the other.

For a ballot profile, $V = (V_1, V_2, \ldots, V_n)$, and any $S \in 2^{[m]}$, define $d(S, V) = \sum_i d(S, V_i)$. Then $d(S, V)$ represents the total distance from S to all the ballots in V. Any committee $S \in 2^{[m]}$ that minimizes $d(S, V)$ must contain every candidate who is supported on more than half the ballots, and cannot contain any candidate who is supported on fewer than half the ballots (Brams et al. 2004). Therefore a Candidate-by-Candidate Majority procedure can be implemented using a total distance minimization criterion, which was called "minisum." But such a procedure does not account for admissibility, so it was not included in Kilgour (2010) and will not be considered here.

It is convenient to define centralization procedures in the space of ballots rather than the space of voters, so we begin by expressing the ballot profile in a way that records the number of times each possible ballot appears in V. Let $W_1, W_2, \ldots, W_{2^m}$ be any enumeration of $[m]$, and for $h = 1, 2, \ldots, 2^m$, define c_h to equal the number of times ballot W_h appears in V. Thus, we use $c = (c_1, c_2, \ldots, c_{2^m})$ to represent V. Any 2^m-vector $w = (w_1, w_2, \ldots, w_{2^m})$ satisfying $w_h \geq 0$ will be called a *ballot weight vector*. For now, we assume that information about V is coded into w; exactly how this is done will be discussed later. (Note that ballot weight vectors are not normalized; normalization would be required if we were to use them to find a weighted mean, but we will use them only for comparison.)

Definition 9. Fix k and $\mathscr{A} \subseteq \mathscr{A}_k$. For any vote profile V and any ballot weight vector w, a winning subset under the *Weighted Minisum* procedure ($MSUM_w$) is any subset in

$$\arg\min_{S \in \mathscr{A}} \sum_{h=1}^{2^m} w_h d(S, W_h),$$

and a winning subset under the *Weighted Minimax* procedure ($MMAX_w$) is any subset in

$$\arg\min_{S \in \mathscr{A}} \max_{h=1,2,\ldots,2^m} w_h d(S, W_h).$$

Note that Minisum and Minimax can be thought of as procedures based on a score, but it is a score that measures unsuitability rather than suitability, and therefore is to be minimized.

Theorem 7. *For any weight vector, Weighted Minisum is upward- and downward-accretive and satisfies composition.*

Proof. Let $S \subseteq [m]$ and suppose that $j_1 \in S$ and $j_2 \notin S$. For any $h = 1, 2, \ldots, 2^m$, consider $\delta(j_1, j_2, h) = d((S - j_1) \cup j_2, W_h) - d(S, W_h)$. It is easy to show that

$$\delta(j_1, j_2, h) = \begin{cases} 0 & \text{if } j_1, j_2 \in W_h \\ 0 & \text{if } j_1, j_2 \notin W_h \\ 2 & \text{if } j_1 \in W_h \text{ and } j_2 \notin W_h \\ -2 & \text{if } j_1 \notin W_h \text{ and } j_2 \in W_h \end{cases}$$

Assume that, when $\mathscr{A} = \mathscr{A}_1$, the Weighted Minisum score of j_1 is strictly less that of j_2. Then

$$\begin{aligned}\sum_{h=1}^{2^m} w_h d(j_1, W_h) < \sum_{h=1}^{2^m} w_h d(j_2, W_h) &= \sum_{h=1}^{2^m} w_h (d(j_1, W_h) + \delta(j_1, j_2, h)) \\ &= \sum_{h=1}^{2^m} w_h d(j_1, W_h) + \sum_{h=1}^{2^m} w_h \delta(j_1, j_2, h),\end{aligned}$$

so that $\sum_{h=1}^{2^m} w_h \delta(j_1, j_2, h) > 0$. It follows that

$$\begin{aligned}\sum_{h=1}^{2^m} w_h d((S - j_1) \cup j_2), W_h) &= \sum_{h=1}^{2^m} w_h (d(S, W_h) + \delta(j_1, j_2, h)) \\ &= \sum_{h=1}^{2^m} w_h d(S, W_h) + \sum_{h=1}^{2^m} w_h \delta(j_1, j_2, h)) \\ &> \sum_{h=1}^{2^m} w_h d(S, W_h).\end{aligned}$$

This is to say that the weighted sum associated with a subset of size k will be increased if any member of the subset is replaced by another candidate whose score when $k = 1$ is greater. Thus, the winning subset at size k must consist of k candidates with the lowest Weighted Minisum scores when $k = 1$. As in Theorems 2 and 4, it follows that Weighted Minisum is accretive and satisfies composition.

□

It remains to define appropriate weights. First, the *count weight* vector is defined by $c = (c_1, c_2, \ldots, c_{2^m})$. One procedure recommended by Brams et al. (2005), Minisum Count ($MSUM_c$), can be based on application of this natural weight vector, which ensures that each voter receives equal treatment. By Theorem 7, $MSUM_c$ is accretive and satisfies composition. As the next result shows, it is also based on approval voting.

Theorem 8. *$MSUM_c$ is based on approval voting.*

Proof. Let $\mathscr{A} = \mathscr{A}_1$. For any vote profile V, the $MSUM_w$ winners are the candidates $j \in [m]$ satisfying

$$\arg\min_{j \in [m]} \sum_{h=1}^{2^m} c_h d(j, W_h).$$

For any ballot W_h and any candidate j,

$$d(j, W_h) = \begin{cases} |W_h| - 1 & \text{if } j \in W_h \\ |W_h| + 1 & \text{if } j \notin W_h \end{cases}$$

Because $\sum_{h=1}^{2^m} c_h|W_h| = \sum_{i=1}^{n}|V_i|$, it follows that

$$\sum_{h=1}^{2^m} c_h d(j, W_h) = \sum_{i=1}^{n} |V_i| + n - 2|\{i : j \in V_i\}|.$$

Clearly, the minimum of this sum occurs for those values of j that maximize $|\{i : j \in V_i\}|$. This set is exactly $AV(V)$. □

It was observed that extreme or isolated voters have substantial influence on the outcome of a $MMAX_w$ election carried out using count weights, and that their influence does not depend on the number of voters whose ballots are near the median (Kilgour et al. 2006). To reduce the importance of ballots cast by extreme voters, *proximity weights* were proposed; they are defined by

$$p_h = \frac{c_h}{\sum_{\ell=1}^{2^m} c_\ell d(W_h, W_\ell)}.$$

Thus, the proximity weight of a ballot is a fraction with the count (the number of times the ballot was cast) in the numerator and the total distance to all other ballots, weighted by their counts, in the denominator. The proximity weights associated with extreme voters are small, because their ballots are farther away from all others; on the other hand, each ballot has proximity weight proportional to the number of times it was cast, so common ballots receive more weight. Because of these features, we recommend only the $MSUM_c$ and $MMAX_p$ procedures.

We illustrate these two centralization procedures using Example 4. The ballots with positive counts are 12, 24, 34, 13, and the respective counts are 2, 1, 1, 2. Since $\sum_{h=1}^{2^4} c_h d(12, W_h) = 10$, it follows that the proximity weight of 12 is $\frac{2}{10}$. The complete set of proximity weights is $\frac{2}{10}, \frac{1}{14}, \frac{1}{14}, \frac{2}{10}$. It is convenient to multiply these weights by 70 to normalize them to integers; they become 14, 5, 5, 14. Then it follows that

	$MSUM_c$	$MMAX_p$
$\mathscr{A} = \mathscr{A}_1$	1 (10)	1 (15)
$\mathscr{A} = \mathscr{A}_2$	12, 13 (10)	12, 13, 14, 23 (28)
$\mathscr{A} = \mathscr{A}_3$	123 (10)	123 (15)

This example shows that $MMAX_p$ is not upward-accretive, since 23 wins at $k = 2$ but neither 2 nor 3 wins at $k = 1$. It is also not downward-accretive, since 14 wins at $k = 2$ but there is no winning set containing 4 at $k = 3$. The $MMAX_p$ procedure also fails composition, as can be seen by setting $B = \{1, 2\}$ and $G = \{3, 4\}$. Then the B-election produces 1 uniquely, the G-election results in a tie between 3 and 4, but the combined election produces a tie, including not only 13 and 14, but also 23. Example 5 shows that $MMAX_p$ is not based on approval voting since the procedure results in a tie between 1 and 2. In contrast, $MSUM_c$ is based on approval voting, accretive, and satisfies composition, although it is not candidate-wise.

12.4.6 Sequential Procedures

The first sequential procedure, now known as Sequential Proportional Approval, was proposed by Thiele (1890). (See also Hallett and Hoag (1926), pp. 453–454.) The essential idea is based on Approval Voting with weighted voters. The idea is to build a committee one member at a time, weighting voters according to the number of candidates they support who are already on the committee. Consequently, a k-member committee requires a sequence of k Approval Voting elections, each with different voter weights.

Our contribution to the development of this voting procedure is to note that Approval Voting is not the only way to conduct a single-winner election using approval ballots. In particular, any procedure not based on Approval Voting may produce different winners; we suggest that both Sequential Approval (the original procedure of Thiele (1890), based on Approval Voting) and Sequential Satisfaction Approval (in which the individual weighted elections are conducted using a weighted Satisfaction score) are both reasonable procedures that deserve further study.

First we define candidates' scores in a single-winner election with weighted voters. Let $w = (w_1, w_2, \ldots, w_n)$ be a voter weight vector, and set

$$f_{w,AV}(j) = \sum_{i=1}^{n} w_i |j \cap V_i| \ ; \ f_{w,SAV}(j) = \sum_{i=1}^{n} w_i \frac{|j \cap V_i|}{|V_i|}$$

for each $j \in [m]$. The winner is any candidate with maximum score. Observe that the approval or satisfaction of higher-weight voters is emphasized in the scores.

For k such that $1 \leq k < m$, a Sequential Procedure for an election with $\mathscr{A} = \mathscr{A}_k$ is the following iterative procedure (modelled on Kilgour (2010)):

- Begin by setting $\mathscr{C}_0 = [m]$ and defining the weight w^1 by setting $w_i^1 = 1$ for all voters i. Find the weighted score $f_w(j)$ for all $j \in \mathscr{C}_0$. The first candidate added to the winning subset is any candidate $j_1 \in \mathscr{C}_0$ that maximizes $f_w(j)$. Now set $\mathscr{C}_1 = \mathscr{C}_0 - j_1$.
- Suppose that $1 < h \leq k$ and that candidates $j_1, j_2, \ldots, j_{h-1}$ have already been added to the winning subset. The set of remaining candidates is $\mathscr{C}_{h-1}$. Reweight the voters so that the weight of voter i is

$$w_i^h = \frac{1}{1 + |V_i \cap \{j_1, j_2, \ldots, j_{h-1}\}|}.$$

  Now find the weighted score $f_w(j)$ for all $j \in \mathscr{C}_{h-1}$. The hth candidate added to the winning subset is any candidate $j_h \in \mathscr{C}_{h-1}$ that maximizes $f_w(j)$. If $h = k$, stop. Otherwise set $\mathscr{C}_h = \mathscr{C}_{h-1} - j_h$ and repeat.

This sequential process is called Sequential Approval (SEQ_{AV}) if the candidate score used is the Approval score, $f_{w,AV}(j)$, and Sequential Satisfaction (SEQ_{SAV}) if the candidate score used is the Satisfaction Approval score, $f_{w,SAV}(j)$.

Note that in any stage, h, of a sequential procedure, there may be a tie for the next candidate to be added to the committee, j_h. If so, any of the tied candidates may be selected, potentially resulting in many different committees. In particular, if $h < k$, the process may "branch"; to find all sequential committees, one must consider all possible resolutions of any tie at any stage, and all of their possible consequences.

We use Example 6 to illustrate that SEQ_{AV} and SEQ_{SAV} are different. In choosing a 2-person committee, SEQ_{AV} selects 1 (score 4) in the first stage, and SEQ_{SAV} selects either 3 or 4 (tied at 2.5). In the second stage, SEQ_{AV} produces a tie between 3 and 4 (second stage scores 2.5), so the 2-committee chosen by SEQ_{AV} is either 13 or 14. In the second stage, SEQ_{SAV} produces 34, whether by starting with 3 and adding 4 (score 2.5) or vice versa. This confirms that SEQ_{AV} and SEQ_{SAV} are different, and that SEQ_{SAV} is not based on approval voting. Procedurally, it seems that SEQ_{AV} and PAV are similar, and they do often produce similar results, but there are examples where they differ (Kilgour 2010).

By construction, any sequential procedure is both upward- and downward-accretive. But one variation is of interest. If there is a tie at stage $h < k$, the standard procedure insists that any resolution of this tie can be used as a basis for future stages. In the variant procedure, stage $h + 1$ includes a comparison of the scores of all candidates in each possible version of C_h, with the winner to be a candidate j with the highest observed value of $f_w(j)$. It can be shown that, with this change, sequential procedures remain upward-accretive, but can fail to be downward-accretive. In other words, this example demonstrates that the two directions of accretion are not equivalent.

12.5 Comparison of Procedures

Table 12.1 summarizes our findings about properties of all of the procedures for k-elections that we have studied. Note that the counterexamples demonstrating that PAV and REP_1 fail Composition are based on Example 2, using $B = \{1, 2\}$ and $G = \{3, 4, 5\}$. It is not known whether SEQ_{AV} or SEQ_{SAV} satisfy Composition. Because they are not scoring procedures, it is inappropriate to ask whether the Sequential procedures are Candidate-wise. The conclusion that $MMAX_p$ is not Based on Approval follows from Example 5.

12.6 Conclusions

This paper has surveyed methods of using approval ballots in k-elections, or elections to choose a k-subset of a set of candidates. The selection of a representative committee, with or without other qualifications on membership, is a common application of approval balloting. Ballot restrictions are also common, but they affect voting strategies but not winner-determination procedures. We have begun

Table 12.1 Procedures and properties

	Based on approval	Candidate-wise	Upward-accretive	Downward-accretive	Composition
AV	Yes	Yes	Yes Thm. 2	Yes Thm. 2	Yes Thm. 4
PAV	Yes	No Ex. 2	No Ex. 2	No Ex. 6	No Ex.2
REP_1	Yes	No Ex. 2	No Ex. 2	No Ex. 2	No Ex.2
REP_2	No Ex. 2	No Thm. 2	No Ex. 7	No Ex. 2	No Ex. 2
MT	Yes	No Thm. 2	No Ex. 4	No Ex. 4	No Ex. 4
SMT	Yes	No Thm. 2	No Ex. 5	No Ex. 5	No Ex. 5
SAV	No Ex. 6	Yes Thm. 5	Yes Thm. 2	Yes Thm. 2	Yes Thm. 4
CSA	Yes	No Ex. 6	Yes Thm. 6	Yes Thm. 6	Yes Thm. 6
MSA	Yes	No Thm. 2	No Ex. 6	No Ex. 6	No Ex. 6
$MAXREP$	Yes	No Thm. 2	No Ex. 4	No Ex. 4	No Ex. 4
$MSUM_c$	Yes	No Ex. 4	Yes Thm. 7	Yes Thm. 7	Yes Thm. 7
$MMAX_p$	No Ex. 5	No Ex. 4	No Ex. 4	No Ex. 4	No Ex. 4
SEQ_{AV}	Yes	–	Yes	Yes	??
SEQ_{SAV}	No Ex. 6	–	Yes	Yes	??

the identification and classification of approval-balloting procedures for k-elections, and the study of properties that are relevant in this context.

The systems listed here are classed as Generalized Approval procedures, Majority Threshold procedures, Satisfaction-related procedures, the Maximum Representation procedure, Centralization procedures and Sequential procedures. Of the 14 specific procedures discussed, two (AV and CSA, were shown (Theorem 5) to produce identical winners, and another, $MSUM_c$, is already known Kilgour (2010) to be identical to the Net Approval procedure, one that is related to AV but not included here because it is difficult to interpret in the context of k-elections. As can be seen from Table 1, there is a good base of knowledge of these procedures, though a few details remain to be filled in.

It is also noteworthy that many of the procedures appearing in Table 12.1 possess identical combinations of properties – at least, of the properties we have studied. But the procedures are all different, which implies that more properties must be identified to characterize the procedures. All of the properties studied here, for instance, depend only on one vote profile, rather than compare the results for two related vote profiles, as is required, for example, to define monotonicity. Nonetheless, the construction of a table such as Table 1 seems the right way to proceed, since it facilitates a direct study of procedures. Moreover, combinations of procedures have been recommended to break ties (Brams et al. 2005), to which many procedures are prone. Nonetheless, it must be more efficient to study procedures singly, in order that the conclusions be as simple and direct as possible.

It is to be hoped that the theoretical analysis of procedures can be mirrored, eventually, by studies of their performance on large-scale data sets, as initiated, for example, by Brams and Kilgour (2011). This kind of study would facilitate assessment of the likely representativeness of the winners under various procedures, and of their computational requirements. For example, would it be desirable to give voters a greater role in determining the outcome, making the size of the committee, as well as the membership, an output of the election, rather than a parameter specified in advance?

Acknowledgements Valuable suggestions from Steven J. Brams and Arkadii Slinko are gratefully acknowledged. Thanks also for helpful comments from a referee and the editors.

References

Arrow, K. J., Sen, A. K., & Suzumura, K. (Eds.). (2002). *Handbook of social choice and welfare* (Vol. 1). Amsterdam: North-Holland.

Brams, S. J. (2008). *Mathematics and democracy: Designing better voting and fair-division procedures*. Princeton, NJ: Princeton University Press.

Brams, S. J. & Fishburn, P. C. (1978). Approval voting. *American Political Science Review, 72*(3), 831–857.

Brams, S. J. & Fishburn, P. C. (1983). *Approval voting*. Cambridge, MA: Birkhäuser Boston.

Brams, S. J. & Fishburn, P. C. (2002). Voting procedures. In K. Arrow, A. Sen, & K. Suzumura (Eds.), *Handbook of social choice and welfare* (pp. 175–236). Amsterdam: Elsevier Science.

Brams, S. J. & Fishburn, P. C. (2005). Going from theory to practice: The mixed success of approval voting. *Social Choice and Welfare, 25*(2–3), 457–474.

Brams, S. J. & Kilgour, D. M. (2001). Fallback bargaining. *Group Decision and Negotiation, 10*(4), 287–316.

Brams, S. J. & Kilgour, D. M. (2011). Satisfaction approval voting. Preprint, New York University.

Brams, S. J., Kilgour, D. M., & Sanver, M. R. (2004). A minimax procedure for negotiating multilateral treaties. In M. Wiberg (Ed.), *Reasoned choices: Essays in honor of Hannu Nurmi* (pp. 108–139). Turku, Finland: Finnish Political Science Association.

Brams, S. J., Kilgour, D. M., & Sanver, M. R. (2005). A minimax procedure for electing committees. *Public Choice, 132*(3–4), 401–420.

Fishburn, P. C. & Pekeč, A. (2004). Approval voting for committees: Threshold approaches. Available at http://dimacs.rutgers.edu/Workshops/DecisionTheory\discretionary-2/PekecFishburn04a.pdf. Cited April 14, 2009.

Hallett, G. H. & Hoag, C. G. (1926). *Proportional representation*. New York: Macmillan.

Kilgour, D. M. (2010). Approval balloting for multi-winner elections. In J. Laslier & M. R. Sanver (Eds.), *Handbook on approval voting* (pp. 105–124). Heidelberg: Springer.

Kilgour, D. M., Brams, S. J., & Sanver, M. R. (2006). How to elect a representative committee using approval balloting. In B. Simeone & F. Pukelsheim (Eds.), *Mathematics and democracy: Recent advances in voting systems and collective choice* (pp. 83–95). Heidelberg: Springer.

Merrill, S. III & Nagel, J. H. (1987). The effect of approval balloting on strategic voting under alternative decision rules. *American Political Science Review, 81*(2), 509–524.

Monroe, B. L. (1995). Fully proportional representation. *American Political Science Review, 89*(4), 925–940.

Nurmi, H. (2002). *Voting procedures under uncertainty*. Berlin: Springer.

Potthoff, R. F. & Brams, S. J. (1998). Proportional representation: Broadening the options. *Journal of Theoretical Politics, 10*(2), 147–178.

Ratliff, T. C. (2003). Some startling inconsistencies when electing committees. *Social Choice and Welfare, 21*, 433–454.

Ratliff, T. C. (2006). Selecting committees. *Public Choice, 126*(3–4), 343–355.

Thiele, T. N. (1890). Sequential proportional approval voting. Available at http://www.nationmaster.com/encyclopedia/Sequential-proportional-approval-voting. Cited May 21, 2009.

Chapter 13
And the Loser Is... Plurality Voting

Jean-François Laslier

13.1 Introduction

Experts have different opinions as to which is the best voting procedure. The Leverhulme Trust sponsored 2010 *Voting Power in Practice* workshop, held at the Chateau du Baffy, Normandy, from 30 July to 2 August 2010, was organized for the purpose of discussing this matter. Participants of the workshop were specialists in voting procedures and, during the wrap-up session at the end of the workshop, it was decided to organize a vote among the participants to elect "the best voting procedure". The present paper reports on this vote. It contains in the Appendix statements by some of the voters/participants about this vote and voting rules in general.

13.2 The Vote

Previous discussion had shown that different voting rules might be advisable under different circumstances, so that a more concrete problem than "What is the best voting rule" should be tackled. The question for the vote was: "What is the best voting rule for your town to use to elect the mayor?"

Even with this phrasing, it was realized afterwards that not all participants had exactly the same thing in mind. In particular, some of them were thinking of a large electorate and some were rather thinking of a committee (the city council) as the electorate. This can be inferred from the participants' comments in the Appendix and is clearly a weakness of this "experiment."

J.-F. Laslier (✉)
CNRS and Department of Economics, Ecole Polytechnique, Palaiseau, France
e-mail: jean-francois.laslier@polytechnique.edu

D.S. Felsenthal and M. Machover, *Electoral Systems*, Studies in Choice and Welfare,
DOI 10.1007/978-3-642-20441-8_13, © Springer-Verlag Berlin Heidelberg 2012

Of course, an interesting feature of this vote is the fact that it was a vote on voting rules *by voting theorists*. So the participants arrived with quite a heavy background of personal knowledge and ideas. But the way the vote was improvised was such that no one had much time to think things over, discuss and coordinate with others, or calculate. Moreover, no candidates were clear common knowledge front-runners, and the final result was apparently not anticipated by most voters.

The possibilities of strategic manipulation were thus quite limited and one can indeed see from the comments that most of these approval votes should be interpreted as the expression of sincere individual opinions. As one referee pointed out: this vote may be the last "naive" vote on voting rules. This adds a particular significance to its result, and also suggests that the experiment should be done again, now that the results are known.[1]

13.2.1 Candidates: The Voting Rules in Question

The set of "candidates," that is the list of considered voting rules was rather informally decided: participants just wrote on the paper board voting rules to be voted upon. Eighteen voting rules were nominated, the definitions of which can be found in Appendix B.

Some rules should really be considered as possible ways to organize elections and will, in usual circumstances, provide indeed a unique winner. Others will often yield not a single winner but a set of possible winners, among which the final choice has to be made by one means or another. The list contains several Condorcet-consistent rules which agree on a unique outcome when the Condorcet winner exists[2] but which differ when there is no Condorcet winner and, in that case, often yield several winners. For instance the Uncovered set is a singleton only if there is a Condorcet winner,[3] the Copeland winner is always in the Uncovered set and the Uncovered set is always included in the Top Cycle. The voting rules also differ as to their informational basis.

1. Some of them require very little information: Plurality voting and Majority voting with a runoff simply ask the voter to provide the name of one (or two) candidates. Approval Voting asks the voter to say "yes" or "no" to each candidate.
2. Most rules require that the voter ranks the candidates: this is the classical framework of Arrowian social choice (Arrow 1951). There is no inter-personal comparisons of alternatives, which means that the ballots are not intended to convey interpretation of the kind "candidate a is better for voter i than for

[1]My guess, based on the theoretical analysis of strategic voting under Approval Voting, is that the result would not be different.

[2]The existence of several Condorcet winners simultaneously is a rare phenomenon.

[3]No randomization scheme was considered. In particular the optimal solutions to the Condorcet paradox studied by Laffond et al. (1993) and Dutta and Laslier (1999) were not on the list of voting procedures.

voter j". The intra-personal structure is purely ordinal, which means that we may know that a is better than b for i, but we cannot know *how much* better.

3. Finally, some of them allow inter-personal comparisons, with intra-personal comparisons being ordinal (Leximin, Majority Judgement) or cardinal (Range Voting).

Most rules extend the majority principle in the sense that, if there are only two candidates, they select the one preferred by a majority of the voters. Range voting, which maximizes the average evaluation, does not fulfill this principle: indeed, according to classical utilitarianism, if a majority of voters slightly prefer a to b while a minority strongly prefers b to a, it may be better to choose b than a, against the majority principle. Therefore, under Range voting, if voters reflect in their vote this pattern of interpersonal comparisons, the minority candidate b may well be elected against a. Such is also the case for the "Majority judgement" system (despite its name) which maximizes the median evaluation and for the Leximin, which maximizes the worst evaluation.

13.2.2 *The Procedure*

13.2.2.1 The Electorate

The 22 voters were the participants of the workshop (one participant abstained). Some of them are advocates of a specific voting rule; for instance, Ken Ritchie and Alessandro Gardini are active in Great Britain in promoting the "Alternative Vote": a system of vote transfers also known as the "Hare" system. Others, like Dan Felsenthal, are advocates of the Condorcet principle and strongly defended this principle during the workshop. But it is fair to say that most of the participants would say that different voting rules have advantages and disadvantages. This might be one of the reasons why no-one objected to the use of Approval voting for this particular vote.

13.2.2.2 The Voting Rule

We used Approval voting for this election. Somebody made the suggestion and there was no counter-proposal. In retrospect, this choice was quite natural: this procedure is fast and easy to use even if the number of candidates is large. Asking voters to rank the 18 candidates was hardly feasible in our case. Approval voting is also advisable when the set of alternatives has been loosely designed and contains very similar candidates.[4] One may nevertheless regret that we lost the occasion to gather, through the vote, more information on the participants' opinions about the different

[4]See the cloning-consistency condition (Tideman 1987) and the composition-consistency property (Laffond et al. 1996).

Table 13.1 Number of approved candidates

Number of approvals	0	1	2	3	4	5	6	7	8	9	10	> 10	Total
Number of ballots	0	2	7	3	5	2	1	1	0	0	1	0	22

Table 13.2 Approval scores

Voting rule		Approvals	Approving percentage
Approval voting	App	15	68.18
Alternative vote	Alt	10	45.45
Copeland	Cop	9	40.91
Kemeny	Kem	8	36.36
Two-round majority	2R	6	27.27
Coombs	Coo	6	27.27
Simpson	Sim	5	22.73
Majority judgement	Bal	5	22.73
Borda	Bor	4	18.18
Black	Bla	3	13.64
Range voting	RV	2	9.09
Nanson	Nan	2	9.09
Leximin	Lex	1	4.54
Top-cycle	TC	1	4.54
Uncovered set	UC	1	4.54
Fishburn		0	0
Untrapped set		0	0
Plurality		0	0

voting rules. Hopefully the next section, where results are presented, will show that we can already learn quite a lot from the analysis of the Approval ballots.

13.3 The Results

13.3.1 Approval Score and Other Indicators

Voters approved on average 3.55 candidates out of 18, with a distribution provided in Table 13.1. This figure is not at odds with what has been observed in other circumstances (Laslier and Sanver 2010).

Table 13.2 provides the scores of the candidates:

Approvals. This is the number of voters who approve the candidate.

Approval score. This is the percentage of the population who approve the candidate. *Approval Voting* is approved by 15 voters out of 22, that is 68.18%.

Table 13.3 Various indicators

	Approvals	Markov	Focus	Central.	Simil.	Satisf.	Dilution
App	15	36.44	12.53	15.65	10.44	5.14	4.07
Alt	10	11.70	9.01	11.90	8.39	2.67	4.50
Cop	9	14.60	6.90	9.86	8.12	3.24	4.22
Kem	8	10.56	6.56	8.16	7.61	2.39	4.00
2R	6	6.91	5.50	7.14	6.86	1.71	4.5
Coo	6	6.28	4.88	7.48	6.89	1.63	4.67
Sim	5	2.82	7.17	8.50	6.52	0.99	6.00
Bal	5	2.26	4.69	7.48	6.62	1.05	5.40
Bor	4	4.42	5.08	5.78	6.16	1.24	5.25
Bla	3	0.86	5.59	6.12	5.83	0.47	7.00
RV	2	1.95	1.60	1.70	5.55	0.70	3.50
Nan	2	0.29	4.35	5.10	5.49	0.24	8.50
Lex	1	0.42	1.67	1.36	5.17	0.20	5.00
TC	1	0.30	1.82	1.70	5.17	0.17	6.00
UC	1	0.21	2.25	2.04	5.16	0.14	7.00

Approval Voting is the winner of the election. It is worth noticing that it is the only candidate approved by more than half of the voters.[5] Three candidates received no vote at all: *Fishburn*, *Untrapped Set*, and *Plurality*.

There are actually many different ways to compute scores and other indicators from a set of Approval ballots. Table 13.3 provides some, which are now defined. The number of voters who approved of both candidates c and c' is called the association of c and c', and is denoted by $as(c, c')$. The number of voters who approved c is denoted by $as(c)$.

Markov score. This score is computed as follows. The candidate "present at date t" is denoted $c(t)$. At date t chose at random one voter v. If v approves $c(t)$, keep this candidate for the next date: $c(t+1) = c(t)$. If not choose $c(t+1)$ at random among the candidates that v approves. This defines a Markov chain over candidates whose stationary distribution is the Markov score. For instance a candidate with Markov score 0.3 is, in the long run of this process, present 30% of the time.

Focus. The focus of candidate c is the sum over all candidates k of the fraction of k-voters who also approved c.

$$f(c) = \sum_k \frac{as(c,k)}{as(k)}.$$

The focus measures the ability of a candidate to attract votes from voters who also voted for others.

[5]One voter wrote on his/her ballot "Approval Voting with a runoff." This procedure was not on the list. This ballot was counted as an approbation of Approval voting.

Centrality. This indicator is based on the following Markov chain. The transition probability from c to c' is $as(c,c')/\sum_{c \neq c'} as(c,c')$. The centrality measure is the associated stationary probability. This is a natural measure of centrality in the multi-graph where there is a link between two candidates each time a voter approves them both.

Similarity. This indicator is based on the following Markov chain. Given the candidate c, one chooses at random a voter v. If v approves c one replaces c by c' chosen at random among the candidates that v approves. If v does not approve c, one replaces c by c' chosen at random among all the candidates. The similarity measure is the associated stationary probability. This means that, given a candidate c, one looks for a candidate c' which is similar to c in the sense that a voter has approved both.

Satisfaction. If v has approved $B(v)$ candidates, count $1/B(v)$ points for each. The total count of candidate c is thus between 0 and the number of voters, and the sum over candidates is the number of voters. (See Kilgour 2010.)

Dilution. This is the average number of candidates approved by the voters who approve a given candidate. Let $as(v,c,k)$ be 1 if voter v approves both c and k, and 0 ifnot. Then:

$$dil(c) = \frac{1}{as(c)} \sum_{v} \sum_{k} as(v,c,k).$$

Notice that this indicator can be computed with a formula somehow dual to the focus:

$$dil(c) = \sum_{k} \frac{as(c,k)}{as(c)}.$$

The dilution thus measures to what extent supporters of a candidate also vote for other candidates. It should not be interpreted as an indicator of the strength of the candidate but as a part of the description of the electorate of the candidate: do these voters give exclusive support (low dilution), or do they support many other candidates (high dilution).

These indicators are all highly correlated with the approval score, except for the dilution (see Table 13.4).

Table 13.4 Correlations among indicators

	Approvals	Markov	Focus	Central.	Simil.	Adjust.	Dilution
Approvals	1	0.935	0.930	0.958	0.999	0.980	−0.547
Markov	0.935	1	0.851	0.981	0.930	0.878	−0.285
Focus	0.930	0.851	1	0.839	0.943	0.966	−0.511
Central	0.958	0.981	0.839	1	0.952	0.897	−0.347
Simil.	0.999	0.930	0.943	0.952	1	0.984	−0.546
Adjust.	0.980	0.878	0.966	0.897	0.984	1	−0.603
Dilution	−0.547	−0.285	−0.511	−0.347	−0.546	−0.603	1

13.3.2 Structure of the Set of Candidates

Table 13.5 shows the number of voters who approved each pair of candidates, and Table 13.6 shows the distribution of these association numbers for each candidate. For instance $7/10 = 70\%$ of the voters who approved the *Alternative Vote* also approved *Approval Voting* while $7/15 = 47\%$ of the voters who approved *Approval Voting* also approved the *Alternative Vote*. It is interesting to note that 83% of the supporters of two-round majority voting also support the *Alternative Vote*, but such is the case of only 22% of the *Copeland* supporters. One may also notice that all the supporters of the *Majority Judgement* are also supporters of *Approval Voting*.

To obtain a more global view, one may compute various distances between candidates. Consider for instance for the similarity index

$$sim(c, c') = \frac{as(c, c')}{as(c)} + \frac{as(c, c')}{as(c')}$$

which ranges from 0 (when the electorates of c and c' are disjoint) to 2 (when they are identical) and define

$$dist(c, c') = 2 - sim(c, c').$$

It turns out that there exists a very good Euclidean representation of the 15 candidates in 3 dimensions, that renders 90% of the sum of square of distances.[6] Figures 13.1 and 13.2 are side views of this representation. *Approval Voting* is in the center. The points on the right are rules which are important in the social choice literature: *Uncovered set*, *Copeland*, *Nanson*, *Kemeny*, *Simpson*, even if they are not very practical. *Borda* is in this group, close to *Nanson*. The points on the left contain three practical solutions to the voting problem: *Two-round majority*, the *Alternative Vote*, and *Black*. *Leximin* is not far from this group. *Coombs* and *Majority Judgement* are close one to the other, with *Range Voting* not far. The *Top-cycle* is isolated.

This structure reflects the vote profile since by definition, two voting rules are represented close one to the other when the same voters approved both.

Studying how candidate rules are associated in the voters' ballots, it appears that the winner is receiving votes associated with all the other candidates. *Approval Voting* can be described as a "centrist" candidate in this vote. Even if one can detect some pattern in the vote profile that differentiates votes for more "theoretical" rules from votes for more "practical" rules, the electorate does not appear to be split.

[6]See Appendix C for more details.

Table 13.5 Association matrix

	App	Alt	Cop	Kem	2R	Coo	Sim	Bal	Bor	Bla	RV	Nan	Lex	TC	UC
App	**15**	7	7	4	3	3	4	5	3	3	2	2	1	1	1
Alt	7	**10**	2	3	5	4	3	3	1	3	1	1	1	1	0
Cop	7	2	**9**	4	1	2	3	4	2	1	0	2	0	0	1
Kem	4	3	4	**8**	1	3	3	1	1	1	0	1	0	1	1
2R	3	5	1	1	**6**	3	1	1	2	2	0	1	1	0	0
Coo	3	4	2	3	3	**6**	1	2	1	1	1	1	0	0	0
Sim	4	3	3	3	1	1	**5**	2	2	2	0	2	0	1	1
Bal	5	3	4	1	1	2	2	**5**	1	1	1	1	0	0	0
Bor	3	1	2	1	2	1	2	1	**4**	1	0	2	0	0	1
Bla	3	3	1	1	2	1	2	1	1	**3**	0	1	1	1	0
RV	2	1	0	0	0	1	0	1	0	0	**2**	0	0	0	0
Nan	2	1	2	1	1	1	2	1	2	1	0	**2**	0	0	1
Lex	1	1	0	0	1	0	0	0	0	1	0	0	**1**	0	0
TC	1	1	0	1	0	0	1	0	0	1	0	0	0	**1**	0
UC	1	0	1	1	0	0	1	0	1	0	0	1	0	0	**1**

Table 13.6 Conditional association matrix

	App	Alt	Cop	Kem	2R	Coo	Sim	Bal	Bor	Bla	RV	Nan	Lex	TC	UC
App	**1**	0.7	0.78	0.5	0.5	0.5	0.8	1	0.75	1	1	1	1	1	1
Alt	0.47	**1**	0.22	0.38	0.83	0.67	0.6	0.6	0.25	1	0.5	0.5	1	1	0
Cop	0.47	0.2	**1**	0.5	0.17	0.33	0.6	0.8	0.50	0.33	0	1	0	0	1
Kem	0.27	0.3	0.44	**1**	0.17	0.5	0.6	0.2	0.25	0.33	0	0.5	0	1	1
2R	0.2	0.5	0.11	0.12	**1**	0.5	0.2	0.2	0.5	0.67	0	0.5	1	0	0
Coo	0.2	0.4	0.22	0.38	0.5	**1**	0.2	0.4	0.25	0.33	0.5	0.5	0	0	0
Sim	0.27	0.3	0.33	0.38	0.17	0.17	**1**	0.4	0.50	0.67	0	1	0	1	1
Bal	0.33	0.3	0.44	0.12	0.17	0.33	0.4	**1**	0.25	0.33	0.5	0.5	0	0	0
Bor	0.20	0.1	0.22	0.12	0.33	0.17	0.4	0.2	**1**	0.33	0	1	0	0	1
Bla	0.20	0.3	0.11	0.12	0.33	0.17	0.4	0.2	0.25	**1**	0	0.5	1	1	0
RV	0.13	0.1	0	0	0	0.17	0	0.2	0	0	**1**	0	0	0	0
Nan	0.13	0.1	0.22	0.12	0.17	0.17	0.4	0.2	0.5	0.33	0	**1**	0	0	1
Lex	0.07	0.1	0	0	0.17	0	0	0	0	0.33	0	0	**1**	0	0
TC	0.07	0.1	0	0.12	0	0	0.2	0	0	0.33	0	0	0	**1**	0
UC	0.07	0	0.11	0.12	0	0	0.2	0	0.25	0	0	0.5	0	0	**1**

Fig. 13.1 Axes 1 and 2

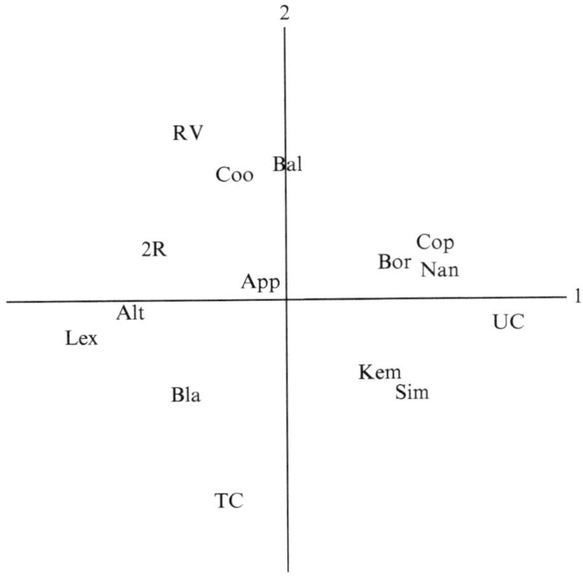

Fig. 13.2 Axes 1 and 3

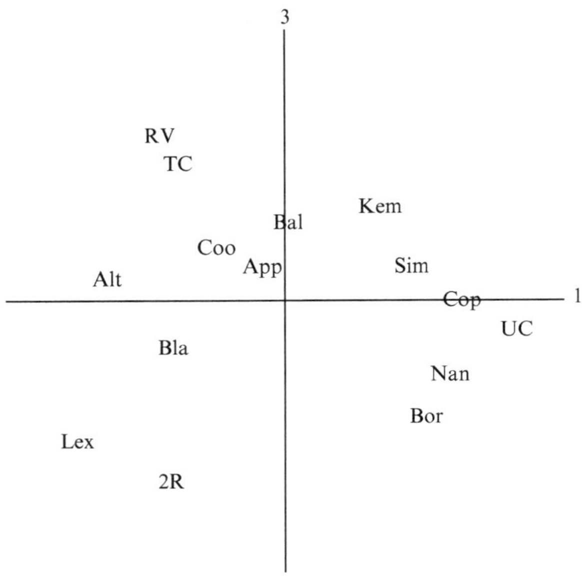

13.4 Conclusion

The analysis of the approval ballots show that *Approval Voting* was a clear winner of this election. This is somehow surprising since *Approval Voting* was not much discussed during the workshop. But this voting rule has already received a lot of attention in the academic literature, and was certainly familiar to the participants.

A striking fact is that *Plurality* rule (First Past the Post) received no approval. The whole ranking of the candidate procedures according to their approval scores seems also robust since alternative ways to count the ballots produce rather similar rankings.

Appendix A: Contributions of the Participants

Participants were ex post invited to write a brief statement that would explain their vote and their view about this "election". One-half of them did it, after having read my own contribution (see below) as an example.

It must be acknowledged that, from a scientific point of view, the experimental protocol was a little loose. The nomination procedure for the set of candidate voting rules was informal and voters did not have all the time required to learn everything about all of them.

> Fuad Aleskerov: "From my point of view our voting for the rules was rather spontaneous. For instance, I know more than 30 rules which can be listed for our voting after reading careful studies, by colleagues, of their properties."

It is also worth remarking that, under Approval Voting, it is not so easy to remember after several days or weeks which candidates you approved out of 18.

> Marc Kilgour: "I really can't remember very well why I voted the way I did. As I recall, the objective was to propose a system to elect the mayor of a town, without any indication of the number of candidates. I think I assumed that there would not be many. I followed the approval strategy of approving everything that seemed to be above average "utility," whatever that would mean. I voted for approval (my actual favorite) and range because they focus on acceptability rather than ranks. I also voted for two or three others that, it seemed to me, were complicated enough to be likely to produce something that would maximize the sum of the utilities but at the same time sophisticated enough to avoid features I don't like such as non-monotonicity. Beyond that, I can't remember much."

This point holds certainly true, as well, for preference-based balloting: it is not so easy to remember how you ranked the whole set of alternatives; it distinguishes these systems from the familiar single-name Plurality and Two-Round Majority voting rules. The set of received contributions, which follow, show well the variety of view points of the experts in the field.

A.1 A. Baujard

The question we have been asked implied to apply the rule in a real political context, involving citizens with their own desires, their own intellectual abilities and the differences among them. Choosing the good rule for a democratic mayor election predicates to pay attention to these features rather than to my own preferences among rules. Through the field-experiments we have conducted, I have learnt that most voters are frustrated by voting rules which gives little scope to the expression of their nuanced preferences, which are sometimes tinted by hesitations, indifference, significant differences between strong vs. weak preferences. These nuances have yet a strong impact on results especially in the case of uninominal voting rules, where a clear cut decision is always required to infer individual preferences. In a democracy, I claim that this should question the legitimacy of the winner in such context. I am therefore not convinced by plurality rules, whatever one or two rounds, Condorcet principle, or any uninominal voting rules in general.

Among plurinominal rules, I focus on the importance of simplicity to explain the rule, and to vote. Above all, I paid attention to transparency, meaning a wide understanding of the process of deriving a result from the ballots, and the ability of citizens to take actively part in the process of counting the votes. These desired properties rule out Alternative Vote, Borda rules and Majority Judgement among others.

Many voters in our experiments spontaneously preferred range voting, begging for the ability of giving negative grades, or a wide range of different grades. Even though this would also be my favorite in an ideal world, I regret how range voting depends on differences in the meaning of grades among people, how it is manipulable – which causes strong inequalities among the voting power of different citizens according to their ability to manipulate. This argument, I admit, may be questionable in a reduced city council, but ruling out range voting seemed cautious in the absence of information on its size and composition.

I have eventually given just one approval in the vote on voting rules: one to approval voting. It is because I had the ability of approving other rules that my choice of giving just one vote was truly meaningful.

A.2 D. Felsenthal

I adhere to the Condorcet Principle as a normative principle when one must elect one out of three or more candidates. This principle prescribes that should a candidate defeat every other candidate in pairwise comparisons (a Condorcet winner), it must be elected, and should a candidate be defeated by every other candidate in pairwise comparisons (a Condorcet loser), it must not be elected. This principle conveys the fundamental idea that the opinion of the majority should prevail, at least when majority comparisons pinpoint an unambiguous winner and/or an unambiguous loser. The Condorcet Principle takes into account only the ordinal preferences of every voter between any pair of alternatives because attempting to take into account also voters' cardinal preferences (as under the Range Voting procedure) would not only imply that a Condorcet winner may not be elected or, worse, that a Condorcet loser may be elected, but also that inter-personal comparisons of utility are possible and acceptable – which they are not!

I rank all the competing procedures for electing one out of m candidates ($m \geq 2$) according to two criteria: First,I prefer all Condorcet-consistent procedures over all procedures that are not Condorcet-consistent. Second, among Condorcet-consistent procedures I prefer those which are not vulnerable to non-monotonicity or to electing a Pareto-dominated candidate when a Condorcet winner does not exist; and among the procedures which are not Condorcet-consistent I prefer those which are not susceptible to one or more of the following four pathologies which I consider as especially serious: non-monotonicity, not electing a candidate who constitutes the top preference of an absolute majority of the voters (aka absolute Condorcet winner), electing a candidate who is a Condorcet loser or is Pareto-dominated.

According to these criteria I approved only Kemeny's and Copeland's procedures because they are both Condorcet-consistent and are not susceptible to any of the above mentioned four pathologies.

My rank-order of the 18 competing procedures is as follows:

Kemeny > Copeland > Black > Nanson > Untrapped Set > Fishburn > Uncovered Set > Top Cycle > Simpson > Borda > Coombs > Alternative Vote > 2-round Majority > Plurality > Majority Judgment > Approval Voting > Leximin > Range Voting.

A.3 W.V. Gehrlein

In all honesty, I do not remember exactly which of the many possible rules that were listed that I voted for during this impromptu exercise. However, my general convictions were expressed on the ballot that I submitted. The first statement on my ballot was: "In a perfect world I would recommend any Condorcet consistent voting rule". Standard arguments against the implementation of majority rule based voting are too heavily focused on one atypical example of something that could conceivably happen to ignore an "almost-majority" minority voting bloc with strong preferences. The obvious question is: What is the likelihood that such a scenario would ever actually exist? We all know that such hypothetical voting situations can always be developed to make any voting rule appear to behave very poorly on some criterion. The only practical way out of this dilemma must therefore be based on the likelihoods that voting rules display such bad behavior. In the context of evaluating voting rules to elect the mayor of a city in a typical situation, my assumption from scenarios that I am familiar with would make the possibility negligible that there would ever be more than four candidates. Since there is a very high probability that a Condorcet winner will exist in such cases, why should we not elect that candidate?

The answer to the immediately preceding question is that Condorcet consistent procedures are not always easy to implement with a larger number of candidates, which led to my second statement on the ballot. "In the real world I would recommend (some elimination rules that I do not recall) and Borda Rule". These rules would give a reasonable probability of electing the Condorcet winner, while also being both explainable to and acceptable to the electorate, without any implication that simplicity should be the only criterion for evaluating voting rules. Arguments about the relatively large probability with which some of these voting rules can be manipulated are typically based on the assumption that one group of voters with similar preferences can manipulate the outcome, while all other voters are completely naive to the situation. When it is further assumed that these other voters are aware of such possibilities and that they can react accordingly, the probability that the winner could actually be changed is significantly reduced. However, it is definitely reasonable to conclude from this exercise that plurality rule is not considered to be acceptable and that Approval Voting is the clear winner when voting is done by Approval Voting. But, it is critical that we must not forget the significant concerns that have been raised about the type of winners that are selected when Approval Voting is employed.

A.4 J.-F. Laslier

I do not adhere to the Condorcet principle as a *normative* principle; if 49% of the population strongly prefer A to B and 51% slightly prefer B to A, I think that A is collectively preferable. My first best decision rule is thus utilitarianism, or "range voting". But I found Approval Voting a very good practical mechanism to approximately achieve the utilitarian outcome. For the practice, I find that Condorcet-consistent procedures advisable, except in the extreme but important case of a society split in two. The best Condorcet procedure to me is the randomized procedure studied by B. Dutta, G. Laffond, M. LeBreton and myself under the name Essential set, but this rule was not proposed. In most cases, the Simpson rule (Minmax procedure) is a good way to select in the Essential set, like Kemeny, Coombs, and others. My preference was:

Range > Approval > various Condorcet methods among which I make little difference > Two round plurality > Alternative vote > Leximin > Majority Judgment > Plurality.

My guess was that, for this election, Approval would win, maybe challenged by Alternative vote (I was right!). Therefore I voted for Approval and Range. Here is my complete ranking, with my sincere utilitarian view scaled on the 0–100 scale:

Range (100) > Approval (99) > Kramer-Simpson (85) > Coombs (84) > Kemeny (83) > Copeland (82) > Nanson (81) > Black (80) > Borda (50) > Fishburn (21) > UncoveredSet (20) > 2-roundMajority (18) > AlternativeVote (17) > UntrappedSet (16) > TopCycle (15) > Leximin (10) > Majority Judgment (1) > Plurality (0)

A.5 M. Machover

I consider that decision about which voting procedure should be used must be governed by some meta-principle. I also consider that an appropriate meta-principle for the present hypothetical case is majority rule. I therefore gave my approval only to Condorcet-consistent procedures, selecting those that have additional desirable properties: Copeland's and Kemeny's procedures.

A.6 V. Merlin

While considering the question "what is the best voting rule that the city council of your town should use to elect a mayor?" my first reaction is that the procedure should be simple and easily understandable by the whole population of the city. The second question to answer is to which degree the Condorcet principle should be implemented. I do not adhere to the Condorcet principle, as a majority of 50% plus epsilon can impose a candidate which is the worst choice of the other voters, without considering compromise candidates. But at least, I do consider that a Condorcet loser should never be elected. Hence, Plurality rule is the worst system in the list.

So, I decided to advise Plurality with two rounds, Alternative Voting and Approval Voting. As long as there is a final duel, any elimination system using the plurality tallies will never elect the Condorcet loser. Plurality with two rounds and Alternative Voting are such systems. They are easy to explain, and have been implemented in different countries (France, Australia), with no major complaints. Moreover, Alternative Voting is hardly manipulable. I also consider that $k+1$ rounds before the final duel are better than k! Though I also voted for Approval Voting, it may be possible for it to select a Condorcet loser, if everybody just reports his first choice. But I think that the risk is quite limited, provided that a sufficiently large part of the population votes sincerely. Experiences show that voters tend also to approve more than one candidate. What would make me rank Approval Voting slightly below the two previous rules, is the fact that it has not been widely used in political elections. I felt that we still need more real life experiences to check that everything goes right with approval voting, but I am ready to give it its chance.

The simplicity argument goes against many Condorcet-consistent rules. Though Kemeny is an extremely elegant solution to the voting problem, it is rather sophisticated. For those who think that the Condorcet criterion should be implemented, I would recommend the Copeland method, which could be easily explained to the voters, as a tournament among the candidates.

At last, I fear that rules like the Borda count or Range voting could lead to undesired outcomes, when a fraction of the voters tries to manipulate it.

A.7 N. Miller

I cast approval votes for Approval Voting and Copeland. My votes did not reflect any general normative principle but rather my sense as to what would be both practical and reasonable for the type of election that Dan Felsenthal stipulated, namely the election of a mayor when a number of candidates are the ballot. A year ago I might have approved of AV/IRV also, but I now think that its problems are quite serious (even in practice, not just in theory). And, as a practical matter, my highest preference would be for Approval Voting, because it is simple to explain to voters, simple to cast votes, and simple to count. Moreover, most voters (in the US at least) would want to see some kind of vote totals in the newspaper the next day, which Copeland does not provide.

While Plurality lost our vote by a landslide, it works perfectly well in most US partisan general elections, since Duverger's Law works so powerfully that there are, literally or effectively, only two candidates in most such elections (the recent Senate contests in Florida and Alaska being notable exceptions). However, Approval Voting might be a definite improvement over Plurality in party primary elections and non-partisan general elections (which is how many mayors are elected), where often three or more candidates are on the ballot.

Finally, voting procedures need to be evaluated not only in terms of their "static" social choice properties (e.g., Condorcet consistency, monotonicity, etc.) but also in terms of their "dynamic" effects, e.g., incentives for candidate entry, candidate ideological positioning, etc., which affect the types of preferences profiles that are most likely to arise.

A.8 H. Nurmi

We were asked to propose voting systems that we could recommend or approve of to be adopted in the mayoral elections of our municipality. Recommend and approve of are two different – albeit related – things, but since we were asked to submit approval ballots, I felt encouraged to suggest more than one system (which I would NOT do if I were asked to recommend "a system"). I proposed Borda, Nanson and probably also Kemeny (someone may have preempted me on the latter, though). Anyway, my ranking is Nanson > Kemeny > Borda > approval voting and these (as far as I now recall) were on my ballot. Nanson and Kemeny are both pretty resistant to misrepresentation of preferences and take into account a great deal of the preference information given by the voters. (One could also point out that they are Condorcet, but I'm not much moved by that property any longer: some systems are vulnerable to adding or removing or cloning alternatives (e.g., Borda) (as shown by Fishburn), others to adding or removing voters with completely tied preferences (Condorcet) (as shown by Saari). Overall, being based on strict majority principle is not a decisive feature in my book. Although it can be argued that it is preferable to be ruled by a majority than by a minority, I think one should also sail clear of the dictatorship of majority. They (Nanson and Kemeny) both do well in terms of several choice theoretic criteria. Borda's advantage is in intuitively plausible metric rationalizability: it looks for the closest (in terms of inversion metric) consensus profile (in terms of the first ranked alternative) and since we are looking for a single winner, this makes sense. Borda count also does well in minority protection (as shown by Nitzan). Approval voting was also on my ballot, not so much because of its choice-theoretic properties, but because of intuitive appeal of its results: it sounds nice to have a mayor who is deemed acceptable by more voters than any other. I must say, though, that the interpretation of "approvability" is not obvious (and this pertains to the interpretation of our balloting result as well). Does the fact that I approve of a candidate mean that I can tolerate him as the mayor without resorting to active resistance or does it mean that I positively support him/her? I think this is what makes the approval voting results hard to interpret, but I guess a mayor that is even tolerated by more voters than any other candidate has at least tolerable prospects.

A.9 F. Plassmann

I view voting as a useful mechanism for making collective decisions when unanimous agreement is not possible. Elections should generally be preceded by discussions about the candidates and the importance that the voters attach to the election. If a minority of voters feels strongly about some candidates while the other voters are almost indifferent between these candidates, then it should be possible for the minority to convince sufficiently many of the others to change their minds prior to the vote-casting process. (I believe that in cases of near-indifference, most people's desire to preserve social harmony trumps rent-seeking.) If it is not possible to change sufficiently many voters' minds, then I would interpret this as evidence that the intensity in preferences between the groups is not as disparate as it might appear. I therefore feel comfortable ignoring voting rules that take account of the intensities of voters' preferences.

I value the Condorcet principle, and I see the main issue as what we should do when there is no Condorcet winner. Apart from the fact that it is not Condorcet consistent, the Borda rule has many attractive properties. Thus my first choice is Black's rule, which seems to be least susceptible, among many popular voting rules, to a wide range of voting paradoxes and which has a very small frequency of ties (as preliminary research with Nic Tideman suggests). The discontinuity of Black's rule also makes strategizing difficult. However, the need to understand two separate evaluation criteria might make Black's rule too complicated for some voters. Voters will accept the outcome of an election only if they understand how the ballots are to be counted. Approval voting is very simple and avoids some of the most egregious shortcomings of the plurality rule. Thus I would endorse approval voting in situations when simplicity is important.

A.10 M. Salles

I voted for Approval Voting and for Borda. I share Jean-François' view regarding the difficulty concerning majority rule. However, I do not go as far as him and would not recommend "range voting". In case there are a sufficient number of candidates, the Borda rule proposes a way to deal somehow with intensity of preferences without going as far as "Range Voting". Also I think that the voting method must be simple enough to be understood by the quasi-totality of the voters, which might not be the case of the alternative vote system or Kemeny's rule.

A.11 N. Tideman

A group of experts on voting theory wanted to learn their collective judgments of a variety of voting rules. They decided (by something like acclimation) to proceed by using approval voting. I thought this was a reasonable way of learning the general level of support for different voting rules, as a prelude to future discussion. I would not have recommended approval voting as a way to make a collective judgment of which voting rule is best. That, I think, requires both more time and a procedure for ranking the options, so that direct paired comparisons can be made.

I am quite startled by the high level of support for approval voting as a way of electing a mayor. What I find particularly distressing about approval voting is that it requires a voter to decide whether to draw a line between generally acceptable and unacceptable candidates, or to leave that task to other voters and instead to draw a line between the very best and the close contenders who are not quite as good. I think that voters for a mayor should not be required to choose between drawing those two types of lines.

The relevant criteria for a voting rule for mayor, in my opinion, are:

- First, the capacity of the rule to gain the trust of voters. This depends on the reasonableness and understandability of the logic of the rule and the ease with which the counting process can be followed. Investigating this requires psychological methods as well as knowledge of the logic of voting procedures.
- Second, the likely statistical success of the rule in identifying the outcome with the greatest aggregate utility, under the assumption that voters vote sincerely. This is something that can be investigated by statistical methods.
- Third, the resistance of the rule to strategic voting. This too can be investigated by statistical methods.

It is my guess that the best rule, by some intuitive averaging of these criteria, is the Simpson rule. But the empirical work that would justify this guess remains to be done.

A.12 W. Zwicker

When I suggested we vote on voting rules and use Approval Voting, I thought the proposal could not pass – we'd surely split over the use of Approval Voting. At the time, however, our "rump session" discussion was stuck and it seemed that a conversational grenade might do more good than harm. I was very surprised that no one objected; some, as one might expect, were enthusiastic. Then I realized the exercise might be constructive if we could collectively endorse the principle that plurality rule was terrible... despite the stated goals of our workshop, I'd never thought it likely that we'd reach even a loose consensus on a single alternative. My own ballot approved a large number of rules, for two reasons: I doubt that the current state-of-the-art allows us confidently to select a small number of best rules, and my genuine indecisiveness was consistent with the best strategy for making plurality look bad. In terms of my specific approvals, it seems like false comfort to rely on any single absolute principle as a guide, when every choice of a voting rule entails trade-offs along many dimensions, about which our understanding is limited. For example, I feel the draw of Condorcet's principle but reject it as an absolute, in part because some recent results suggest trade-offs between that principle and any reasonable degree of decisiveness. I've come to view decisiveness as an under-valued trait – very important, though not decisively so of course. I did approve some Condorcet extensions, but not top-cycle, because of its striking indecisiveness. Mathematically, Kemeny is beautiful whereas Black is plug-ugly, but I swallowed hard, approved Black, and disapproved Kemeny (because Kemeny winner are rankings, not individual candidates, and I can imagine what would happen the first time some real world election yielded a tie among several rankings).

Appendix B: 18 Voting Rules

In what follows, the "majority tournament" is the binary relation among candidates: "More than half of the voters prefer *a* to *b*". In that case we say that *a beats b* (according to pair-wise majority rule).

B.1 Approval Voting [App]

Each voter approves as many candidates as she wishes. The candidate with the most approval is elected. See Brams and Fishburn (1983), Laslier and Sanver (2010).

B.2 Alternative Vote [Alt]

Each voter submits a ranking (possibly incomplete) of the candidates. One first counts the number of times each candidate appears as top-ranked (his plurality score). The candidate with the lowest plurality score is eliminated. In a second count, the votes for this candidate are transferred to the second-ranked candidate (if any) on these ballots. The process is then repeated again and again until one candidate is ranked first by an absolute majority of the votes (original or transferred) is elected. See Farrell (2001), Farrell and McAllister (2006). Other names for this procedure or its variants: "Hare" system, "Single Transferable Vote", "Instant runoff".

B.3 *Copeland [Cop]*

Each voter submits a ranking of the candidates. For each candidate one computes his pairwise comparison score, that is the number of challengers this candidate beats under pair-wise majority rule. The candidates with the largest score are chosen. This Condorcet-consistent rule does not specify how ties (which are common when there is no Condorcet winner) are broken. See Laslier (1997). Other name: Tournament score.

B.4 *Kemeny [Kem]*

Each voter submits a ranking of the candidates. The rule defines a summary ranking as follows. For any ranking R of candidates one computes the sum, over all pairs (a, b) of candidates of the number of voters who agree with how R ranks a and b. Then R^* is chosen to maximize this total number of agreements. The elected candidate is the top-ranked candidate according to R^*. This procedure is Condorcet-consistent. See Young and Levenglick (1978), Young (1988). Other name: Median ranking.

B.5 *Two-Round Majority [2R]*

Each voter votes for one candidate. If a candidate obtains an absolute majority, he is elected. If not, a runoff election takes place among the two candidates who obtained the most votes. This rule is the most common rule throughout the world for direct elections, but it has seldom retained the attention of social choice theorists. See Lijphart (1994), Blais et al. (1997,2010), Taagera (2007). Other name: Plurality with a run-off.

B.6 *Coombs [Coo]*

Similar to the Alternative Vote but, at each round, if no candidate is ranked first by an absolute majority of the ballots, the eliminated candidate is the one who is most often ranked last. This procedure is Condorcet-consistent in the single peaked domain. Coombs (1964).[7]

B.7 Majority Judgement [Bal]

Each voter grades each candidate according to some pre-specified finite grading scale expressed in verbal terms. For each candidate one computes his median grade. Among the candidates with the highest median grade, a linear approximation scheme (described in Balinski and Laraki 2007) is used in order to choose the elected candidate. See Basset and Persky (1999), Gerlein and Lepelley (2003), Felsenthal and Machover (2008), Laslier (2011). Other names for this procedure or its variants: "Robust voting", "Best median".

B.8 Simpson [Sim]

Each voter submits a ranking of the candidates. The pair-wise vote matrix is computed. Then the chosen candidate is the one against which the smallest majority (in favor of another candidate) can be gathered. See Simpson (1969). Other names: "Minimax procedure", "Simpson-Kramer rule".

B.9 Borda [Bor]

Each voter submits a ranking of the candidates. For K candidates, each one receives $K-1$ points each time he is ranked first, $K-2$ points each time he is ranked second, etc. The elected candidate is the one who receives the largest number of points.

B.10 Black [Bla]

Choose the Condorcet winner if it exists and the Borda winner if not. Suggested by Black (1958).

B.11 Nanson [Nan]

Each voter submits a ranking of the candidates. The Borda score is computed. Candidates with Borda score equal to or below the average are eliminated. Then

[7] Thanks to Dan Felsenthal for pointing to me details of this definition.

a new Borda count is computed, on the reduced profile and the process is iterated. This procedure is Condorcet-consistent. See Nanson (1883).

B.12 Range Voting [RV]

Each voter gives to each candidate as many points as she wishes between zero and, say, 10 points. The elected candidate is the one who receives the largest number of points. Range Voting is not often considered in the voting rule literature since, from the theoretical point of view, it is essentially plain utilitarianism. See Arrow et al. (2002), Dhillon and Mertens (1999), Baujard and Igersheim (2010) or the `rangevoting.org` web site. Other names for this procedure or its variants: "Utilitarianism", "Point voting" and in French: "vote par note", which just means "voting by grading."

B.13 Top Cycle [TC]

Each voter submits a ranking of the candidates. The majority tournament is computed. The Top-Cycle is the smallest set of candidates such that all candidates in this set beat all candidates outside this set. This Condorcet-consistent rule does not specify how ties (which occur when there is no Condorcet winner) are broken. See Schwartz (1972), Laslier (1997).

B.14 Uncovered Set [UC]

Each voter submits a ranking of the candidates. The majority tournament is computed. A candidate a belongs to the Uncovered set if and only if, for any other candidate b, either a beats b or a beats some c who beats b. This Condorcet-consistent rule does not specify how ties (which occur when there is no Condorcet winner) are broken. See Miller (1980), McKelvey (1986), Laslier (1997). Other name (in the graph-theory literature): "Kings procedure".

B.15 Leximin [Lex]

Each voter grades each candidate according to some pre-specified grading scale. Each candidate k is evaluated according to the worst grade he received, say $g(k) = \min_v g(k,v)$. The elected candidate is the one with the best evaluation $g^* = \max_k g(k)$. If several candidates have the same evaluation g^*, the elected candidate is the one who receives g^* the least often. This rule is an important benchmark for normative economics. See Arrow et al. (2002).

B.16 Fishburn

This choice correspondence is a variant of the Uncovered set which is useful when the majority relation contains ties (exactly as many voters prefer a to b than b to a). See Aleskerov and Kurbanov (1999).

B.17 Untrapped Set

This choice correspondence defined by Duggan (2007) is a variant of the Top-Cycle which is useful when the majority relation contains ties (exactly as many voters prefer a to b than b to a).

B.18 Plurality

Each voter votes for one candidate. The candidate with the most votes is elected. This is the most common voting rule in the Anglo-saxon world and the literature is very large. Other name: First Past the Post.

Appendix C: Statistical Significance of the 3D Representation

The method for spatial representation of data sets is derived from multivariate factor analysis. Given is a symmetric matrix of positive numbers, intended to measure the distances between the items, say $dist(c, c')$. If each item c is represented by a point $\phi(c)$ in the Euclidean space of dimension d one can compute the sum of the squares of the distances between the items:

$$\sum_{c,c'} dist^2(c, c'),$$

called the total variance, and compare this sum to the sum of squares of the distances between the corresponding points:

$$\sum_{c,c'} (\phi(c) - \phi(c'))^2,$$

called the explained variance. The best representation with d dimensions can be computed numerically using linear algebra. The quality of the representation is measured by the ratio between explained and total variance. This technique was used for Approval Voting data by Laslier and Van der Straeten (2004) and by Laslier (2006).

Of course the quality of the representation can only increase with the number of dimensions. In the text I show a 3D representation that explains about 90% of the the variance. In order to check whether this figure should be considered as large, I replicated the same computation on randomly generated data. Recall that, with the real data the explained percentages are, respectively 39, 66 and 90 for 1, 2 and 3 dimensions.

In a first test, suppose that each voter approves of each candidate independently with a probability p that corresponds to the average approval rate (here: $p = 78/(15 * 22) \simeq 0.236$). Running 10.000 simulations I find that the observed figures $39\%, 66\%, 90\%$) are respectively attained with probability 0.016, 0.005 and 0.0005. It is thus clear that our data set has much more structure than a totally random one, in which all candidates are alike, up to random fluctuations.

In a second test, suppose that we set the expected number of approval votes received by each candidate c to its actual value. So suppose that each voter independently approves of each candidate c with a probability $p(c)$ equal to the actual approving percentage of this candidate. For instance for the candidate *Approval Voting*, $p(\texttt{App}) = 15/22 \simeq 0.6818$. We thus keep trace that some candidates are good and some are not, but we lose the correlation among candidates. In that case, I find that that the observed figures $(39\%, 66\%, 90\%)$ are respectively attained with probability 0.07, 0.07 and 0.03. Again one can conclude from this statistical test that it is not by chance that the real data set provides such large figures.

References

Aleskerov, F. & Kubanov, E. (1999). A degree of manipulability of known social choice procedures. In S. Aliprantis, A. Alkan, & N. Yannelis (Eds.), *Studies in economic theory* (Vol. 8), *Current trends in economics: Theory and applications* (pp. 13–28). Berlin: Springer.

Arrow, K. J. (1951). *Social choices and individual values*. New York: Wiley.

Arrow, K. J., Sen, A. K., & Suzumura, K. (2002). *Handbook of social choice and welfare* (Vol. 1). Amsterdam: Elsevier.

Balinski, M. & Laraki, R. (2007). A theory of measuring, electing, and ranking. *PNAS, 104*, 8720–8725.

Basset, G. W. & Persky, J. (1999). Robust voting. *Public Choice, 99*, 299–310.

Baujard, A. & Igersheim, H. (2010). Framed-field experiments on approval voting in the political context. In J.-F. Laslier & R. Sanver (Eds.), *Handbook on approval voting* (pp. 357–395). Heidelberg: Springer.

Black, D. (1958). *The theory of committees and elections*. Cambridge: Cambridge University Press.

Blais, A., Massicotte, L., & Dobrzynska, A. (1997). Direct presidential elections: A world summary. *Electoral Studies, 16*, 441–455.

Blais, A., Labbé-St-Vincent, S., Laslier, J.-F., Sauger, N., & Van der Straeten, K. (2010). Strategic vote choice in one round and two round elections. *Political Research Quarterly, 20*, 1–9.
Brams, S. J. & Fishburn, P. C. (1983). *Approval voting*. Boston: Birkhäuser.
Coombs, C. H. (1964). *Theory of data*. New York: Wiley.
Dhillon, A. & Mertens, J.-F. (1999). Relative utilitarianism. *Econometrica, 67*, 471–498.
Duggan, J. (2007). A systematic approach to the construction of non-empty choice sets. *Social Choice and Welfare, 28*, 491–506.
Dutta, B. & Laslier, J.-F. (1999). Comparison functions and choice correspondences. *Social Choice and Welfare, 16*, 513–532.
Farrell, D. M. (2001). *Electoral systems: A comparative introduction*. New York: Palgrave.
Farrell, D. M. & McAllister, I. (2006). *The australian electoral system: Origins, variations, and consequences*. Sydney: University of New South Wales Press.
Felsenthal, D. & Machover, M. (2008). The majority judgment voting procedure: A critical evaluation. *Homo Oeconomicus, 25*, 319–334.
Gehrlein, W. V. & Lepelley, D. (2003). On some limitations of the median voting rule. *Public Choice, 117*, 177–190.
Kilgour, M. K. (2010). Approval balloting for multi-winner elections. In J.-F. Laslier & R. Sanver (Eds.), *Handbook on approval voting* (pp. 357–395). Heidelberg: Springer.
Laffond, G., Laslier, J.-F., & LeBreton, M. (1993). The Bipartisan set of a tournament game. *Games and Economic Behavior, 5*, 182–201.
Laffond, G., Lainé, J., & Laslier, J.-F. (1996). Composition-consistent tournament solutions and social choice functions. *Social Choice and Welfare, 13*, 75–93.
Laslier, J.-F. (1997). *Tournament solutions and majority voting*. In *Series studies in economic theory* (Vol. 7). Heidelberg: Springer.
Laslier, J.-F. (2006). Spatial approval voting. *Political Analysis, 14*, 160–185.
Laslier, J.-F. (2011). On choosing the alternative with the best median evaluation. *Public Choice*, forthcoming.
Laslier, J.-F. & Sanver, R. M. (Eds.). (2010). *Handbook on approval voting*. Heildelberg: Springer.
Laslier, J.-F. & Van der Straeten, K. (2004). Election présidentielle: une expérience pour un autre mode de scrutin. *Revue Française de Science Politique, 54*, 99–130.
Lijphart, A. (1994). *Electoral systems and party systems: A study of twenty-seven democracies, 1945–1990*. Oxford: Oxford University Press.
McKelvey, R. D. (1986). Covering, dominance, and institution-free properties of social choice. *American Journal of Political Science, 30*, 283–314.
Miller, N. R. (1980). A new solution for tournaments and majority voting. *American Journal of Political Science, 24*, 68–96.
Nanson, E. J. (1883). Methods of elections. *Transactions and Proceedings of the Royal Society of Victoria, 19*, 197–240.
Schwartz, T. (1972). Rationality and the myth of maximum. *Noûs, 6*, 97–117.
Simpson, P. (1969). On defining areas of voter choice. *Quarterly Journal of Economics, 83*, 478–490.
Taagerera, R. (2007). Electoral systems. In C. B. Boix & S. Stokes (Eds.), *The oxford handbook of comparative politics*. Oxford, Oxford University Press.
Tideman, T. N. (1987). Independence of clones as a criterion for voting rules. *Social Choice and Welfare, 4*, 185–206.
Young, H. P. (1988). Condorcet's theory of voting. *American Political Science Review, 82*, 1231–1244.
Young, H. P. & Levenglick, A. (1978). A consistent extension of Condorcet's election principle. *SIAM Journal on Applied Mathematics, 35*, 285–300.

FSC
www.fsc.org
MIX
Papier aus verantwortungsvollen Quellen
Paper from responsible sources
FSC® C105338

If you have any concerns about our products,
you can contact us on
ProductSafety@springernature.com

In case Publisher is established outside the EU,
the EU authorized representative is:
Springer Nature Customer Service Center GmbH
Europaplatz 3, 69115 Heidelberg, Germany

Printed by Libri Plureos GmbH
in Hamburg, Germany